Biological Functions
of Microtubules
and Related Structures

Academic Press Rapid Manuscript Reproduction

Proceedings of the 13th Oji International Seminar on Biological Functions of Microtubules and Related Structures, November 30–December 4, 1981, Tokyo, Japan, Held under the Sponsorship of the Japan Society for the Promotion of Science and the Fujihara Foundation of Science

Biological Functions
of Microtubules
and Related Structures

Edited by

HIKOICHI SAKAI
Department of Biophysics and Biochemistry
Faculty of Science
The University of Tokyo
Tokyo, Japan

HIDEO MOHRI
Department of Biology
College of General Education
The University of Tokyo
Tokyo, Japan

GARY G. BORISY
Laboratory of Molecular Biology
University of Wisconsin
Madison, Wisconsin

1982

ACADEMIC PRESS
A Subsidiary of Harcourt Brace Jovanovich, Publishers
Tokyo   New York
London   Paris   San Diego   San Francisco   São Paulo   Sydney   Toronto

ACADEMIC PRESS, INC.
111 Fifth Avenue, New York, New York 10003

United Kingdom Edition published by
ACADEMIC PRESS, INC. (LONDON) LTD.
24/28 Oval Road, London NW1 7DX

Library of Congress Cataloging in Publication Data

Oji International Seminar on Biological Functions of
    Microtubules and Related Structures (13th : 1981 :
    Tokyo, Japan)
    Biological functions of microtubules and related
structures.

    Proceedings of the 13th Oji International Seminar on
Biological Functions of Microtubules and Related Struc-
tures, held at Komaba Eminence in Tokyo, Nov. 30-Dec. 4,
1981.
    Bibliography: p.
    Includes index.
    1. Microtubules--Congresses. I. Sakai, Hikoichi,
date   . II. Mohri, Hideo. III. Borisy, Gary G.
IV. Title.
QH603.M44O38 1981   574.87'34   82-11609
ISBN 0-12-615080-X

PRINTED IN THE UNITED STATES OF AMERICA

82 83 84 85   9 8 7 6 5 4 3 2 1

Contents

I TUBULIN AND MICROTUBULES; BIOCHEMISTRY, MOLECULAR BIOLOGY, AND REGULATION OF ASSEMBLY

II DYNEIN, MICROTUBULES, AND OTHER PROTEINS IN CELL MOTILITY

III MICROTUBULES AND RELATED PROTEINS IN MITOSIS

IV   INTERACTIONS OF CYTOSKELETAL COMPONENTS

V   CYTOSKELETONS

VI MICROTUBULES IN MEMBRANE FUNCTIONS AND TRANSPORT

Contributors

Numbers in parentheses indicate the pages on which the authors' contributions begin.

CHRISTOPHER W. BELL (137), *Pacific Biomedical Research Center, University of Hawaii, Honolulu, Hawaii*

R. D. BERLIN (405), *Department of Physiology, University of Connecticut Health Center, Farmington, Connecticut*

GARY G. BORISY (233), *Laboratory of Molecular Biology, University of Wisconsin, Madison, Wisconsin*

J. M. CARON (405), *Department of Physiology, University of Connecticut Health Center, Farmington, Connecticut*

THEODORE G. CLARK (433), *Department of Biology, Kline Biology Tower, Yale University, New Haven, Connecticut*

WILLIAM L. DENTLER (297), *Department of Physiology and Cell Biology, Center for Biomedical Research, McCollum Laboratories, University of Kansas, Lawrence, Kansas*

SACHIKO ENDO (391), *Department of Biophysics and Biochemistry, Faculty of Science, The University of Tokyo, Tokyo, Japan*

LINDA M. GRIFFITH (311), *Department of Structural Biology, Sherman Fairchild Center, Stanford University School of Medicine, Stanford, California*

YUKIHISA HAMAGUCHI (199), *Biological Laboratory, Tokyo Institute of Technology, Tokyo, Japan*

FUMIKO HARADA (355), *The Tokyo Metropolitan Institute of Medical Science, Tokyo, Japan*

THOMAS S. HAYS (227), *Department of Zoology, Duke University, Durham, North Carolina*

YUKIO HIRAMOTO (247), *Biological Laboratory, Tokyo Institute of Technology, Tokyo, Japan*

R. HUNTLEY (405), *Department of Physiology, University of Connecticut Health Center, Farmington, Connecticut*

G. ISENBERG (275), *Max-Planck-Institute for Psychiatry, Munich, Federal Republic of Germany*

KOICHI ISHIGURO (163), *Department of Biophysics and Biochemistry, Faculty of Science, The University of Tokyo, Tokyo, Japan*

HARUNORI ISHIKAWA (343, 377), *Department of Anatomy, Faculty of Medicine, The University of Tokyo, Tokyo, Japan*

TOMOHIKO ITO (211), *Sugashima Marine Biological Laboratory, Nagoya University, Toba, Mie, Japan*

FUYUKI IWASA (199), *Department of Biology, College of General Education, The University of Tokyo, Tokyo, Japan*

B. M. JOCKUSCH (275), *Department of Developmental Biology, University of Bielefeld, Bielefeld, Federal Republic of Germany*

MARY ANN JORDAN (61), *Department of Biological Sciences, University of California, Santa Barbara, California*

SHIRO KAKIUCHI (83), *Department of Neurochemistry, Institute of Higher Nervous Activity, Osaka University Medical School, Osaka, Japan*

SHINJI KAMIMURA (177), *Zoological Institute, Faculty of Science, The University of Tokyo, Tokyo, Japan*

RITSU KAMIYA (189), *Institute of Molecular Biology, Nagoya University, Nagoya, Japan*

TIMOTHY L. KARR (49), *Department of Chemistry, University of California, Santa Barbara, California*

TOYOKI KATO (211), *Sugashima Marine Biological Laboratory, Nagoya University, Toba, Mie, Japan*

YOSHITO KAZIRO (11), *Institute of Medical Science, The University of Tokyo, Tokyo, Japan*

CHARLES H. KEITH (365), *Department of Pharmacology, New York University School of Medicine, New York, New York*

TAKAAKI KOBAYASHI (23), *Department of Biochemistry, The Jikei University School of Medicine, Tokyo, Japan*

SUSUMU KOTANI (285), *Department of Biophysics and Biochemistry, Faculty of Science, The University of Tokyo, Tokyo, Japan*

DAVID KRISTOFFERSON (49), *Department of Chemistry, University of California, Santa Barbara, California*

PAUL J. KRONEBUSCH (233), *Laboratory of Molecular Biology, University of Wisconsin, Madison, Wisconsin*

DONNA F. KUBAI (227), *Department of Zoology, Duke University, Durham, North Carolina*

HIROMICHI KUMAGAI (73), *Department of Biophysics and Biochemistry, Faculty of Science, The University of Tokyo, Tokyo, Japan*

MASANORI KUROKAWA (321, 425), *Department of Biochemistry, Institute of Brain Research, Faculty of Medicine, The University of Tokyo, Tokyo, Japan*

TOMOYUKI KUWAKI (285), *Department of Biophysics and Biochemistry, Faculty of Science, The University of Tokyo, Tokyo, Japan*

RAYMOND J. LASEK (329), *Neurobiology Center and Department of Anatomy, School of Medicine, Case Western Reserve University, Cleveland, Ohio*

SUN-HEE LEE (49), *Department of Chemistry, University of California, Santa Barbara, California*

ISSEI MABUCHI (261), *Department of Biology, College of General Education, The University of Tokyo, Tokyo, Japan*

TADAKAZU MAEDA (41), *Mitsubishi-Kasei Institute of Life Sciences, Tokyo, Japan*

ROBERT L. MARGOLIS (61), *The Hutchinson Cancer Research Center, Seattle, Washington*

GEN MATSUMOTO (391), *Electrotechnical Laboratory, Tsukuba, Ibaraki, Japan*

R. N. MELMED (405), *Department of Physiology, University of Connecticut Health Center, Farmington, Connecticut*

TAIKO MIKI-NOUMURA (41), *Department of Biology, Ochanomizu University, Tokyo, Japan*

KOICHI MIZUNO (1), *Department of Biology, Faculty of Science, Osaka University, Toyonaka, Osaka, Japan*

HIDEO MOHRI (125), *Department of Biology, College of General Education, The University, of Tokyo, Tokyo, Japan*

HIROSHI MORI (321), *Department of Biochemistry, Institute of Brain Research, Faculty of Medicine, The University of Tokyo, Tokyo, Japan*

MASAAKI MORISAWA (151), *Ocean Research Institute, The University of Tokyo, Tokyo, Japan*

J. R. MORRIS (329), *Cancer Research Center, Massachusetts Institute of Technology, Cambridge, Massachusetts*

AILEEN MORSE (61), *Department of Biological Sciences, University of California, Santa Barbara, California*

HIROMU MUROFUSHI (163, 391), *Department of Biophysics and Biochemistry, Faculty of Science, The University of Tokyo, Tokyo, Japan*

REIKO NAGAI (189), *Department of Biology, Faculty of Science, Osaka University, Toyonaka, Osaka, Japan*

SHOGO NAKAMURA (189), *Department of Biology, Toyama University, Toyama, Japan*

R. BRUCE NICKLAS (227), *Department of Zoology, Duke University, Durham, North Carolina*

EISUKE NISHIDA (73, 285), *Department of Biophysics and Biochemistry, Faculty of Science, The University of Tokyo, Tokyo, Japan*

KAZUO OHNISHI (115), *Institute of Biological Sciences, The University of Tsukuba, Ibaraki, Japan*

MAKOTO OKUNO (151), *Department of Biology, College of General Education, The University of Tokyo, Tokyo, Japan*

J. M. OLIVER (405), *Department of Physiology, University of Connecticut Health Center, Farmington, Connecticut*

THOMAS D. POLLARD (311), *Department of Cell Biology and Anatomy, The Johns Hopkins University School of Medicine, Baltimore, Maryland*

DANIEL L. PURICH (49), *Department of Chemistry, University of California, Santa Barbara, California*

JOEL L. ROSENBAUM (433), *Department of Biology, Kline Biology Tower, Yale University, New Haven, Connecticut*

JUNKO SAKAGUCHI (33), *Institute of Medical Science, The University of Tokyo, Tokyo, Japan*

HIKOICHI SAKAI (73, 163, 285, 391), *Department of Biophysics and Biochemistry, Faculty of Science, The University of Tokyo, Tokyo, Japan*

HIDEMI SATO (211), *Sugashima Marine Biological Laboratory, Nagoya University, Toba, Mie, Japan*

RICHARD F. SATTILARO (297), *Department of Physiology and Cell Biology, Center for Biomedical Research, McCollum Laboratories, University of Kansas, Lawrence, Kansas*

ANDREA CIMINO-SAUCIER (49), *Department of Chemistry, University of California, Santa Barbara, California*

S. CHARLES SELDEN (311), *Department of Cell Biology and Anatomy, The Johns Hopkins University School of Medicine, Baltimore, Maryland*

MICHAEL L. SHELANSKI (365), *Department of Pharmacology, New York University School of Medicine, New York, New York*

HIROH SHIBAOKA (1), *Department of Biology, Faculty of Science, Osaka University, Toyonaka, Osaka, Japan*

YOSHINOBU SHIGENAKA (91, 105), *Department of Information and Behavior Science, Faculty of Integrated Arts and Sciences, Hiroshima University, Hiroshima, Japan*

YÔKO SHÔJI (247), *Biological Laboratory, Tokyo Institute of Technology, Tokyo, Japan*

TOSHINOBU SUZAKI (91, 105), *Zoological Laboratory, Faculty of Science, Hiroshima University, Hiroshima, Japan*

KEIICHI TAKAHASHI (177), *Zoological Institute, Faculty of Science, The University of Tokyo, Tokyo, Japan*

T. CHOKU TAKAHASHI (211), *Sugashima Marine Biological Laboratory, Nagoya University, Toba, Mie, Japan*

TOMOKO TASHIRO (425), *Department of Biochemistry, Institute of Brain Research, Faculty of Medicine, The University of Tokyo, Tokyo, Japan*

BRIAN J. TERRY (49), *Department of Chemistry, University of California, Santa Barbara, California*

SHIN-ICHI TOMINAGA (11), *Institute of Medical Science, The University of Tokyo, Tokyo, Japan*

CHIKASHI TOYOSHIMA (125), *Department of Physics, Faculty of Science, The University of Tokyo, Tokyo, Japan*

SACHIKO TSUKITA (377), *Department of Anatomy, Faculty of Medicine, The University of Tokyo, Tokyo, Japan*

SHOICHIRO TSUKITA (343, 377), *Department of Anatomy, Faculty of Medicine, The University of Tokyo, Tokyo, Japan*

TAKEYUKI WAKABAYASHI (125), *Department of Physics, Faculty of Science, The University of Tokyo, Tokyo, Japan*

YOSHIO WATANABE (115), *Institute of Biological Sciences, The University of Tsukuba, Ibaraki, Japan*

LESLIE WILSON (61), *Department of Biological Sciences, University of California, Santa Barbara, California*

ICHIRO YAHARA (355), *The Tokyo Metropolitan Institute of Medical Science, Tokyo, Japan*

MASAYUKI YAMAMOTO (33), *Institute of Medical Science, The University of Tokyo, Tokyo, Japan*

SHOKO YAMAZAKI (41), *Department of Biology, Ochanomizu University, Tokyo, Japan*

KAZUHIDE YANO (105), *Department of Information and Behavior Science, Faculty of Integrated Arts and Sciences, Hiroshima University, Hiroshima, Japan*

YOKO YANO (125), *Department of Biology, College of General Education, The University of Tokyo, Tokyo, Japan*

REIKO YOGOSAWA (105), *Department of Information and Behavior Science, Faculty of Integrated Arts and Sciences, Hiroshima University, Hiroshima, Japan*

KAZUHITO YOKOYAMA (321), *Department of Biochemistry, Institute of Brain Research, Faculty of Medicine, The University of Tokyo, Tokyo, Japan*

Preface

The 13th Oji International Seminar on The Biological Functions of Microtubules and Related Structures was held at Komaba Eminence in Tokyo from November 30 through December 4, 1981. The organizing committee divided the Seminar into six sessions—the biochemistry and molecular biology of tubulin, including regulation of microtubule assembly, microtubule–dynein systems and other proteins in cell motility, microtubules and related proteins in mitosis, the interactions of cytoskeletal components, the cytoskeleton, and microtubules in membrane functions and transport.

In studies on microtubules before 1972, electron microscopic observations identified microtubules in a variety of cells and tissues. Biochemical analyses of flagellar microtubules clarified some properties of the unit protein, and studies of the colchicine-binding protein from brain tissues led to the purification and characterization of "tubulun," which was the name originally used by Mohri for the flagellar microtubule protein. Investigations of the dynamic nature of microtubules were made possible by the establishment of appropriate conditions for microtubule assembly *in vitro*. This was the beginning of the second stage of microtubule research, which focused on the mechanism of microtubule assembly and its regulation. Understanding the importance of multifunctional properties of the tubulin molecule and its role in cellular functions has become one of the central objectives of cell biology.

Early studies of the tubulin molecule were followed by studies of microtubule-associated proteins and microtubule-organizing centers, the demonstration of cytoskeletal networks using immunofluorescence techniques, and studies on the possible involvement of microtubules in many cellular processes including motility, mitosis, transport, membrane excitation, and so on. More recently, the important concept of microtubule subunit treadmilling has expanded our knowledge of microtubule dynamics.

If we assume that we are now at the end of the second stage of microtubule research, we must also recognize that many intriguing aspects of research are just beginning in such areas as the molecular genetics of tubulin and the dynamics of cytoskeletal networks (including microtubules, intermediate filaments, and microfilaments). Determination of the primary structure of tubulin is now providing new protein chemical and structural approaches for its study and the monoclonal antibody technique is providing a useful tool for further insight.

The objectives of this meeting were to consolidate our knowledge of the molecular function of tubulin in various biological processes or events and to discuss possible future developments and directions in which we see unique ideas growing. This meeting, we hope, was held at an appropriate time for promoting collaboration, and for pointing the way to the next age of microtubule research.

The Organizing Committee
of the Thirteenth Oji International Seminar
> *Hikoichi Sakai*
> *Hideo Mohri*
> *Gary G. Borisy*
> *Yoshito Kaziro*
> *Masanori Kurokawa*
> *Yukio Hiramoto*
> *March 1982*

Acknowledgments

We are very grateful to the Japan Society for the Promotion of Science and the Fujihara Foundation of Science for extending to us their cordial assistance in organizing this seminar, and especially to Mr. Y. Ichikawa, the director of the Fujihara Foundation, for his encouragement. Our thanks are also due to Mrs. Masako Takahashi, the secretary of this meeting, for her work before, during, and after the seminar.

Photographs by Hiromichi Kumagai, Shohei Maekawa, and Shinichi Hisanaga. First row: Berlin, Sakai, and Shelanski; Mohri; Bell and Wilson. Second row: Mizuno; Yamamoto; Hatano. Third row: Tsukita; Dentler; Yahara. Fourth row: Murofushi; Kurokawa; Borisy, Rosenbaum, and Nicklas.

First row: Ishikawa and Lasek; Watanabe; Kaziro. Second row: Rosenbaum; Takahashi and Sakai; Shigenaka. Third row: Tashiro; Hiramoto; Kakiuchi and Kirschner. Fourth row: Purich; Hamaguchi; Nicklas and Okuno; Kamiya.

First row: Borisy and Mabuchi; Kobayashi; Pollard. Second row: Kumagai; Shelanski, Sato, and Nicklas; Miki-Noumura. Third row: Suzaki; Matsumoto; Wilson, Kirschner, and Dentler. Fourth row: Nicklas; Isenberg and Nishida; Yokoyama.

CHAPTER 1

TUBULIN AND TUBULIN ASSOCIATED CDR-LIKE PROTEIN FROM AZUKI BEAN EPICOTYLS

Koichi Mizuno

Hiroh Shibaoka

Department of Biology
Osaka University
Toyonaka, Osaka

Microtubules play important roles in plant cell morphogenesis. In interphase cells, microtubules run close to the wall of the cell and are considered to control the direction of cellulose microfibrils in the wall (1), which directly control the direction of cell expansion (2). In unisodiametrically expanding cells, both cellulose microfibrils and microtubules near the cell wall run perpendicular to the long axis of the cell. With cells treated with colchicine or other microtubule disrupting agents, neither microtubules nor perpendicularly arranged cellulose microfibrils are observed and the cells no longer expand unisodiametrically; microfibrils in a wide spread of direction are observed and the cells expand isodiametrically.

Thus, before we try to answer the question of how plant cells control the direction of cell expansion, we have to answer the question of how plant cells control the arrangement of microtubules near the cell wall.

As the first step toward answering this question, we attempted to isolate tubulin and tubulin associated proteins from higher plant materials. In isolating tubulin from higher plant materials, use was made of our previous finding that ethyl phenylcarbamate, a kind of herbicide, disrupted microtubules in azuki bean stem cells (3). We prepared ethyl phenyl carbamate-Sepharose and carried out an affinity chromatography (4).

Biological Functions of Microtubules
and Related Structures

MATERIALS AND METHODS

Preparation of Azuki Bean Stem Extract

The epicotyl segment of azuki bean *(Vigna angularis)* seedlings grown for 5 days at 27°C in the dark were homogenized
with extraction buffer [50 mM MES, 1 mM $MgCl_2$, 1 mM GTP,
1 mM dithiothreitol, 10% glycerol, 1 mM EGTA (EGTA extraction
buffer) or $CaCl_2$ (Ca^{2+} extraction buffer)] and the homogenate
was centrifuged for 1 hr at 100,000 x g at 4°C. The supernatant was retained and used for isolation of tubulin and
tubulin associated proteins.

Preparation of Ethyl Phenylcarbamate(EPC)-Sepharose

Aminoethyl-Sepharose was prepared according to Cuatracasas
(5). 500 mg of carboxyl ethyl phenylcarbamate in ethanol was
mixed with 50 ml of aminoethyl-Sepharose and to the mixture
200 µl of water soluble carbodiimide was added and the
reaction was allowed to proceed at room temperature for 18 hr.

Gel Electrophoresis

SDS-polyacrylamide gel electrophoresis was performed
according to Laemmli (6) by using 10% and 12.5% gels.
Samples were run on gels in the presence of 0.1 mM $CaCl_2$ or
EGTA. Two dimensional electrophoresis was performed according
to O'Farrell (7). Gels were stained with Coomassie Brilliant
Blue R-250.

Colchicine Binding Assay

250 µg of purified tubulin sample was incubated with 0.5
µCi of [^{14}C]colchicine (57.08 mCi/mmol) at 30°C for 45 min,
then applied to a column of Sephadex G-50 (0.6 x 25 cm). The
column was eluted with the EGTA extraction buffer and 0.5 ml
fractions were collected. The fractions were assayed for
protein content and for radioactivity by Lowry et al's method
(8) and by the liquid scintillation method, respectively.

Assay of NAD Kinase

NAD kinase and its activator were prepared according to Anderson and Cormier (9) from the leaves of pea *(Pisum sativum)* seedlings grown for 8 days in the light at 27°C. 30 g of leaves were homogenized with 50 mM Tris-HCl (pH 8.0) containing 2% polyvinylpyrrolidone. The homogenate was filtered through 4 sheets of gauze and centrifuged at 14,000 x g for 30 min. To the supernatant solution was added solid ammonium sulfate to bring 50% saturation. The precipitate was dissolved in the homogenizing buffer. The protein solution was desalted by gel filtration with a Sephadex G-25 column and the desalted solution was applied to a DEAE-cellulose column. NAD kinase was recovered in the DEAE-cellulose non-binding fraction. The NAD kinase activator, which was absorbed onto the column, was eluted with 50 mM Tris-HCl (pH 8.0) containing 0.5 M NaCl and 0.1 mM EGTA. The eluate was heated to 90°C for 2 min and resulted precipitate was removed. The supernatant was used as the activator of NAD kinase. NAD kinase was assayed by the method of Muto and Miyachi (10).

Assay of Dithiothreitol Oxidation

Enzyme sample was prepared from leaves of light-grown 5 day old azuki bean seedlings by the same method as that employed for the preparation of NAD kinase. 200 µl of enzyme sample was mixed with 1 ml of 50 mM MES buffer (pH 6.5) containing 1 mM $MgCl_2$, 0.1 mM dithiothreitol, with or without CDR-like protein and incubated at 36°C. After 30 min incubation, the reaction was terminated by adding 1 N HCl and the remaining dithiothreitol was determined with phenazinemethosulphate and 2,6-dichlorophenolindophenol as an electron carrier and an acceptor, respectively.

RESULTS

Isolation of Azuki Bean Tubulin

Azuki bean epicotyls were homogenized with the EGTA extraction buffer and the homogenate was centrifuged at 100,000 x g. The supernatant was applied to a column of EPC-Sepharose (3.5 x 15 cm). After the column was washed with the EGTA elution buffer (the EGTA extraction buffer minus GTP), the proteins absorbed onto the column were eluted with the same buffer containing 100 mM NaCl (Fig. 1). Tubulin absorbed

onto the EPC-Sepharose column was further purified by chromatography on DEAE-Sephadex A-50 and Sephadex G-200. The behavior of purified azuki bean tubulin in gel filtration on
Sephadex G-200 column indicated that this protein had a molecular weight of about 100,000. The SDS-polyacrylamide gel of
azuki bean tubulin showed two bands having the same mobilities
as α- and β-subunits of rabbit brain tubulin. The purified
tubulin was analyzed also by two dimensional electrophoresis.
The subunits of azuki bean tubulin migrated to approximately
the same positions as those of rabbit brain tubulin. But, the
gels showed a distinct difference between azuki bean tubulin
and rabbit brain tubulin; the slow moving subunit in SDS
electrophoresis showed a more acidic isoelectric point than
that the fast moving subunit in azuki bean, whereas the slow
moving subunit showed a more basic isoelectric point than the
fast moving subunit in rabbit.

Colchicine Binding Activity of Azuki Bean Tubulin

Gel filtration of the mixture of the purified azuki bean
tubulin and [^{14}C]colchicine revealed that the former had an
ability to bind the latter. Colchicine binding activity
coincided with the protein peak which appeared in the void
volume of column. One mole of azuki bean tubulin prepared by
the procedure described above bound only 0.1 mole colchicine.
The colchicine binding activity of azuki bean tubulin was
found to be labile; about 50% of the activity was lost within
20 hr at 0°C. Dithiothreitol should be included in the buffer
used for the isolation of tubulin. If a buffer containing no
dithiothreitol was used in any one step in isolating and
purifying tubulin, tubulin showed no colchicine binding
activity.

Isolation of Calcium-Dependent Regulator (CDR)-Like Protein

The presence of the CDR-like protein in azuki bean stem
extract was easily demonstrated by analyzing the protein
absorbed onto EPC-Sepharose column by SDS-polyacrylamide gel
electrophoresis in the presence and absence of Ca^{2+}. The CDR-
like protein moved faster in the presence of Ca^{2+} than in its
absence. In the absence of Ca^{2+}, the CDR-like protein migrated
to the position where the protein having a molecular weight of
about 19,000 would be expected (Fig. 2). As calmodulins from
animal and plant sources, azuki bean CDR-like protein was heat
stable. The mobility of this protein in SDS-polyacrylamide gel
electrophoresis was not changed by heat (90°C, 5 min) treat-

ment. As shown in Fig. 2, the heat treatment effectively removed other proteins than the CDR-like protein from the crude CDR-like protein preparation. The EGTA elution buffer containing 100 mM NaCl eluted both tubulin and CDR-like protein from EPC-Sepharose column loaded with the supernatant of the homogenate prepared with the EGTA extraction buffer, but the same buffer containing 50 mM NaCl eluted CDR-like protein leaving tubulin in the column (Fig. 1 and 2), i.e. tubulin and CDR-like protein was eluted separately from the column by stepwise elution.

Interaction of CDR-Like Protein with Tubulin

The heat (90°C, 2 min) treated proteins in the peak 1 in Fig. 1 were mixed with the protein in the peak 2 in Fig. 1 and the mixture was dialyzed against the Ca^{2+} elution buffer overnight. The mixture was rechromatographed on EPC-Sepharose column equilibrated with the Ca^{2+} elution buffer. The column was eluted with the Ca^{2+} elution buffer containing 50 mM NaCl and then with the buffer containing 100 mM NaCl. CDR-like protein was not detected in the fraction eluted with 50 mM NaCl, but was detected in the fraction eluted with 100 mM NaCl in which tubulin was also detected, suggesting that CDR-like protein has an ability to interact with tubulin in the presence of Ca^{2+} (Fig. 3).

The supernatant of azuki bean stem homogenate prepared with the Ca^{2+} extraction buffer was applied to EPC-Sepharose column equilibrated with the Ca^{2+} elution buffer (Ca^{2+} extraction buffer minus GTP). The column was eluted with the Ca^{2+} elution buffer containing 50 mM NaCl and then with the same buffer containing 100 mM NaCl. In this experiment, a majority of CDR-like protein was recovered from the fraction eluted with the buffer containing 50 mM NaCl.

The yield of CDR-like protein was greater in the experiments in which the EGTA buffer was used than in that in which the Ca^{2+} buffer was used.

Inability of CDR-Like Protein to Activate NAD Kinase

As plant calmodulin has been known to activate NAD kinase, the effect of azuki bean CDR-like protein on NAD kinase was examined. CDR-like protein used in this experiment was prepared with the Ca^{2+} buffer. As Table 1 shows, azuki bean CDR-like protein did not activate pea NAD kinase, while the pea activator did activate it, suggesting that azuki bean CDR-like protein is not identical with calmodulin.

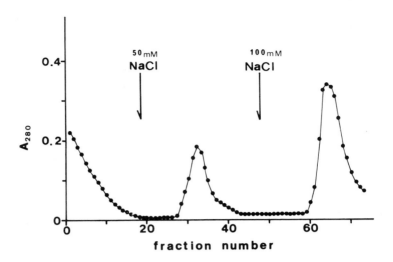

FIGURE 1. *Purification of azuki bean CDR-like protein by chromatography on EPC-Sepharose. 100,000 × g supernatant of azuki bean stem homogenate was applied to the column. The column was eluted with the EGTA elution buffer containing 50 mM NaCl and then with the same buffer containing 100 mM NaCl.*

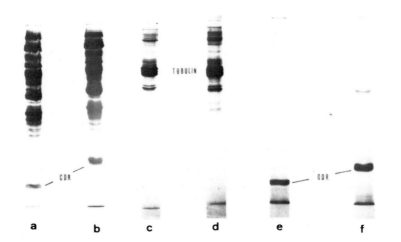

FIGURE 2. *SDS-polyacrylamide gel of purification steps of azuki bean CDR-like protein. a, peak 1 in Fig. 1 + CaCl₂ (1 mM); b, peak 1 in Fig. 1 + EGTA (1 mM); c, peak 2 in Fig. 1 + CaCl₂; d, peak 2 in Fig. 1 + EGTA; e, heat (90°C, 2 min)-treated peak 1 + CaCl₂; f, heat-treated peak 1 + EGTA.*

Activation of Dithiothreitol Oxidation by CDR-Like Protein

As the presence of Ca^{2+} in the extraction buffer seemed to accelerate the oxidation of dithiothreitol in the buffer, the effect of CDR-like protein on the oxidation of dithiothreitol was examined in the presence and absence of Ca^{2+}. As shown in Table 2, CDR-like protein accelerated the oxidation of dithiothreitol and the accelerating effect of CDR-like protein was Ca^{2+}-dependent. Trifluoperazine, a well known calmodulin inhibitor, suppressed the accelerating effect of CDR-like protein.

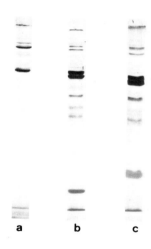

a   b   c

FIGURE 3.   Interaction of CDR-like protein with tubulin.
Protein in peak 1 in Fig. 1 was heat (90°C, 2 min)-treated
and were mixed with proteins in peak 2 in Fig. 1 and the
mixture was dialyzed against the Ca^{2+} elution buffer.
Then, the mixture was applied to EPC-Sepharose column. The
column was eluted with the Ca^{2+} elution buffer containing 50
mM NaCl and then with the same buffer containing 100 mM NaCl.
Eluates thus obtained were analyzed by SDS-polyacrylamide gel
electrophoresis. a, eluate with the buffer containing 50 mM
NaCl; b and c, eluate with the buffer containing 100 mM NaCl;
b, + $CaCl_2$ (1 mM); c, + EGTA (1 mM).

TABLE 1. *Effect of Azuki Bean CDR-Like Protein in the Activity of Pea NAD Kinase*

	NAD kinase activity (units[*])
Enzyme	0.4
" + pea activator + Ca^{2+}	20.0
" + " + EGTA	0.6
" + azuki bean CDR-like protein + Ca^{2+}	0.4

Ca^{2+}; 1 mM $CaCl_2$, EGTA; 1 mM.
*A unit of enzyme activity was defined as the 0.001 decrease of absorbance at 600 nm per 1 minute

TABLE 2. *Effect of Azuki Bean CDR-Like Protein on Oxidation of Dithiothreitol (DTT) by Azuki Bean Enzyme*

	DTT oxidation (units[*])
Enzyme + Ca^{2+}	22
" + EGTA	16
" + CDR-like protein + Ca^{2+}	137
" + " + EGTA	3
" + " + Ca^{2+} + TFP	38

Ca^{2+}; 1 mM $CaCl_2$, EGTA; 1 mM, TFP; 5 µM trifluoperazine.
* A unit was defined as the 0.005 decrease of absorbance at 600 nm. DTT oxidation activity was shown as the difference in the amount of remaining DTT between blank and test.

DISCUSSION

EPC-Sepharose affinity chromatography was demonstrated to be useful in isolating tubulin from azuki bean stem tissues. Using this method, we succeeded to isolate a protein which consisted of two subunits possessing the same mobilities as α- and β-subunits of rabbit brain tubulin on SDS-polyacrylamide gel electrophoresis and showed colchicine binding activity. Owing to the lack of reliable means of isolating tubulin from higher plants, we have scarce information about the biochemical nature of plant tubulin. EPC-Sepharose affinity chromatography will enable us to obtain such information and, thereby, will allow us to discuss about the nature and function of plant microtubules.

EPC-Sepharose chromatography was useful also in isolating
calcium-dependent regulator (CDR)- or calmodulin-like protein
from azuki bean stem tissues. A protein bearing several similarities to calmodulin was isolated from azuki bean stems by
chromatography with EPC-Sepharose.
The similarities are as follows:
(1) The azuki bean CDR-like protein possesses the same mobility on SDS-polyacrylamide gel to that of calmodulin.
(2) The mobility of the protein on SDS-polyacrylamide gel is
greater in the presence of Ca^{2+} than in its absence.
(3) The protein requires Ca^{2+} when it activates some enzymes.
(4) The protein does not show its promoting effect on enzymes
in the presence of trifluoperazine, a well known calmodulin
inhibitor.
(5) The protein is heat stable.
    Notwithstanding the above mentioned similarities, the
azuki bean CDR-like protein is not identical to plant
calmodulin, because the former does not promote pea NAD kinase
whereas the latter does.
    The difference in the yield of CDR-like protein between
the experiment using the EGTA buffer and that using the Ca^{2+}
buffer seems to indicate that a part of CDR-like proteins
absorbed onto the EPC-Sepharose column in the absence of Ca^{2+}
is not absorbed onto the column in the presence of Ca^{2+}.
In other words, this difference seems to suggest the presence
in the azuki bean extract of, at least, two kinds of CDR-like
proteins, i.e. one which passes through the column in the
presence of Ca^{2+}, but is absorbed onto the column in the
absence of Ca^{2+} and another which is absorbed onto the column
even in the presence of Ca^{2+}. Therefore, although the CDR-like
protein absorbed onto the column in the presence of Ca^{2+} did
not activate NAD kinase, there still remains the possibility
that another kind of CDR-like protein does activate the
enzyme.
    The well known fact that Ca^{2+} disrupts microtubules and
the present result that the azuki bean CDR-like protein purified by the heat treatment interacts with azuki bean tubulin
lead us to speculate that the CDR-like protein has an important role in the regulation of polymerization and depolymerization of microtubules in plant cells. But, the result that
a majority of CDR-like protein was eluted from EPC-Sepharose
column separately from tubulin even in the presence of Ca^{2+}
suggests that a majority of CDR-like protein is not interacting with tubulin in cells. To understand the role of CDR-
like protein in cells, the discrepancy between these two
results, i.e. the difference between the heat-untreated CDR-
like protein and the heat-treated protein, should be
clarified.

REFERENCES

1. Hepler, P. K. and Palevitz, B. A., *Ann. Rev. Plant Physiol.* 25, 309 (1974).
2. Preston, R. O., *in* "The Physical Biology of Plant Cell Walls" Chapman and Hall, London, (1974).
3. Shibaoka, H. and Hogetsu, T., *Bot. Mag. Tokyo 90*, 317 (1977).
4. Mizuno, K., Koyama, M. and Shibaoka, H., *J. Biochem. 89*, 329 (1981).
5. Cuatracasas, P., *J. Biol. Chem. 245*, 3059 (1970).
6. Laemmli, U. K., *Nature 227*, 680 (1970).
7. O'Farrell, P. H., *J. Biol. Chem. 250*, 4007 (1975).
8. Lowry, O. H., Rosebrough, N. J., Farr, A. L. and Randall, R. J., *J. Biol. Chem. 193*, 265 (1951).
9. Anderson, J. M. and Cormier, M. J., *Biochem. Biophys. Res. Comm. 84*, 595 (1978).
10. Muto, S. and Miyachi, S., *Plant Physiol. 59*, 55 (1977).

CHAPTER 2

PURIFICATION AND PROPERTIES OF MICROTUBULE-
ASSOCIATED ATPASES FROM BOVINE BRAIN

Shin-ichi Tominaga

Yoshito Kaziro

Institute of Medical Science
University of Tokyo
Takanawa, Minato-ku, Tokyo

I. INTRODUCTION

Microtubules are involved in various cellular functions
including chromosomal movements, axonal transport, hormonal
secretion, and motility of cilia and flagella (see Ref. 1 for
a review). These processes presumably require the expenditure
of energy which may be supplied through hydrolysis of ATP. In
fact, it is established that the motive force of cilia and
flagella depends on the presence of dynein ATPase associated
with the outer doublet microtubules (see Ref. 2 for a review).

On the other hand, little is known about the ATPase of
cytoplasmic microtubules. Although the presence of an ATPase
in brain microtubules had been reported (3,4), it was not
clear whether the ATPase is specifically associated with a
component of cytoplasmic microtubules.

In 1978, Gelfand *et al*. (5) reported that a microtubule
preparation isolated from bovine brain by cycles of assembly
and disassembly contained an ATPase activity which is not
sensitive to membrane ATPase inhibitors. About the same time,
Ihara *et al*. (6) in our laboratory found the presence of a
similar ATPase activity in the MAPs fraction of rat brain
microtubules, and furthermore, demonstrated that the ATPase
activity is markedly stimulated by the addition of purified
6S tubulin.

The virtually complete dependency of the ATPase activity
on tubulin suggested that the ATPase may be a component of

brain microtubules and may participate in axoplasmic transport. In this report, we studied the purification and properties of the ATPase to clarify its function in brain microtubules.

II. PARTIAL PURIFICATION OF MICROTUBULE-ASSOCIATED ATPASES FROM BOVINE BRAIN

A. Association of the ATPase Activity with Microtubules

Before starting purification, we examined whether this
ATPase activity is physically associated with rat brain
microtubules. In this experiment, ATPase activity was measured
using $[\gamma-^{32}P]ATP$ as a substrate, and the activity to hydrolyze 1 nmol of ATP per minute at 37°C was defined as 1 unit.
During the temperature-dependent assembly and disassembly,
ATPase activity was found to be closely associated with
tubulin (Fig. 1). There was a considerable amount of ATPase
activity in the first supernatant (Sup-1), probably due to the
presence of non-specific ATPases in the crude fraction. How-

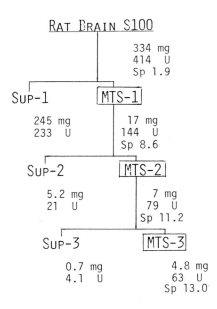

RAT BRAIN S100
334 mg
414 U
Sp 1.9

SUP-1   MTS-1
245 mg   17 mg
233 U   144 U
            Sp 8.6

SUP-2   MTS-2
5.2 mg   7 mg
21 U   79 U
            Sp 11.2

SUP-3   MTS-3
0.7 mg   4.8 mg
4.1 U   63 U
            Sp 13.0

FIGURE 1. Association of ATPase activity with rat brain
microtubules. Microtubules were purified by cycles of successive assembly and disassembly (6) and the ATPase activity of
the supernatant and precipitate fractions was measured.

ever, there was little ATPase activity in the second and third supernatants (Sup-2 and Sup-3), and the specific activity of the ATPase in the microtubules which had been reassembled twice and three times (MTS-2 and MTS-3) remained almost constant. From this result, we concluded that the ATPase was firmly associated with brain microtubules.

B. Partial Purification of Two Kinds of ATPases from Bovine Brain

Figure 2 shows the outline of the purification procedure of ATPases from bovine brain microtubules. Twice-cycled microtubules were applied to a phosphocellulose column to separate tubulin and the MAPs fraction. The fraction eluted with 0.6 M KCl was further purified on a DEAE-cellulose column. Two distinct ATPases were separated by this column chromatography.

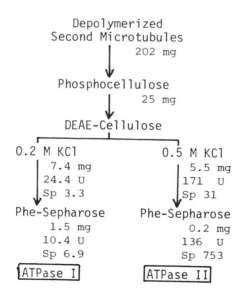

FIGURE 2. Outline of the purification of microtubule-associated ATPases from bovine brain. Total proteins, total activity, and the specific activity of two ATPases are shown. The standard assay for these ATPases contained in 50 µl; 20 mM MES-K (pH 6.5 at 20°C), 5 mM 2-mercaptoethanol, 0.4 mM magnesium acetate, 5 %(v/v) glycerol, 0.3 mM $[\gamma-^{32}P]ATP$ and 1.1 µM tubulin. Further additions were 10 mM $CaCl_2$ for ATPase I and 20 mM KCl for ATPase II, respectively. After incubation for 20 min at 37°C, liberated inorganic phosphate was determined.

One eluted with 0.2 M KCl was designated as ATPase I, and the
other eluted with 0.5 M KCl, as ATPase II. Each of them was
purified separately by phenyl-sepharose column chromatography
(data not shown). The final yield of activity was 10.4 units
for ATPase I and 136 units for ATPase II. The details of the
purification procedure will be described elsewhere (Tominaga
and Kaziro, manuscript in preparation).

III. SOME PROPERTIES OF ATPases I AND II

A. *Heat-Stability*

    The heat stability of ATPases I and II is shown in Fig. 3.
After incubation for 10 min at the indicated temperature, the
remaining ATPase activity was measured. As can be seen,
ATPase I was exceedingly unstable; its activity being inac-
tivated almost completely after incubation for 10 min at 37°C.
On the other hand, the activity of ATPase II was stable up to
50°C under the same conditions. Therefore, we first attempted
to stabilize ATPase I.

B. *Stabilization of ATPase I*

    The addition of glycerol at high concentrations was found
to be effective for stabilization of ATPase I. In the presence of 40 to 50 %(v/v) glycerol, little activity was lost
after incubation for 20 min at 33°C whereas about 50 % inactivation was observed when the glycerol concentration was
lowered to 20 %. The effect of several proteins and nucleotides to protect ATPase I against heat-inactivation was also
studied. As expected, a marked stabilization effect was observed when 0.3 mg/ml native tubulin was present. On the
other hand, chicken gizzard F-actin (0.6 mg/ml) was only partially effective, and bovine serum albumin (0.4 mg/ml) had no
effect. ATP and its non-hydrolyzable analogue, AMP-PNP, at
0.1 mM, had little effect, but 0.1 mM ADP had a marked stabilizing effect.

C. *Effect of Divalent and Monovalent Cations*

    We studied the effect of Ca^{2+} ions on the activity of
both ATPases I and II (Fig. 4). As shown in Fig. 4A, ATPase I
was markedly dependent on tubulin in the presence of 10 mM
Ca^{2+}, *i.e.*, at unphysiologically high concentrations. Other

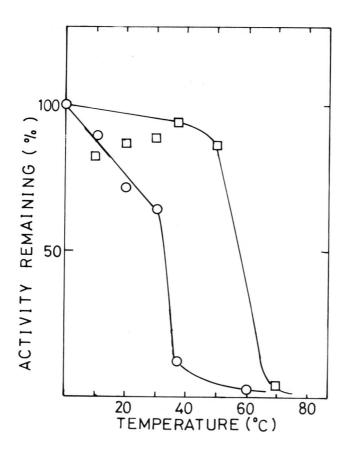

FIGURE 3.   Heat-stability of ATPases I and II. The enzyme preparations after phenyl-sepharose column chromatography were incubated for 10 min at indicated temperatures in a buffer containing 20 mM Tris-HCl (pH 7.5 at 10°C), 0.5 mM Mg^{2+}, 5 mM 2-mercaptoethanol, 10 % (v/v) glycerol. The protein concentrations were 1.4 mg/ml and 0.1 mg/ml, for ATPase I (○) and ATPase II (□), respectively.

divalent cations such as Sr^{2+}, Ba^{2+}, Zn^{2+}, Ni^{2+} and Co^{2+} could not replace Ca^{2+}. On the other hand, ATPase II, which is partially dependent on added tubulin was rather inhibited by Ca^{2+} ions.

As for the effect of KCl, ATPase I was markedly inhibited by KCl in the presence of 10 mM Ca^{2+} (Fig. 5A). ATPase II was

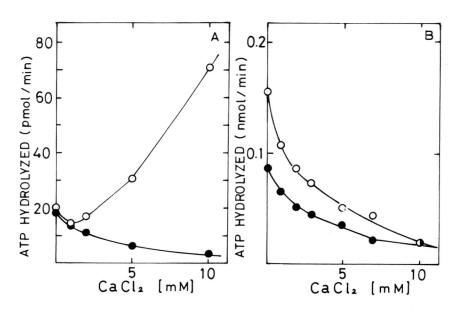

FIGURE 4. Effect of Ca^{2+} ions on ATPases I (A) and II (B). ATPase activity was assayed as described in the legend to Fig. 2 with varying concentrations of Ca^{2+} in the presence (O) or in the absence (●) of 1.1 μM tubulin.

most active in the presence of 20-40 mM KCl but was inhibited at higher concentrations of KCl (Fig. 5B).

D. Nucleotide Specificity

Nucleotide specificity of both ATPases I and II was studied using ^{3}H-labelled NTPs as substrates, and analyzing the reaction products by thin layer chromatography. As shown in TABLE I, GTP was hydrolyzed almost equally as ATP by ATPase I, while ATPase II was found to be specific to ATP.

Km values for ATP were determined as 40 μM for ATPase I and 140 μM for ATPase II.

E. Activation of ATPases I and II by the Addition of Tubulin

Both ATPases were found to be activated by the addition

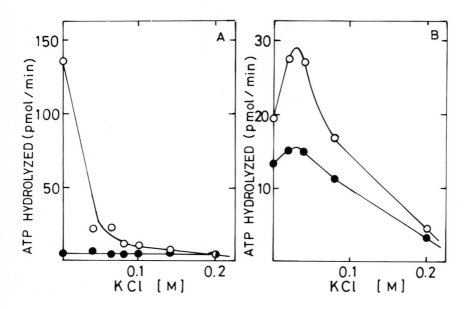

FIGURE 5.   *Effect of KCl on ATPases I (A) and II (B). ATPase activity was assayed in the presence (O) or in the absence (●) of 1.1 µM tubulin. The assay conditions were as described in Fig. 2 except for KCl concentrations as indicated.*

TABLE I.   *Nucleotide specificity of both ATPases I and II. ^{3}H-labelled NTPs (4 µM) were used as substrates and the ATPase activity was assayed in the presence of tubulin. Other conditions were the same as in Fig. 2.*

NTP	NTP hydrolyzed (pmol)
ATPase I	
ATP	4.5
GTP	4.9
CTP	2.2
UTP	2.0
ATPase II	
ATP	7.7
GTP	0.6
CTP	0.2
UTP	0.2

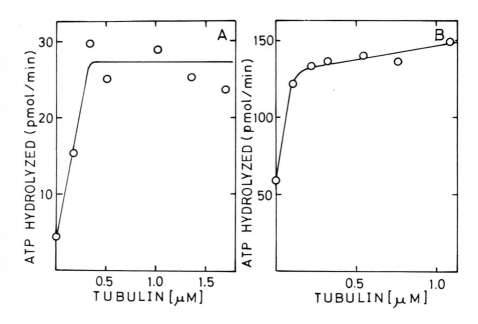

*FIGURE 6. Activation of ATPases I (A) and II (B) by
native tubulin. ATPase activity was assayed as in Fig. 2.
Tubulin concentrations were assayed according to Arai et al.
(7).*

of tubulin. As shown in Fig. 6, ATPase I was almost completely
dependent on tubulin in the presence of Ca^{2+} ions, whereas the
activity of ATPase II was stimulated by tubulin, but only partially. The stimulation of the activity of ATPase I was observed only with native tubulin; tubulin preparations denatured either by heating for 1 min at 90°C, or more mildly by
dialysis in the absence of glycerol being completely ineffective. The addition of chicken gizzard F-actin was also found
to be ineffective.

F. Inhibitors

To see the possibility whether or not these ATPases play
any function in axonal transport, we tested the effect of
vanadate which is known to be a specific inhibitor for reactions catalyzed by dynein of *Tetrahymena* cilia and sea urchin

sperm flagella (8,9). However, the results indicated that ATPase I was inhibited only weakly (Ki was about 100 μM), and ATPase II was insensitive to vanadate up to the concentration of 3×10^{-4} M.

We examined also the effect on ATPase activities of about 20 different compounds which are known to suppress neurons. They include mitotic inhibitors such as colchicine and nocodazole. However, none of them inhibited the activity of ATPase-I. On the other hand, as shown in Fig. 7, ATPase II was found to be inhibited by chlorpromazine, trifluoperazine and

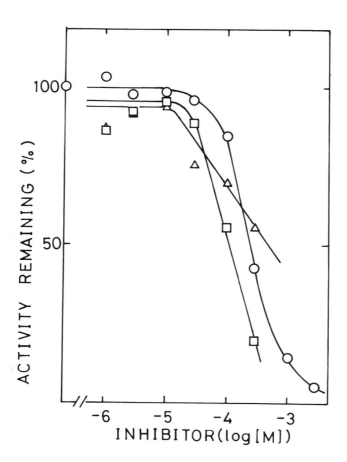

FIGURE 7.   Effect of chlorpromazine (○), trifluoperazine (□) and nicardipin (△) on ATPase II. The reaction was carried out as described in Fig. 2, and the inhibitors were added to the indicated concentrations.

nicardipin. Chlorpromazine was reported to inhibit axonal transport at a concentration of about 10^{-3}-10^{-4} M (10,11). Trifluoperazine has a similar structure as chlorpromazine and both of them are known to bind calmodulin specifically (12). The relationship between the function of ATPase II and the inhibitory effect of these compounds are not very clear, but they may be useful for investigation of the functions of the ATPases in brain microtubules *in vivo*.

IV. DISCUSSION

The difficulty encountered in purifying ATPases I and II from bovine brain was that they occur only as minor components of microtubule associated proteins. Contrary to our expectation, they did not correspond to any of the major components of MAPs, *e.g.*, MAP1, MAP2, and *tau*. Nevertheless, the fact that these two ATPases are found not only in bovine brain but also in rat brain suggests their wide distribution in mammalian brain microtubules.

Since neither of the two ATPases are purified to homogeneity, their approximate molecular sizes were estimated by glycerol density gradient centrifugation. The *s* values for ATPase I and ATPase II determined by this procedure was 5S and 27S, respectively.

The total activity of ATPase I was about one-tenth of that of ATPase II indicating that the latter represents the major ATPase activity associated in brain microtubules. On the other hand, the dependency on tubulin of ATPase activity was more pronounced in ATPase I than in ATPase II. However, the tubulin-dependent ATPase activity of ATPase I required unphysiologically high concentrations of Ca^{2+} ions (5 to 10 mM), and the addition of calmodulin from porcine brain did not lower the optimum Ca^{2+} concentration.

A large *s* value and the hydrophobic nature of ATPase II suggest the possibility that it could be a membrane-bound ATPase. In fact, when ATPase II was run on a discontinuous sucrose density gradient centrifugaion, the most of the activity was recovered in the border of 0.5 and 1.5 M sucrose after centrifugation for 19 hrs at 230,000 x g. The inhibition by chlorpromazine and trifluoperazine which are known to inhibit Ca^{2+}, Mg^{2+}-ATPase of sarcoplasmic reticulum through interaction with membrane phospholipids (13) may also point to the vesicular nature of ATPase II.

Although the localizaiton and physiological function of both ATPases I and II are remained to be studied, copurification with brain microtubules, protection of ATPase I with tubulin, and stimulation of both ATPase activities by the

native form of tubulin strongly suggest the presence of close
structural and functional association of both ATPases with
brain microtubules.

V. SUMMARY

Two distinct ATPases, ATPase I and ATPase II, were partially purified from bovine brain microtubules. ATPase I was
dependent on tubulin in the presence of 10 mM Ca^{2+} while
ATPase II was stimulated only several fold by tubulin and was
inhibited by Ca^{2+} ions. ATPase I was unstable but could be
stabilized by 50 % glycerol, 0.1 mM ADP or 0.3 mg/ml tubulin.
Km for ATP of ATPase I and ATPase II was 40 µM and 140 µM,
respectively. ATPase II was specific to ATP while ATPase I
was active with both ATP and GTP. The activity of ATPase II
was inhibited by chlorpromazine, trifluoperazine and
nicardipin. Localization and physiological function of both
ATPases are unknown.

REFERENCES

1.   Olmsted, J.B. and Borisy, G.G. *Ann. Rev. Biochem. 42,*
      507 (1973).
2.   Warner, F.D. and Mitchell, D.R. *International Review of
      Cytology 66,* 1 (1980).
3.   Gaskin, F., Kramer, S.B., Cantor, C.R., Adelstein, R.
      and Shelanski, M.L. *FEBS Lett. 40,* 281 (1974).
4.   Burns, R.C. and Pollard, T.D. *FEBS Lett. 40,* 274 (1974).
5.   Gelfand, V.I., Gyoeva, F.K., Rosenblat, V.A. and
      Shanina, N.A. *FEBS Lett. 88,* 197 (1978).
6.   Ihara, Y., Fujii, T., Arai, T., Tanaka, R. and Kaziro,
      Y. *J. Biochem. 86,* 587 (1979).
7.   Arai, T., Ihara, Y., Arai, K. and Kaziro, Y. *J. Biochem.
      77,* 647 (1975).
8.   Kobayashi, T., Martensen, T., Nath, J. and Flavin, M.
      Biochem. Biophys. Res. Comm. 81, 1313 (1978).
9.   Gibbons, I.R., Cosson, M.P., Evans, J.A., Gibbons, B.H.,
      Houck, B., Martinson, K.H., Sale, W.S. and Tang, W.-J.Y.
      Proc. Natl. Acad. Sci. USA 75, 2220 (1978).
10.   Edström, A., Hansson, H.A. and Norström, A.
      Z. Zellforsch. 143, 53 (1973).
11.   Edström, A., Hansson, H.A. and Norström, A.
      Z. Zellforsch. 143, 71 (1973).
12.   Levin, R.M. and Weiss, B. *Biochim. Biophys. Acta 540,*
      197 (1978).

13.   Hasselbach, W., *in "The Enzymes"* (Boyer, P.D., ed.)
       Vol. 10, p. 431. Academic Press, New York (1974).

CHAPTER 3

FRACTIONATION AND ELECTROPHORETIC ANALYSIS
OF MICROTUBULE PROTEINS FROM RAT BRAIN

Takaaki Kobayashi

Department of Biochemistry
The Jikei University School of Medicine
Minato-ku, Tokyo

I. INTRODUCTION

Mammalian brain microtubules consist of tubulin and other proteins---microtubule-associated protein 1 (MAP 1), MAP 2 and τ proteins (1-3). High molecular weight MAPs are found in neurons but not in glial or other non-neuronal cells (4-6), though they are contained in the mitotic spindle (7), and it seems that the variety of microtubule-associated proteins is at least partly responsible for the variety of functions of microtubules.

Here I report a rapid and systematic method for fractionation of microtubule proteins and demonstrate that MAP 1 and MAP 2 contain 2 and 3 polypeptides, respectively.

II. FRACTIONATION AND ELECTROPHORETIC ANALYSIS OF MICROTUBULE PROTEINS

A. *Fractionation of Microtubule Proteins by DEAE-Cellulose Column Chromatography*

In order to avoid proteolysis, all the experiments were done in the presence of protease inhibitors (100 μM PMSF, 10 μM TLCK, 10 μg/ml each Pepstatin and Leupeptin).

Microtubule proteins were purified from rat brain by two cycles of assembly and disassembly (8) and were then fractionated by DEAE-cellulose column chromatography (Fig. 1).

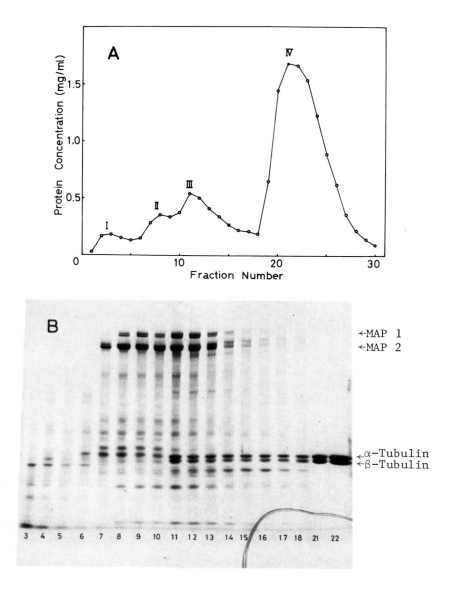

Fig. 1. Fractionation of rat brain microtubule proteins by DEAE-cellulose column chromatography. (A) The column elution profile. (B) Each fraction eluted from the column as shown in (A) was analyzed by electrophoresis with a 3-12 % polyacryl-amide gel containing 8 M urea and 0.1 % SDS. Fraction numbers are shown.

Fig. 2 shows a high resolution electrophoretic separation of the high molecular weight components of several MAP samples. There are 2 polypeptides in the MAP 1 region. In the case of MAP 2, one sample contains 3 polypeptides, another contains 2, and the others seem to contain 2 or 3, though it is unclear.

Since all of these samples were incubated at 37°C during preparation, there may have been some degradation by protease. It is, therefore, interesting to analyze samples which have been prepared without any heating. Such a sample can be easily obtained by gel filtration of a crude extract of brain, which gives a preparation of "rings"(1,10).

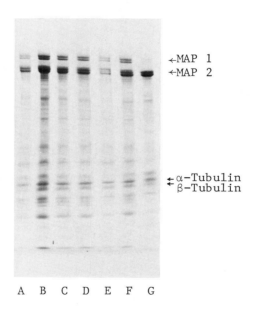

Fig. 2. SDS-urea polyacrylamide gel electrophoresis of microtubule-associated proteins. Samples (A)-(F) were prepared by DEAE-cellulose column chromatography. Sample (G) was prepared by heating microtubule proteins in boiling water and subsequent centrifugation (9).

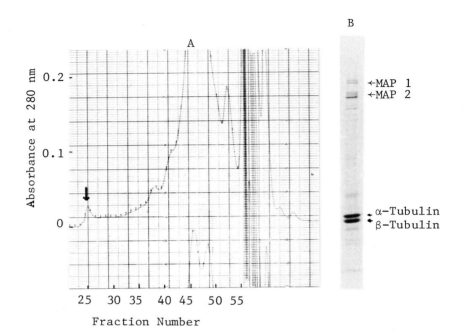

Fig. 3. Preparation and SDS-urea polyacrylamide gel electrophoresis of crude "ring" fraction. (A) Rat brain crude extract was fractionated by filtration through TSK-GEL G4000SW. (B) The flow-through fraction, indicated by arrow in (A), was analyzed by SDS-urea polyacrylamide gel electrophoresis.

B. High Molecular Weight Components of "Rings"

A crude extract of rat brain was gel-filtered through TSK-GEL G4000SW   The flow-through fraction which contained "rings" was analyzed by SDS-urea polyacrylamide gel electrophoresis (Fig. 3). It is obvious that there are 2 and 3 polypeptides in the MAP 1 and MAP 2 regions, respectively.

C. Fractionation of Microtubule-Associated Proteins by Gel Filtration

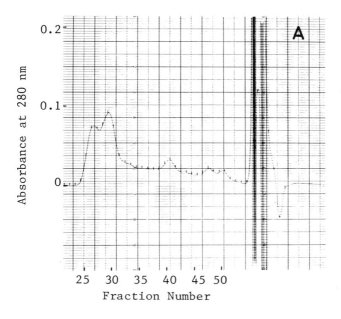

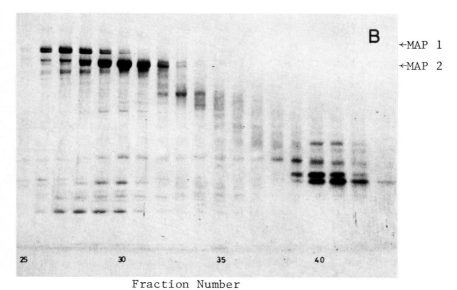

Fig. 4. Fractionation of microtubule-associated proteins by gel filtration through TSK-GEL G4000SW. (A) Elution profile. (B) SDS-urea polyacrylamide gel electrophoresis of each fraction .

   Microtubule-associated proteins were then further fractionated with TSK-GEL G4000SW into MAP 1, MAP 2, τ-factors
and other proteins (Fig. 4A). SDS-urea polyacrylamide gel
electrophoresis reveals that there are some low molecular
weight components in the high molecular weight region of the
column effluent. Some may be bound specifically to high molecular weight components, but some may be degradative products generated during and after fractionation.

   Protease inhibitors were fairly effective but never per-
fect. The best way to prepare good samples, therefore, was to
fractionate them as rapidly as possible and finish the experiment before degradation could occur.

   Fig. 5 shows the result of another experiment which is
better than that shown in Fig. 4 with respect to proteolysis.
The relatively higher content of high molecular weight components suggests less extensive degradation.

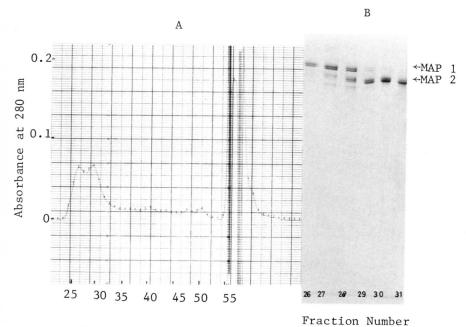

   Fig. 5. Fractionation of microtubule-associated proteins
by gel filtration through TSK-GEL G4000SW. (A) Elution pro-
file. (B) Gel electrophoretic profile.

D. Summary

Fig. 6 shows a summary of the fractionation of microtubule proteins from rat brain.

III. DISCUSSION

There have been many reports concerning the significance
of microtubule-associated proteins.

Purified MAP 2 or τ proteins can induce polymerization of
pure tubulin (11, 12), while MAP 1 is not implicated in the
polymerization (13).

Map 2 is also responsible for the interaction of microtubules with actin (14) or neurofilaments (15).

Phosphorylation of MAP 2 (16) inhibits microtubule assem-

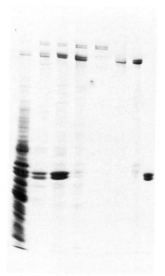

A B C D E F G H

*Fig. 6. SDS-urea polyacrylamide gel electrophoresis of
microtubule proteins. (A) Crude extract of rat brain. (B)
Crude "ring" fraction. (C) Microtubule proteins. (D) Microtubule-associated proteins. (E) Microtubule-associated protein 1. (F) Microtubule-associated protein 2. (G) Microtubule-
associated protein 2 prepared by heating. (H) Tubulin prepared
by DEAE-cellulose column chromatography.*

bly (17) and actin bundle formation (18).

Microtubule assembly is also controlled by calcium and calmodulin (19, 20), and the calmodulin-binding protein in microtubules is supposed to be tubulin (21) or τ protein (22).

The results reported here should be very useful in the investigation of the roles of microtubule-associated proteins, and the function of microtubules.

REFERENCES

1. Weingarten, M. D., Lockwood, A. H., Hwo, S. Y., & Kirschner, M. W., *Proc. Natl. Acad. Sci. U. S. A. 72*, 1858 (1975).
2. Murphy, D. G., & Borisy, G. G., *Proc. Natl. Acad. Sci. U. S. A. 72*, 2696 (1975).
3. Dentler, W. L., Granett, S., & Rosenbaum, J. L.,*J. Cell Biol. 65*, 237 (1975).
4. Connolly, J. A., & Kalnins, V. I., *Exptl. Cell Res. 127*, 341 (1980).
5. Izant, J. G., & McIntosh, J. R., *Proc. Natl. Acad. Sci. U. S. A. 77*, 4741 (1980).
6. Matus, A., Bernhardt, R., & Hugh-Jones, T., *Proc. Natl. Acad. Sci. U. S. A. 78*, 3010 (1981).
7. Sherline, P., & Schiavone, K., *J. Cell Biol. 77*, R9 (1978)
8. Shelanski, M.L., Gaskin, F., & Cantor, C.R., *Pooc. Natl. Acad. Sci. U. S. A. 70*, 765 (1973).
9. Fellous, A., Francon, J., Lennon, A. M., & Nunez, J., *Eur. J. Biochem. 78*, 167 (1977).
10. Borisy, G. G., & Olmsted, J. B., *Science 177*, 1196 (1972).
11. Cleveland, D. W., Hwo, S. Y., & Kirschner, M. W., *J. Mol. Biol. 116*, 207 (1977).
12. Herzog, W., & Weber, K., *Eur. J. Biochem. 92*, 1 (1978).
13. Villasante, A., Torre, J., Manso-Martinez, R., & Avila,J., *Eur. J. Biochem. 112*, 611 (1980).
14. Griffith, L. M., & Pollard, T. D., *J. Cell Biol. 78*, 958 (1978).
15. Runge, M. S., Lave, T. M., Yphantis, D. A., Lifsics, M. R. , Saito, A., Altin, M., Reinke, K., & Williams, R.C., *Proc. Natl. Acad. Sci. U. S. A. 78*, 1438 (1981).
16. Sloboda, R. D., Rudolph, S.A., Rosenbaum, J. L., & Greengard. P., *Proc. Natl. Acad. Sci. U. S. A. 72*,177 (1975).
17. Jameson, L., Frey, T., Zeeberg, B., Dalldorf, F., & Caplow, M., *Biochemistry 19*, 2472 (1980).
18. Nishida, E., Kuwaki, T., & Sakai, H., *J. Biochem. 90*, 575 (1981).

19. Weisenberg, R. C., *Science 177*, 1104 (1972).
20. Marcum, J. M., Dedman, J. R., Brinkley, B. R., & Means, A. R., *Proc. Natl. Acad. Sci. U. S. A. 75*, 3771 (1978).
21. Kumagai, H., & Nishida, E., *J. Biochem. 85*, 1267 (1979).
22. Sobue, K., Fujita, M., Muramoto, Y., & Kakiuchi, S., *FEBS Lett. 132*, 137 (1981).

CHAPTER 4

DETECTION OF TUBULIN GENES IN YEASTS

Masayuki Yamamoto
Junko Sakaguchi

Institute of Medical Science
University of Tokyo
Minato-ku, Tokyo

I. INTRODUCTION

There are many teleological merits in employing those
organisms which have a nerve system or a motile system like
cilia or flagella as the material to study nature of tubulin
genes and their expression. Purification of the gene product,
tubulin, and its mRNA is relatively easier in such organisms.
They exhibit a wide variety of interesting biological phenomena in which tubulin participates. In fact, molecular-genetic
analysis of tubulin genes has recently started with the aid
of molecular cloning methods in chicken (1), *Chlamydomonas* (2)
and some other eukaryotes (1,3).

We believe, however, that the study of tubulin and tubulin
genes in simpler lower eukaryotes is equally important and
benefited by definite advantages. Among various organisms that
have tubulin, only lower eukaryotes, especially yeasts, can be
handled genetically with ease. It will be possible to isolate
various kind of tubulin mutants in these organisms once a gene
for tubulin has been identified. Although the number of microtubules in yeast or fungal cells is small, their participation
in chromosome separation and in primitive cytoskeltal function
appears certain. Furthermore, it is possible that the tubulin
in these microbes conserves the most primeval characteristics
of this protein.

There are already many studies that have focused on tubulin or tubulin genes of lower eukaryotes. Tubulin was detected
(4) and purified (5) in *Saccharomyces*. Microtubules in this
yeast were visualized electron microscopically and their close
association with spindle pole bodies, or their participation
in nuclear division and bud formation was pointed out (6,7).

In *Aspergillus nidulans*, mutants resistant to benzimidazole derivatives were isolated and genetically analyzed (8).
Subsequently some of them were shown to have mutations in a
β-tubulin gene (9). Through analysis of suppressor mutations,

Biological Functions of Microtubules
and Related Structures

33

a gene for α-tubulin of this fungus was also determined (10).

In the fission yeast *Schizosaccharomyces pombe*, which will be the main subject of analysis in this short report, colcemid resistant mutants were isolated more than ten years ago (11). Biochemical or genetical analysis of these mutants, however, is yet incomplete to prove them to be tubulin mutants.

We have isolated and genetically analyzed *S. pombe* mutants resistant to benzimidazole derivatives (12), the compounds which had been successfully used to select β-tubulin mutants in *Aspergillus* (8,9). We found three loci on the chromosome of *S. pombe* that were responsible for the drug resistance. Mutations in one locus (*ben1*) made the cell more resistant than the other two (*ben2, ben3*). Precise mapping of *ben1* and *ben2* was done (12). The results of genetic analysis of benzimidazole derivative resistant mutants in *S. pombe* were very similar to those obtained in *Aspergillus*. By analogy, we infer that *ben1* may code for β-tubulin in *S. pombe*, although biochemical evidence is insufficient for the conclusion.

In this article, we will describe our recent approach to detect tubulin genes in *S. pombe*. We will show that *Chlamydomonas* α- and β-tubulin cDNAs hybridize to *S. pombe* DNA in a specific manner, thereby giving information about the copy number and the arrangement of tubulin genes in this yeast. Our observation also proves that *Chlamydomonas* tubulin cDNAs are useful as the probes to screen *S. pombe* tubulin clones. Similar analysis of tubulin genes in *Saccharomyces cerevisiae* will be also presented.

II. MATERIALS AND METHODS

Yeast DNA

DNA was prepared from *Schizosaccharomyces pombe* L972 (h⁻ auxotroph) and *Saccharomyces cerevisiae* MT13 (homothallic diploid *his4 ura3 lys1*) essentially according to Cryer et al. (13).

Chlamydomonas *tubulin cDNA*

Chlamydomonas α-tubulin cDNA clone α253 and β-tubulin cDNA clone β37 were described (2) and kindly supplied by Drs. C. Silflow and J. Rosenbaum. Plasmids were prepared from the bacteria harboring them. They were digested with a restriction endonuclease PstI which cut out tubulin cDNA segments from the vector pBR322. The digests were separated in an agarose gel. Two α-tubulin cDNA pieces together (380bp and 320bp), or one β-tubulin cDNA piece (1110bp) was eluted electrophoretically and repurified as described (14).

Hybridization

Yeast DNA was digested with several restriction endonucleases and electrophoretically run in an 1% agarose gel. After the run, DNA fragments in the gel were irradiated with UV to reduce their sizes, denatured, renatured, and transferred to a nitrocellulose filter according to Southern (15). The filter was processed for DNA-DNA hybridization as described (16). *Chlamydomonas* α- and β-tubulin cDNAs were labeled with α-^{32}P dCTP by nick-translation (17). Hybridization was carried out in 3xSSC with supplements described by Denhardt (18,16) for 20hrs at 65°C. After hybridization, the filter was washed in 6xSSC for 15min x 5 times at 60-62°C. It was then air-dried and exposed to an X-ray film (Fuji Rx-s) at -70°C with a Dupont Lightning-Plus intensifying screen.

III. RESULTS AND DISCUSSION

Schizosaccharomyces pombe DNA was digested completely with one of six different restriction endonucleases, run in an agarose gel and transfered to a nitrocellulose filter. If ^{32}P-labeled *Chlamydomonas* α-tubulin cDNA was hybridized to this filter, only one hybridization band appeared in each digestion (Fig. 1). On the other hand, if *Chlamydomonas* β-tubulin cDNA was used as the hybridization probe, two bands (BamHI, HindIII, PstI or PvuII digestion) or two main bands plus one weak band (EcoRI digestion) could be seen (Fig. 2). Two bands seemed to overlap in the case of SalI digestion.

Two control experiments were necessary to verify that those bands described above indeed reflected specific hybridization of *Chlamydomonas* tubulin cDNA to *S. pombe* DNA. Hybridization of ^{32}P-labeled pBR322 to *S. pombe* DNA was examined, since there was some contaminant pBR322 DNA in our purified cDNA fractions. Hybridization of ^{32}P-labeled poly(dG)·poly(dC) to *S. pombe* DNA was also examined, since *Chlamydomonas* tubulin cDNA clones used here were constructed using GC tailing methods (2) and purified cDNA fragments bore some GC stretch at their termini. As the results, no cross hybridization was detected between pBR322 and *S. pombe* DNA (data not shown). Under our experimental conditions, poly(dG)·poly(dC) turned out to be very sticky to a nitrocellulose filter, but it never hybridized to *S. pombe* DNA specifically (data not shown). Therefore, we conclude that those bands seen in Fig. 1 and

FIGURES 1 (LEFT) and 2 (RIGHT).

Fig. 1: *Hybridization pattern of* Chlamydomonas α-*tubulin* cDNA *to* Schizosaccharomyces pombe *DNA.* S. pombe *DNA was completely digested with one of the following restriction endonucleases and transfered to a nitrocellulose filter as described in* Materials and Methods. *About 3μg DNA was loaded in each lane. Input radioactive counts for hybridization were 4.5 x 10^6 cpm. Exposure was for 430hrs.*

Lanes 1: BamHI, *2:* EcoRI, *3:* HindIII, *4:* PstI, *5:* PvuII, *6:* SalI. *The arrow indicates the position of a 4.4kb size marker.*

Fig. 2: *Hybridization pattern of* Chlamydomonas β-*tubulin* cDNA *to* S. pombe *DNA. The experimental details are same as Fig. 1 except for the hybridization probe and the exposure span (100hrs).*

Fig. 2 are the results of specific hybridization of *Chlamydomonas* tubulin cDNA sequences to *S. pombe* DNA.

From the comparison of Fig. 1 and Fig. 2, we can tell that segments of *S. pombe* DNA which hybridize to *Chlamydomonas* α-tubulin cDNA can also hybridize to β-tubulin cDNA even more intensively. This observation is interesting since Silflow and Rosenbaum reported no cross hybridization between these two *Chlamydomonas* tubulin cDNAs (2). No cross hybridization between α- and β-tubulin cDNAs was reported in chicken, either (1). Therefore, if we may assume that the segment of *S. pombe*

DNA which hybridizes to *Chlamydomonas* α-tubulin cDNA carries
the unique α-tubulin gene in *S. pombe*, the situation will be
interpreted as follows. There are two tubulin genes on the
haplontic genome of *S. pombe*, one coding for α-tubulin and the
other coding for β-tubulin. Since both of them can hybridize
to *Chlamydomonas* β-tubulin cDNA, they are likely to share some
common nucleotide sequences. Probably they have not diverged
much from their ancestor in the genic evolution. Apparently

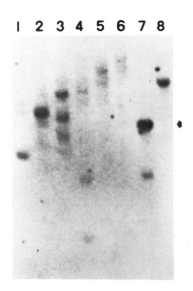
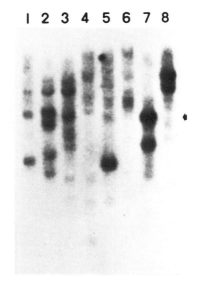

FIGURES 3 (LEFT) and 4(RIGHT).
   Fig. 3: Hybridization pattern of Chlamydomonas *α-tubulin*
cDNA to Saccharomyces serevisiae *DNA (lane 1-6) and to* S.
pombe *DNA shown for comparison (lane 7,8). The enzyme used for*
each lane is shown below. Each lane received about 3μg DNA.
Input radioactive counts for hybridization were 1.6 x 10⁷ cpm.
Exposure was for 60hrs.
   Lanes 1: BamHI, 2: EcoRI, 3: HindIII, 4: PstI, 5: PvuII,
6: SalI, 7: HindIII, 8: PvuII. Tha arrow indicates the posi-
tion of a 4.4kb size marker. This figure has better resolution
than Fig. 1 and a minor band can be seen in lane 7 which could
not be detected in Fig. 1 lane 3.
   Fig. 4: Hybridization pattern of Chlamydomonas *β-tubulin*
cDNA to S. cerevisiae *DNA (lane 1-6) and to* S. pombe *DNA (lane*
7,8). The experimental details are same as Fig. 3 except for
the hybridization probe, input radioactivity (1.4 x 10⁷ cpm)
and the exposure span (48hrs).

genes for β-tubulin in higher organisms preserve proximity to
them. Genes for α-tubulin seem to be more versatile and have
lost homology with those for β-tubulin in higher organisms.

Similar hybridization experiments were carried out with
Saccharomyces cerevisiae DNA. The hybridization patterns of
Chlamydomonas tubulin cDNAs to this DNA are shown in Fig. 3
and in Fig. 4. Again, control experiments were done to demonstrate that observed bands were specific to tubulin cDNAs
(data not shown).

The obtained results were difficult to interpret. Compared
to *S. pombe* DNA (Fig. 3 & 4, lanes 7,8), the hybridization to
S. cerevisiae DNA was weaker and multiple bands with heteroge-
neous intensity appeared in most lanes (Fig. 3 & 4, lanes 1-
6). We suspect that *Chlamydomonas* tubulin cDNAs may not have
much homology with tubulin genes in *S. cerevisiae*.

In one case, *Chlamydomonas* α-tubulin cDNA hybridized to
three bands, either of which was much larger in size than that
cDNA (Fig. 3 lane 3). It is unlikely that these bands were the
products of incomplete digestion, since much excess nucleases
were used. Therefore, *S. cerevisiae* seems to have at least two
chromosomal sites that have some homology with *Chlamydomonas*
α-tubulin cDNA.

The results of hybridization of the β-tubulin cDNA appeared too complex to draw any solid conclusion (Fig. 4). However,
one obvious thing was that most bands which hybridized to
α-cDNA were hybridizable to β-cDNA, as was the case with *S.
pombe*. So, probably α- and β-tubulin genes in *S. cerevisiae*
resemble each other to some extent.

In *Saccharomyces cerevisiae*, as was found in *Chlamydomonas*
(2), there may be two genes for α-tubulin and two genes for
β-tubulin. Alternatively, it may have only one gene for α-
tubulin, one gene for β-tubulin, and possibly a few pseudo-
genes. Before these speculations, we must make it clear whether every hybridization band observed with *S. cerevisiae* DNA
is indeed related to tubulin genes, though.

It will be striking if *S. pombe* and *S. cerevisiae*, both
being *Ascomycetes*, have a different number of genes for
tubulin. Cloning of tubulin genes from these two yeasts and
their precise comparison will provide the answer to this
point.

ACKNOWLEDGMENTS

We thank Dr. H. Uchida for critical reading of the manuscript. We also thank Drs. C. Silflow and J. Rosenbaum for
their generous supply of *Chlamydomonas* tubulin cDNA clones,
which made this investigation possible.

REFERENCES

1. Cleveland, D. W., Lopta, M. A., MacDonald, R. J., Cowan,
    N. J., Rutter, W. J., and Kirschner, M. W., *Cell 20*, 95
    (1980).
2. Silflow, C. D., and Rosenbaum, J. L., *Cell 24*, 81 (1981).
3. Sánchez, F., Natzle, J. E., Cleveland, D. W., Kirschner,
    M. W., and McCarthy, B. J., *Cell 22*, 845 (1980).
4. Baum, P., Thorner, J., and Honig, L., *Proc. Natl. Acad.
    Sci. USA 75*, 4962 (1978).
5. Kilmartin, J. V., *Biochemistry 20*, 3629 (1981).
6. Byers, B., and Goetsch, L., *J. Bacteriol. 124*, 511
    (1975).
7. Peterson, J. B., and Ris, H., *J. Cell Sci. 22*, 219
    (1976).
8. Tuyl, J. M. van, *Med. Fac. Landbouww. Rijksuniv. Gent 40*,
    691 (1975).
9. Sheir-Neiss, G., Lai, M. H., and Morris, N. R., *Cell 15*,
    639 (1978).
10. Morris, N. R., Lai, M. H., and Oakley, C. E., *Cell 16*,
    437 (1979).
11. Lederberg, S., and Stetten, G., *Science 168*, 485 (1970).
12. Yamamoto, M., *Mol. Gen. Genet. 180*, 231 (1980).
13. Cryer, D. R., Eccleshall, R., and Marmur, J., *in* "Methods
    in Cell Biology, vol. 12" (D. M. Prescott, ed.), p. 39.
    Academic Press, New York, (1975).
14. Yang, R. C.-A., Lis, J., and Wu, R., *in* "Methods in Enzy-
    mology, vol. 68" (R. Wu, ed.), p. 176. Academic Press,
    New York, (1979).
15. Southern, E. M., *J. Mol. Biol. 98*, 503 (1975).
16. Holland, M. J., Holland, J. P., and Jackson, K. A., *in*
    "Methods in Enzymology, vol. 68" (R. Wu, ed.), p. 408.
    Academic Press, New York, (1979).
17. Rigby, P. W. J., Dieckmann, M., Rhodes, C., and Berg, P.,
    J. Mol. Biol. 113, 237 (1977).
18. Denhardt, D. T., *Biochem. Biophys. Res. Commun. 23*, 641
    (1966).

CHAPTER 5

FLEXURAL RIGIDITY OF SINGLET MICROTUBULES
ESTIMATED FROM
STATISTICAL ANALYSIS OF FLUCTUATING IMAGES

Shoko Yamazaki

Department of Biology
Ochanomizu University
Tokyo

Tadakazu Maeda

Mitsubishi-Kasei Institute of Life Sciences
Tokyo

Taiko Miki-Noumura[1]

Department of Biology
Ochanomizu University
Tokyo

I. INTRODUCTION

It has recently been demonstrated that a cytoskeleton,
a framework within an intact cell, consists of F-actin
filaments, intermediate filaments and microtubules. This
framework is considered to mechanically support the shape
of a cell (1-3). Though it is worth while to study the
mechanical properties of cytoskeleton, only those of F-actin
filaments have been reported (4,5).

[1]*This study was in part supported by grants extended to
Miki-Noumura by the Japan Ministry of Education, Science
and Culture.*

Biological Functions of Microtubules
and Related Structures

Our purpose in this research has been to measure the
flexural rigidity of singlet microtubules. The principles of
the measurement are as follows: (i) Singlet microtubules in
solution can be observed under a dark-field microscope
equipped with a high-brightness light source (6). (ii) When
observed, these microtubules show incessant Brownian movement
such as translational, rotational and flexing motion. If
their images are recorded on photographs, the contour length,
L, and end-to-end distance, R, of each microtubule can be
measured. If an image of a microtubule is continuously
recorded on videotapes or cinefilms, its L and R become
measurable from frame to frame of the recorded images. (iii)
From the statistical analysis of L and R of filamentous
polymers, we can estimate their flexural rigidity (7,8).

In our previous study, a large number of microtubules
were photographed under a dark-field microscope, the values
of L and R of their images were ensemble-averaged, and the
flexural rigidity of singlet microtubules was estimated (9).
In the present study, we continuously recorded images of
microtubules on videotapes with a dark-field microscope and
a high-sensitivity TV camera. From the value of L and the
time-averaged value of R of each microtubule, we estimated
the flexural rigidity of singlet microtubules. We compared
the estimated value of the flexural rigidity in the present
study with that in the previous study. We also compared the
mechanical properties of singlet microtubules with those of
F-actin filaments.

II. MATERIALS AND METHODS

Partially purified tubulin was prepared from porcine brain
by the method of Shelanski et al (10) with some modifications.
Tubulin polymerization at 35°C and its depolymerization at 0°C
were repeated three times to purify a tubulin fraction.
Following the final centrifugation at 0°C, the supernatant
was used as the tubulin fraction in our experiment. SDS-gel
electrophoresis showed that about 85% of the protein in this
fraction was tubulin. Tubulin was polymerized to singlet
microtubules in an assembly buffer solution containing 50 mM
KCl, 0.5 mM MgSO$_4$, 1 mM EGTA, 1 mM ATP and 10 mM MES-NaOH
buffer (pH 6.8) at 25°C. The viscosity of the tubulin
solution while in the process of polymerization was repeatedly
measured with an Ostwald-type capillary viscometer. It was
confirmed with electron-microscopy that, after completion of
the increase in solution viscosity, there were singlet microtubules in the solution.

A drop of the reassembled microtubule solution was placed on a glass-slide and covered with a coverslip. The glass-slide was placed on a stage of a dark-field microscope equipped with a high-brightness light source and a high-sensitivity TV camera. Images of microtubules in the solution on the glass-slide were displayed on a TV screen, and were continuously recorded on videotapes. The focus of the microscope was continuously adjusted during observation, and a microtubule could be observed in focus along its length for 15 to 50 sec. Observation and recording were carried out in a room maintained at $25 \pm 1°C$.

Recorded images of microtubules on videotapes were copied on 35-mm film with a video-motion-analyzer (Sony SV-1100) and a motor-driven 35-mm camera (Nikon FL). The copied images were projected onto a graphic-digitizer (Kontron, West Germany). The total magnification of the microtubule images was 3000× on the digitizer. It was possible to measure image length within an accuracy of 0.1 mm on the digitizer.

The end-to-end distance, r, and contour length, l, of the filament images belonging to the same microtubule were measured from frame to frame of the 35-mm film on the digitizer. In some frames, the middle part of the filament was out of focus, with both ends in focus. In such a case, only the end-to-end distance, r, could be measured. The contour length, L, of a microtubule was assumed to be equal to the largest l value of the images divided by the total magnification. The mean-squared end-to-end distance, $<R^2>$, of the microtubule was assumed to be equal to the squared average of the end-to-end distances, r, of the images divided by the total magnification[1].

The mean-squared end-to-end distance, $<R^2>$, of a semiflexible rod is related to its contour length, L, as,

$$<R^2> = (\exp(-2\lambda L) - 1 + 2\lambda L)/(2\lambda^2) \tag{1}$$

where λ is the inverse of the statistical length of the rod (7)[2]. The parameter λ is inversely proportional to the

---

[1]*If the observation time of a microtubule is much longer than ours (15-50 sec for each microtubule), the mean-squared end-to-end distance, $<R^2>$, obtained here will be equal to the time-averaged value of square of the end-to-end distance of the microtubule.*

[2]*Qualitatively speaking, the parameter λ is a measure of rod stiffness or flexibility; that is, the more flexible the rod, the larger the parameter λ.*

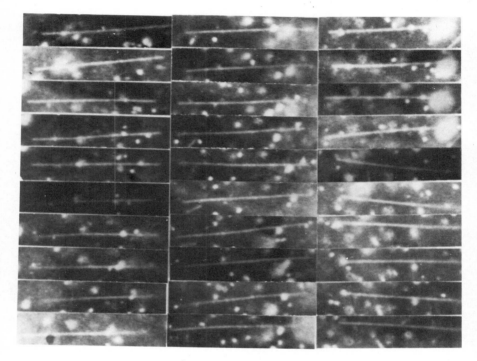

10 μm

FIGURE 1. *Recorded images of a singlet microtubule taken from a videotape in sequence. The time interval between each image is 1/3.5 sec. Magnification: 1500×.*

elastic modulus, ε, for the bending of a rod, and proportional to the thermal energy; that is,

$$\varepsilon = 3k_B T / (4\lambda) \qquad\qquad (2)$$

where k_B and T are the Boltzmann constant and the absolute temperature, respectively (8).

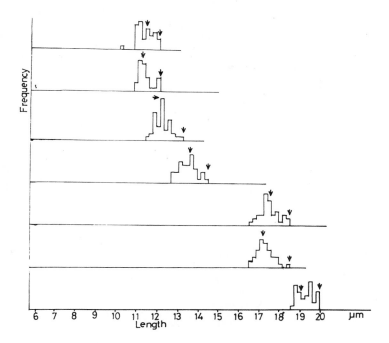

*Figure 2.   Histograms of the end-to-end distance of
seven microtubules.   The arrow on the left side of each
histogram shows the root-mean-squared end-to-end distance
of each microtubule, and the arrow on the right shows the
contour length.*

III. RESULTS AND DISCUSSION

Microtubules observed under a dark-field microscope
showed incessant Brownian motion.   Microtubules were not
straight in all cases but slightly curved from time to time.
This indicates that microtubules in solution are not rigid
but slightly flexible.   Figure 1 shows recorded images of
a microtubule taken from a videotape in sequence.   The
time-interval between each image is 1/3.5 sec.   Figure 2
shows histograms of the end-to-end distance, r, of seven
microtubules.   The arrow on the left side of each histogram
in Fig. 2 points to the root-mean-squared end-to-end
distance, $\sqrt{<R^2>}$, of each microtubule, and the other arrow on
the right points to the contour length, L.   The values of the
mean-squared end-to-end distance, $<R^2>$, and contour length,

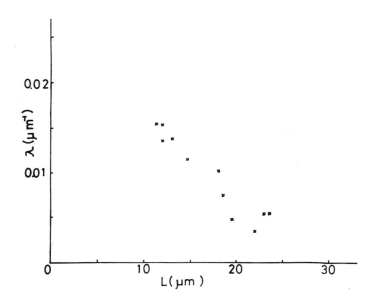

FIGURE 3. The value of the inverse of the statistical length, λ, of 11 microtubules. Ordinate: λ. Abscissa: contour length.

L, of each microtubule were substituted into eq. (1), and the most probable value of the parameter λ was estimated for each microtubule.

Figure 3 shows the estimated values of λ for 11 microtubules. Figure 3 seems at first to suggest that the value of λ depends on the contour length of a microtubule, but this should be carefully interpreted on the basis of systematic and statistical errors inevitably involved in measurement.

The error arises from two sources, primarily. One is geometrical and the other, statistical. Geometrical error results from the fact that the depth of the focus, d, of the microscope is not zero but finite. Consider a length-L rod, whose axis is so slightly tilted from the horizontal plane that the rod is in focus along its length. Its end-to-end distance measured under a microscope will thus be less than its true value (by as much as $L(1 - \cos\theta)$, where θ is the tilt angle). This geometrical error is by no means negligible and its effect is systematically larger in the case of shorter filaments. Thus the values of λ for shorter filaments in Fig. 3 should be considered as upper limits of the true value of the parameter λ.

Statistical error arises from the fact that the observation time for a filament is not much longer than the relaxation time, τ, of the flexing motion of the filament. Since τ is roughly proportional to the fourth power of L for a slightly flexible filament (11), the ratio of the observation time to the relaxation time becomes smaller for longer filaments, and the effect of statistical error becomes larger for longer filaments. The observed values of λ for longer filaments, however, will be distributed about the true value of the parameter λ, since the Brownian movements of filaments are statistically independent of each other.

The average value of λ of the 11 microtubules in Fig. 3 was computed to be $\lambda = (9.8 \pm 4.5) \times 10^{-3}$ μm^{-1}, and that of the 6 microtubules longer than 15 μm in Fig. 3 was $\lambda = (6.3 \pm 2.4) \times 10^{-3}$ μm^{-1}. The former value of λ is the upper limit of the true value of λ. In our previous study, we photographed a large number of microtubules spontaneously stuck to the under surface of a coverslip (9). From the ensemble-averaged values of L and R of the images, we estimated the value of the parameter λ to be $\lambda = (6.8 \pm 0.2) \times 10^{-3}$ μm^{-1}. Our present study on the flexibility of microtubules gave roughly the same value of the parameter λ as our previous one.

By substituting the value of λ into eq. (2), we calculated the elastic modulus for bending, ε, to be $\varepsilon = \sim 10^{-16}$ dyne cm^2. Comparing this value with that of F-actin filaments (4,5), we concluded that a singlet microtubule was more resistant toward bending than an F-actin filament.

ACKNOWLEDGMENT

The authors should like to thank Dr. S. Fujime of Mitsubishi-Kasei Institute of Life Sciences for his valuable comments on filament flexibility. The authors are indebted to Miss Kaoru Shibuya for typing the manuscript.

REFERENCES

1. Lazarides, E. and Weber, K., *Proc. Natl. Acad. Sci. 71*, 2268 (1974).
2. Weber, K., Pollack, R. and Bibring, T., *Proc. Natl. Acad. Sci. 72*, 459 (1975).
3. Lazarides, E., *Nature 283*, 249 (1980).
4. Takebayashi, T., Morita, Y. and Oosawa, F., *Biochim. Biophys. Acta 492*, 357 (1977).

5. Nagashima, H. and Asakura, S., *J. Mol. Biol. 136*, 169 (1980).
6. Miki-Noumura, T. and Kamiya, R., *Exptl. Cell Res. 97*, 451 (1976).
7. Kratky, O. and Porod, G., *Rec. Trav. Chim. 68*, 1106 (1949).
8. Harris, R.A. and Hearst, J.E., *J. Chem. Phys. 44*, 2595 (1966).
9. Mizushima, J., Maeda, T. and Miki-Noumura, T., in preparation.
10. Shelanski, M.W., Gaskin, F. and Cantor, C.R., *Proc. Natl. Acad. Sci. 70*, 765 (1973).
11. Maeda, T. and Fujime, S., *Macromolecules 14*, 809 (1981).

CHAPTER 6

EVALUATION OF POTENTIAL ROLES FOR GDP
IN MICROTUBULE ASSEMBLY AND DISASSEMBLY

David Kristofferson
Sun-Hee Lee
Brian J. Terry
Andrea Cimino Saucier
Timothy L. Karr
Daniel L. Purich

Department of Chemistry
University of California
Santa Barbara, California

I. INTRODUCTION

Microtubule assembly is generally conceptualized in terms
of the condensation polymerization model of Oosawa and Kasai
(1). In this respect, the involvement of GTP interactions and
hydrolysis in tubulin polymerization can be analyzed as effects
on the nucleation (or initiation) phase, the elongation phase,
and the conversion of the initial kinetically-controlled polymer length distribution to an equilibrium distribution. With
the observation by Margolis and Wilson (2) of the steady state
tubulin flux (also termed "treadmilling" or steady-state head-
to-tail polymerization), there is yet another demonstrable process involving GTP hydrolysis (3-5). The involvement and properties of the intrinsic GTPase reactions of tubulin during the
polymerization events have been characterized largely on the
basis of studies in Japanese and American laboratories (6-9).
On the other hand, there are still a number of important unanswered questions or incomplete notions regarding guanine nucleotide participation. For example, one may ask how the guanine nucleotide interactions alter the ability of tubulin to
self-assemble. Likewise, it remains to understand fully the
chemical mechanisms of polar growth of microtubules at steady-
state, and we lack an answer to why GTP, and not ATP, plays a
role in these assembly reactions. Finally, there needs to be

Biological Functions of Microtubules
and Related Structures

49

an effort to reconcile many of the seemingly disparate observations on GDP action in brain microtubule assembly processes.
The principal focus of this chapter is this latter question
with an aim toward analyzing GDP roles in microtubule assembly
and regulation. From what is known about the role of nucleoside diphosphate kinase in adjusting the cellular GTP/GDP ratio
relative to the fluctuating ATP/ADP ratio, one may anticipate
that GDP effects may be of significance in metabolic control of
microtubule functions.

II. RESULTS AND DISCUSSION

A. *Observations on Elongation Properties of Tubulin-GDP*

Potential for tubulin-GDP (TbGDP) elongation (see Fig. 1A)
was first noted by MacNeal and Purich (8) who observed the same
extent of microtubule (MT) assembly even when up to nearly 40%
of the tubulin (Tb) was bound to GDP. This experiment used
stoichiometric levels (1:1) of Tb and exchangeable site (E-site)
nucleotide. Further studies on TbGDP elongation (Fig. 1B) were
-----

FIGURE 1. Polymerization Properties of TbGDP.

*A, Rate and extent of polymerization of TbGDP (prepared by
the acetate kinase regenerating method) with varying amounts of
TbGDP (designated "untreated tubulin") at $37^{O}C$. MT protein conc.
is 1.7 mg/ml. For other details, see Reference 8.*

*B, Effect of GDP addition on GTP-supported assembly of su-
crose tubulin. Assay conditions: prot. (1.15 mg/ml), GTP (0.2
mM) at $30^{O}C$; GDP (2 mM) added with 2.33 mM $MgSO_4$. Similar behavior at $37^{O}C$, higher protein conc., and without added divalent
cation. For other details, see Reference 10.*

*C, Effect of GDP addition on GTP-supported assembly of so-
called Shelanski MT protein. Assembly conditions: protein (1.6
mg/ml), GTP (0.2 mM) at $30^{O}C$; GDP (2 mM) added at various times.
At $37^{O}C$, the final plateau values are closer but not identical.*

*D, GDP-induced disassembly. Each curve is the average de-
crease in turbidity for three determinations and corrected for
dilution of protein concentration by using a buffer addition
control. Protein (Shelanski preparation) is 2 mg/ml, and 2 mM
GDP is added with or without 1.6 mM $MgSO_4$.*

*E. Elongation of MT fragments with TbGDP (Shelanski prepar-
ation). Protein (1.4 mg/ml, charged with GTP (8), assembled, and
sheared with a 22 gauge needle) was combined with two volumes
Tb (1.8 mg/ml, preincubated with 1 mM GDP 40 min, 37° C). MT
length data was obtained by standard electron microscopy.*

*F. Critical concentration with and without GDP. Protein
(Shelanski preparation) was centrifuged at 190,000xg at $4^{O}C$ to
reduce GTPase. Fit with GDP excludes high concentration point.*

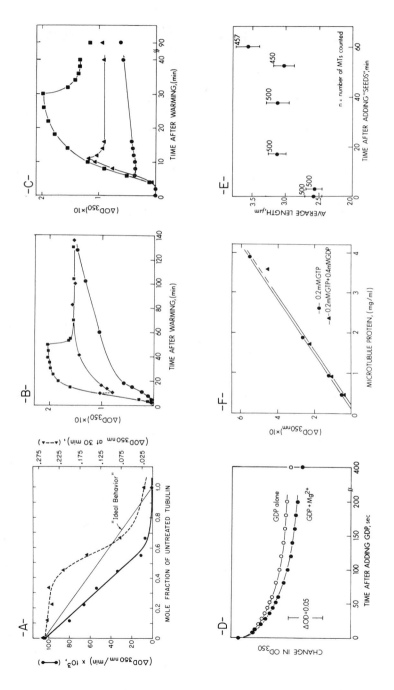

Figure 1

performed by Karr *et al.*(10). Addition of 2 mM GDP to MT protein initially polymerized with 0.2 mM GTP to different extents
of polymerization resulted in the same final assembly plateau.
At a 10:1 molar excess of GDP over GTP, most of the tubulin is
complexed with GDP; thus, they concluded that TbGDP was able to
support assembly but not MT nucleation. The lower assembly
plateau was thought to reflect a higher TbGDP critical concentration, C_C (10).

Weisenberg *et al.*(11) found altered MT behavior upon addition of GDP than that noted above. They observed that GDP,
added at different assembly extents, led to progressively higher final plateau values. The Karr *et al.*(10) experiments used
MT protein prepared by a sucrose purification method (12), and
we recently verified that the purification method (12,13) does
affect the outcome of the GDP experiments (see Fig. 1B-C).

GDP addition at the highest extent of MT assembly (Fig.1B)
results in MT depolymerization. If this effect were due to an
increase in the apparent critical concentration, one would expect to see a retarded depolymerization because release of GTP
from the Tb E-site occurs with a half-life of 2.2 min (14). GDP
addition would thereby result in MT disassembly only after an
initial lag period allowing for GTP $\rightarrow$ GDP exchange. Careful
studies of this GDP-induced depolymerization reveals no such
lag (Fig. 1D). Furthermore, GDP-induced disassembly was not a
result of Mg^{2+} sequestration; if anything, addition of both GDP
and magnesium ion slightly enhanced the extent of disassembly.

In view of the questions raised by these intriguing findings from several laboratories, we have sought to reevaluate
the earlier observations. Our reanalysis is summarized below.

While the Fig. 1A results support the hypothesis of TbGDP
elongation, the experiment in Fig. 1B still allows the possibility that elongation after GDP addition is caused by TbGTP
only. This objection would not hold for Fig. 1A because only
stoichiometric nucleotide levels are used, but in Fig. 1B there
is an initial 19-fold molar excess of GTP over Tb.

Jameson and Caplow (15) claim that TbGDP does not elongate
microtubules; instead, GDP forms an "inactive pool" in their
model. They could not observe any appreciable turbidity increase upon adding MT fragments to disassembled TbGDP. The GDP-
induced increase in C_C was also explained by the formation of a
TbGDP pool that is inactive toward elongation. Zackroff *et al.*
(16) also recently claimed that GDP "inactivates" Tb and that
the extent of inactivation depends on the level of microtubule
associated proteins (MAPs). On the other hand, MTs assembled
first with 0.1 mM GTP did not depolymerize when diluted into
unassembled Tb containing 1 mM GDP. (Turbidity plateaus were
stable for at least one hour.) Thus, they would argue that
TbGDP stabilizes MTs against depolymerization but also does not
permit elongation. Using a GTP-trap to convert GTP to GDP with

phosphofructokinase, Margolis (4) came to a similar conclusion.
Total conversion resulted in a stable turbidity at any extent
of microtubule assembly. This suggested that various metasta-
ble states could be attained but no elongation occurred. It is
interesting that Karr *et al.*(17) determined that bovine brain
MTs depolymerize end-wise with a combined off-rate constant of
113 sec^{-1}, and a similar value was obtained for porcine MT protein (18). Because MTs contain TbGDP, these rate constants
should apply to the TbGDP release from tubules. One may reason
that TbGDP stabilization must involve both addition and release
reactions for TbGDP at each end, and it is difficult to claim
that TbGDP does not add. Thus, the question is not whether it
adds; rather, one must ask more about the rate and extent of
TbGDP elongation.

B. *Several Models for Tubulin·GDP Interactions*

Basically, there are three models to consider: a, an elongation model; b, an inactive pool model; and c, a stabilization
model. These models, their implicit theoretical behavior, and
some additional experimental findings are described here.

One would expect that the elongation model is easily tested
by use of tubulin fragments and TbGDP to observe either elongation, depolymerization, or maintenance of a constant length.
Unfortunately, earlier attempts at this seemingly straight forward experiment are flawed. The Jameson and Caplow (15) work,
for example, involved a 40-50 fold lower MT number concentration than typically observed in spontaneous MT assembly, and
their failure to observe a turbidity increase is readily understood in terms of the anticipated 40-50 fold lower level of
elongating ends. Karr *et al.*(10) also used a MT fragment or
"seeded" assembly protocol, but their experiment did not eliminate the 0.11 mM GTP initially present during assembly. The
Zackroff *et al.*(16) elongation experiments with fragments and
TbGDP used electron microscopy to observe directly the changes
in MT length. Their MT fragments grew to eight times their
initial length in about 50 min, and their solution of fragments
contained 50 μM GTP which was reduced to 2.3 μM upon addition
of the TbGDP. We recently performed a similar experiment (Fig.
1E) with a final GTP level of 0.7 μM by using the acetate kinase protocol (8). The average length increase using electron
microscopic measurements was not as prominent as that found by
Zackroff *et al.*(16), but the semiquantitative study did clearly
reveal elongation.

Ideally, one would like to directly measure the incorporation of GDP into microtubules. Unfortunately, even radioactivity studies with labeled GDP can be ambiguous with the presence of endogenous nucleoside diphosphate kinase (NDPK). Here,
labeled GDP and GTP would become scrambled by transphosphoryl-

ation, and our attempts to inhibit the NDPK activity by site-directed reagents led to modification and inhibition of tubulin assembly.

The inactive pool model is strictly eliminated by the above mentioned MT fragment experiments, especially because one expects fragment disassembly upon addition of TbGDP. However, Jameson and Caplow (15) interpreted the GDP-induced increase in C_C on the basis of this model, and we have obtained similar C_C results (see Fig. 1F). Such results were interpreted to mean that GDP addition to MTs causes disassembly by forming inactive TbGDP complex. MTs will stabilize at a new plateau only when sufficient TbGTP is present to balance the on- and off-rates. By definition, this concentration of TbGTP is the normal C_C in the absence of TbGDP (designated C_C^o) because the inactive pool model assumes that TbGDP does not interfere with the TbGTP equilibria at MT ends. The observed critical concentration in the presence of GDP (designated C_C^{obs}) is given as:

$$C_C^{obs} = [TbGTP] + [TbGDP] \qquad \text{(Eqn. 1)}$$

If K_D and K_T (dissociation constants for TbGDP and TbGTP) are known, one may calculate [TbGDP] as follows:

$$[TbGDP] = (\delta/\tau)[TbGTP] \qquad \text{(Eqn. 2)}$$

where δ and τ are reduced concentrations, $[GDP]/K_D$ and $[GTP]/K_T$, respectively. Thus combining Eqns. 1 and 2 leads to:

$$C_C^{obs} = C_C^o (1 + \delta/\tau) \qquad \text{(Eqn. 3)}$$

This last equation predicts that C_C will increase by a constant amount for any GDP-GTP ratio (or, equivalently, the amount of polymer formed at any protein concentration will drop by a constant amount). Thus, a C_C plot with GDP present should be parallel to a plot without GDP. This is observed in Fig. 1F, where the least-squares fit gives slopes of 0.1475 and 0.1476 with and without GDP, respectively.

We recently determined the values of K_T and K_D at 37^oC for phosphocellulose-purified bovine brain Tb. By the Hummel-Dreyer method (19), K_D is 1.7×10^{-7} M. From the Penefsky centrifuge assay method (20), K_T/K_D is 1/2.1, giving a K_T of 8.0×10^{-8} M. Equation 3 gives a C_C^{obs} of 0.22 mg/ml using these values and the data in Fig. 1F ($C_C^o = 0.11$ mg/ml, [GDP] = 400 µM and [GTP] = 200 µM), suggesting very good agreement with the experimentally determined value.

Actually, one should not make too much of this agreement with the inactive pool model, because there are at least four sources of potential ambiguity. First, the GDP-GTP ratio will vary with protein concentration because GTP is hydrolyzed to

GDP on a stoichiometric basis relative to tubulin assembled to
MTs. This is a rather minor effect depending on the [GDP]/[Tb]
concentration ratio; in Fig. 1F, C_C increases from 0.22 to 0.23
mg/ml at the highest Tb concentration. Second, at higher nucleotide concentrations, the slope of the C_C plots decrease.
Jameson and Caplow (15) proposed the existence of another
guanine nucleotide binding site with a dissociation constant in
the mM range, and this effector site may complicate interpretation of the quantitative analysis of C_C plots. Indeed, we see
a similar effect with bovine brain tubulin: repeating the experiment in Fig. 1B (0.2 mM GTP & 2 mM GDP) at lower nucleotide
levels (0.1 & 1 mM, respectively) yields higher extents of MT
assembly. Equation 3 does not account for such behavior. [Yet,
the alternative binding site is helpful in explaining the rapid
depolymerization of MTs upon GDP addition (Fig. 1D).] Third,
the contaminating GTPase found in many MT protein preparations
may alter the GTP and GDP levels. More GTPase is present at
higher MT protein concentrations, and the abundance of the
GTPase depends on method of purification (21). This may explain
the differential behavior seen in Fig. 1B-C. Even a high speed
cold centrifugation of MT protein can reduce GTPase activity,
and this was used with Shelanski MT protein in Fig. 1F. (For
that matter, concentration-dependent changes in any MAP component may affect their ability to influence the extent of MT
polymerization.)

   Perhaps the greatest limitation in the use of the inactive
pool model is that parallel-line behavior in C_C plots and increased C_C^{obs} values with GDP may also be explained by the
elongation model. The inactive pool model provides the upper
limit on the value for C_C. Equations 1 and 2 still hold for
the elongation model, but the difference between the two models
is in the calculation of [TbGTP]. In the inactive pool model,
C_C^O is [TbGTP]; however, if TbGDP elongates, then the following
equation results from the equality of on- and off-rates at
steady state:

$$k_1 [\text{TbGTP}] + k_2 [\text{TbGDP}] = k_3 \qquad (\text{Eqn.4})$$

where k_1 and k_2 are the on-rate constants at a MT end for Tb-
GTP and TbGDP, respectively, and k_3 is the off-rate at a MT
end. We assume that TbGTP is hydrolyzed to TbGDP upon adding
to the MT end, so that a single value of k_3 applies. Using
Eqn. 2 to substitute for [TbGDP] in Eqn. 4, and solving for
[TbGTP], one obtains:

$$[\text{TbGTP}] = k_3/(k_1 + k_2 \delta/\tau) \qquad (\text{Eqn.5})$$

In the inactive pool model, k_2 is zero and [TbGTP] is k_3/k_1 (or
C_C^O). As k_2 approaches zero, [TbGTP] approaches a maximal
value of C_C^O. Thus, the C_C in the presence of GDP in the elong-

ation model is given by an equation analogous to Eqn. 3, but we use Eqn. 5 for [TbGTP] instead of C_c^o.

$$C_c^{obs} = [k_3(1 + \delta/\tau)]/[k_1 + k_2\delta/\tau] \qquad (Eqn.6)$$

If the rate of TbGDP elongation equals that of TbGTP (*i.e.*, $k_2 = k_1$), then C_o^{obs} equals C_c^o, as expected. The inactive pool model explanation for parallel C_c plots at different GDP/GTP ratios also holds true here. Equation 6, however, offers a means of measuring k_2. Solving for k_2 and using k_3/k_1 equivalent with C_c^o, we get,

$$k_2 = \frac{k_3}{(\delta/\tau)} \left(\frac{(1 + \delta/\tau)}{C_c^{obs}} - \frac{1}{C_c^o} \right) \qquad (Eqn.7)$$

We are currently refining our data in an attempt to measure k_2. The crucial parameter is the difference term, and this can be very small or negative as a result of experimental error. The value for k_3 has been determined (17) as well as the K_T and K_D values for estimating δ/τ. At this point, it is fair to conclude that distinction between the inactive pool model and the slow elongation model may be a difficult task, but the "seeded" assembly experiments alluded to above militate against a strict inactive pool model. Similarly, determining k_2 may pose great challenges on experimental precision and accuracy.

Finally, there is the TbGDP stabilization model to consider. The original experiments of Weisenberg *et al.*(11) gave results similar to those in Fig. 1C, leading them to suggest that GDP does not promote assembly but does stabilize MTs at different levels. The results presented by Karr *et al.*(10), however, were obtained with MT protein by the sucrose extraction method (12), and elongation was evidently proceeding at nearly the GTP rate. (It will be of interest to use the GTP trapping method of Margolis (4) with the sucrose MT protein to learn more about the influence of preparative methodology.) In any case, one may envision a limited polymerization model for stabilizing a MT end, but careful analysis leads one to the realization that most of these models will yield depolymerization. For example, suppose that binding of a single TbGDP protomer could block a MT end. Assuming all free Tb is TbGDP, after one reaction event (*i.e.*, gain or loss of a protomer), half of the ends are blocked. After the next event, the inactive ends (one-half of the total) will again equipartition. Half will lose a protomer; the other half will be unable to gain one (assuming elongation occurs only when the previous protomer to add hydrolyzed bound GTP). Thus, while the active ends do not contribute to a change in the average length (because they both gain and lose protomers), the inactive ends of MT polymers lead to a decrease in polymer length. As shown above, if one TbGDP blocks an end,

on the average half of the ends are blocked and one-quarter of
the ends lose one protomer without a corresponding gain in the
other one-quarter of blocked ends. Thus, one reaction event
decreases the average length by one-quarter of a protomer. In
general, if n TbGDP protomers add before blocking an end, on
the average $(1/2)^n$ of the ends are blocked, and the average MT
length decreases by $(1/2)^{n+1}$ of a protomer per event. If MTs
irreversibly depolymerize, the average length decreases by one
protomer per event for an off-rate of k_-. Thus, the depolymerization rate for a limited polymerization model is $[(1/2)^{n+1}k_-]$
by this reasoning. If only one TbGDP blocks an end, the rate
of disassembly will be one-forth the irreversible rate. So,
one must assume a high value of n to stabilize a microtubule;
for example, if n is 13, corresponding to the number of protofilaments, the rate will be $k_-/16384$. It should be evident,
however, that the seeded assembly experiments require more extensive binding of TbGDP protomers than only a single turn of
the microtubule helix.

Another type of stabilization model might involve copolymerization of TbGDP and TbGTP to a limited extent. Such models
would possibly help to define the action of MT assembly poisons
such as colchicine and podophyllotoxin, but this is uncertain.

C. Influence of Tubulin·GDP on Microtubule "Treadmilling" Flux

As noted earlier, assembled singlet microtubules are known
to demonstrate opposite-end assembly/disassembly exchange reactions (2). Maintenance of this steady-state flux requires
the expenditure of free energy, supplied in the case of MTs by
GTP hydrolysis (3-5). Interestingly, while the rates of TbGDP
elongation or copolymerization are relatively slow on the time
scale of MT assembly (10-20 min), they may be of consequence on
the time scale of the "treadmilling" flux (1-4 hours). Work in
our laboratory has demonstrated that the mole fraction of Tb
occupancy by GDP and GTP can be experimentally manipulated by
use of UDP and UTP (in large molar excess with respect to the
guanine nucleotide) and added NDPK to equilibrate these components in the following manner:

$$[GTP]/[GDP] = [UTP]/[UDP] \qquad (Eqn.8)$$

From a knowledge of K_T and K_D, one may calculate the levels of
UTP and UDP required to achieve desired mole fractions of TbGDP
and TbGTP. By using low levels of guanine nucleotides, one can
also greatly minimize the activity of the high K_m GTPase (21),
and the high UTP and UDP levels serve as a "phosphorylation
buffer" and maintain the [GTP]/[GDP] ratio.

There are several potential schemes for TbGDP influencing
the kinetics of the microtubule steady-state flux. First, the

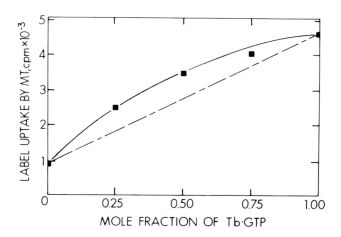

FIGURE 2. *Dependence of the microtubule "treadmilling"*
flux on the mole fraction of tubulin exchangeable-site
liganded by GTP. Acetate kinase-charged tubulin (8)
was assembled with 30 μM additional GTP in the presen-
ce of 2 IU NDPK. After a stable assembly plateau was
attained, appropriate concentrations of UTP and UDP
(1 mM total) were added and a new plateau was acheived.
Incorporation of tritiated guanine nucleotide was then
determined by rapid filtration methods after addition
of labeled nucleotide (itself incubated at an identical
UTP and UDP concentration to approximate the proper
mole fraction values plotted) and incubation for 30 min.
For the mole fractions plotted, the GTP/GDP concentra-
tion ratio was set at 0; 0.16; 0.48; 1.43; and 1.0 to
correct for the differences in GDP and GTP affinities.

TbGDP complex may form an inactive pool of protomers; then, the
rate of treadmilling would only depend on the level of TbGTP.
Second, the TbGDP could act as a competitive inhibitor of TbGTP
addition to microtubule ends; this would cause the so-called
treadmilling flux to be inhibited at low TbGTP mole fractions.
Third, there could be efficient copolymerization of both TbGTP
and TbGDP leading to the very intriguing possibility of modulation of the treadmilling flux by changes in the cellular GTP/
GDP concentration ratio; depending on the copolymer properties.
    To learn more about the potential for GDP action on the
steady-state tubulin flux, the experiment presented in Fig. 2
was carried out. From the protocol, it should be clear that
the total radioactivity of labeled GDP and GTP incorporated into microtubules is plotted on the vertical axis. While these
data suggest that there is a change in the rate of uptake of
radioactivity relative to the TbGTP mole fraction, a detailed
mechanistic interpretation must await additional experiments.

Nonetheless, there is clearly a need to consider the role of
GDP in such interactions, especially because the alternative
site interactions of GDP (15) occur at guanine nucleotide con-
centrations far above the likely physiologic levels of GDP and
GTP.   On the other hand, the GDP and GTP concentrations used in
Fig. 2 are well below the physiologic guanine nucleotide level.

III. CONCLUDING REMARKS

The exact mechanism(s) of GDP action in microtubule assembly and disassembly remain as a challenge for those interested
in the dynamic properties of microtubules. Likewise, the significance of such interactions in modulating cellular microtubule assembly and disassembly has been rather elusive. While
it would appear that the literature is full of disagreement about
the molecular action of GDP, one must consider the possibility
that there are several (or even many) guanine nucleotide interactions which depend on changes in the GTP and GDP levels. We
have already begun to delineate some of these in terms of a
nonexchangeable nucleotide site, an exchangeable nucleotide
site, and more recently a low-affinity alternative nucleotide
site. How these sites serve to achieve coordinated control of
tubule-related functions remains as a central issue, but this
report summarizes several new approaches for characterizing the
in vitro polymerization and depolymerization behavior of bovine
brain microtubule protein.

ACKNOWLEDGMENTS

The research described in this report was supported by a
grant from the National Institutes of Health. D.L.P. is also
pleased to acknowledge support from the same agency in the form
of a Research Career Development Award. We join to thank Mr.
Mel Alderman of the Globe Packing Co. for his assistance in
providing the bovine brain tissue used in our research.

REFERENCES

1. Oosawa, F. and Kasai, M. (1962) *J. Mol. Biol. 4,* 10-21.
2. Margolis, R. L. and Wilson, L. (1978) *Cell 13,* 1-8.
3. Terry, B. J. and Purich, D. L. (1980) *J. Biol. Chem. 255,*
   10532-10536.
4. Margolis, R. L. (1981) *Proc. Natl. Acad. Sci. U.S.A. 78,*

1586-1590.

5. Cote, R. H. and Borisy, G. G. (1981) *J. Mol. Biol. 150,* 577-602.

6. Kobayashi, T. (1975) *J. Biochem. (Tokyo) 77,* 1193-1197.

7. Arai, T. and Kaziro, Y. (1977) *J. Biochem. (Tokyo) 82,* 1063.

8. MacNeal, R. K. and Purich, D. L. (1978) *J. Biol. Chem. 253,* 4683-4687.

9. Andreu, J. S. and Timasheff, S. N. (1981) *Arch. Biochem. Biophys. 211,* 151-157.

10. Karr, T. L., Podrasky, A. E., and Purich, D. L. (1979) *Proc. Natl. Acad. Sci. U.S.A. 76,* 5475-5479.

11. Weisenberg, R. C., Deery, W. J., and Dickinson, P. J. (1976) *Biochemistry 15,* 4248-4254.

12. Karr, T. L., White, H. D., and Purich, D. L. (1979) *J. Biol. Chem. 254,* 6107-6111.

13. Shelanski, M. L., Gaskin, F., and Cantor, R. C. (1973) *Proc. Natl. Acad. Sci. U.S.A. 70,* 765-768.

14. Terry, B. J. and Purich, D. L. (1979) *J. Biol. Chem. 254,* 9469-9476.

15. Jameson, L. and Caplow, M. (1980) *J. Biol. Chem. 255,* 2284-2292.

16. Zackroff, R. V., Weisenberg, R. C., and Deery, W. J. (1980) *J. Mol. Biol. 139,* 641-677.

17. Karr, T. L., Kristofferson, D., and Purich, D. L. (1980) *J. Biol. Chem. 255,* 8560-8566.

18. Zeeberg, B., Reid, R., and Caplow, M. (1980) *J. Biol. Chem. 255,* 9891-9899.

19. Hummel, J. P. and Dreyer, W. J. (1962) *Biochim. Biophys. Acta 63,* 530-532.

20. Penefsky, H. S. (1977) *J. Biol. Chem. 252,* 2891-2899.

21. White, H. D., Coughlin, B. A., and Purich, D. L. (1980) *J. Biol. Chem. 255,* 486-491.

CHAPTER 7

INTERACTION OF VINBLASTINE WITH STEADY-STATE
MICROTUBULES IN VITRO: MECHANISM OF INHIBITION
OF NET TUBULIN ADDITION TO ASSEMBLY ENDS

Leslie Wilson[1]
Aileen Morse[1,2]
Mary Ann Jordan[1]

Department of Biological Sciences
University of California
Santa Barbara, California

Robert L. Margolis[3]

The Hutchinson Cancer Research Center
1124 Columbia Street
Seattle, Washington

I. INTRODUCTION

The vinca alkaloid, vinblastine, is a potent mitotic inhibitor, and appears to prevent cell growth by disruption of
microtubule function (e.g., 1-7). Though much has been learned
about the interaction of the vinca alkaloids with tubulin and
microtubules, the precise mechanism of action of this class of
drugs still remains to be elucidated. In addition to inhibiting the formation of microtubules and to destroying preformed ones, vinblastine possesses the ability to induce the for-

[1]Supported by American Cancer Society grant CD-3G and NIH
grant NS13560.
[2]Present address: Marine Science Institute, University of
California, Santa Barbara, California.
[3]Supported by NIH grant GM28189.

mation of highly birefringent uniaxial crystals in cells, composed of tubulin complexed with vinblastine (8-12). Vinblastine can induce the self-association of tubulin in vitro (13, 14), and at very high concentrations (e.g., 10^{-3}M) the drug is able to precipitate tubulin and a number of other proteins including actin and neurofilament proteins from solution (15-18). This latter activity appears to be nonspecific, and due to the strong cationic character of vinblastine and may result from the binding of the drug to large numbers of nonspecific ionic sites on the surfaces of various proteins.

It seems clear that there are two specific vinblastine binding sites per molecule of tubulin, but there has been disagreement about the magnitude of the binding affinities for the sites (4,12,19-21). Values have ranged from a high of approximately 8×10^6 liters/mole at 37° C to a low of 2.3×10^4 liters/mole at 25° C. Na and Timasheff have suggested that cooperativity between vinblastine binding to tubulin and tubulin self-association could account at least in part for the different vinblastine-tubulin binding constants which have been observed (13,14).

The vinca alkaloids are capable of inhibiting microtubule polymerization at concentrations that are considerably below the concentration of free tubulin (5-6, 22,23). This phenomenon has been called "nonstoichiometric" or "substoichiometric" poisoning. These results have indicated that vinblastine is not acting by complexing with soluble tubulin and inactivating the tubulin so that it cannot participate in the assembly reaction, but rather, at low concentrations, it may be acting by binding itself or as a complex with tubulin to the growing ends of microtubules, thereby blocking assembly.

In addition to blocking the formation of microtubules at low concentrations, vinblastine appears able to disassemble microtubules directly by causing splaying and peeling of protofilament strands at the ends of the microtubules (23-25). This activity is exerted in a higher concentration range than that which produces substoichiometric poisoning. The mechanistic relationship, if any, between the ability of vinblastine at low concentrations to block the formation of microtubules, and the ability of the drug at high concentrations to depolymerize microtubules at their ends by protofilament peeling remains to be determined.

Our approach toward understanding the mechanism of action of the vinca alkaloids has been to investigate the binding of 3H-vinblastine to steady-state bovine brain microtubules at different drug concentrations, with the purpose of correlating the affinity and stoichiometry of drug binding to the microtubules with the ability of the drug to block microtubule polymerization substoichiometrically, and with the ability of vinblastine to induce the unravelling of protofilaments at microtubule ends (23). In this report we focus on the interaction of ^{3}H-vinblastine with steady-state microtubules in the vinblastine concentration range which produces substoichiometric poisoning. Our results indicate that vinblastine prevents the net addition of tubulin to steady-state bovine brain microtubules in vitro by binding rapidly, reversibly, and with high affinity to a very limited number of molecules of tubulin at the net assembly ends of the microtubules.

II. MATERIALS AND METHODS

Bovine brain microtubules were purified by three cycles of assembly-disassembly without glycerol (26). All experiments were begun with microtubule suspensions at steady-state in 100 mM MES, 1.0 mM EGTA, 1.0 mM MgSO$_4$, 0.08 mM GTP, pH 6.75, at 30°C with or without a GTP regenerating system as described in detail elsewhere (23). Determination of the rate of net tubulin addition at the assembly ends of steady-state microtubules was accomplished by addition of ^{3}H-GTP and measurement of the rate of labeled guanine nucleotide incorporation into the microtubules for 1 hour. For assay of vinblastine binding to microtubules, sheared steady-state microtubules were incubated with ^{3}H-vinblastine for 30 minutes. Microtubules were collected for analysis of labeled nucleotide incorporation or vinblastine binding by centrifugation through a stabilizing buffer containing 50% sucrose. Stoichiometries of the quantity of vinblastine binding per microtubule were calculated on the basis of 8 nm for the longitudinal axis of a tubulin molecule in a microtubule and 13 protofilaments per microtubule (27,28). Microtubule length distributions and methods for precise quantitation of tubulin for calculating binding stoichiometries are described in detail elsewhere (23,29).

III. RESULTS

A. Inhibition of Tubulin Addition at the Assembly Ends of Steady-State Microtubules In Vitro

In steady-state conditions in vitro, net tubulin addition onto a microtubule occurs at one end of the polymer (called the assembly or A end) and net tubulin loss occurs at the opposite end (disassembly or D end) (reviewed in 30). The rate of tubulin addition at the A end of a microtubule is a measurement of the rate of flux of tubulin from one end of a steady-state microtubule to the other end. The addition of

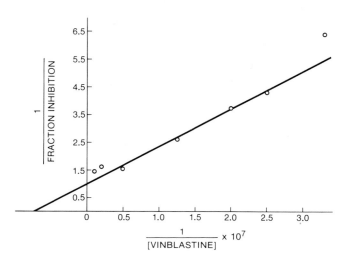

Figure 1.   Inhibition of tubulin addition to the assembly ends of steady-state microtubules by vinblastine. Aliquots of a microtubule suspension at steady-state (3.13 mg/ml total microtubule protein; mean microtubule length, 6 μm) were incubated with different concentrations of vinblastine for 10 min, then pulsed with 5 μCi ^{3}H-GTP for 1 hr in order to measure the rate of tubulin addition to microtubule assembly ends. Initial rates of ^{3}H-GTP uptake in drug-free and drug-containing samples were determined from the slopes of the lines, and the fraction inhibition (percent inhibition) was calculated.

vinblastine to steady-state bovine brain microtubules in vitro inhibits the rate of tubulin addition at the A ends (and, therefore, steady-state tubulin flux) in a concentration-dependent manner, with half-maximal inhibition of tubulin addition occurring at 1.38 x 10^{-7} M added drug (determined from the X intercept of the double-reciprocal plot of the data; Figure 1). The soluble tubulin concentration at steady-state with this preparation of microtubules is 1.8 μM (26), and the dissociation constant for the binding of vinblastine to pure 6S bovine brain tubulin is 5 x 10^{-6} M (19). When this constant is used to calculate the quantity of vinblastine actually bound to soluble tubulin under conditions of half-maximal assembly inhibition, a ratio of vinblastine-bound soluble tubulin to soluble free tubulin of 2:100 is found. These data indicate that vinblastine inhibits the addition of tubulin to the assembly ends of steady-state microtubules by interacting either by itself, or as a complex with tubulin, at the assembly ends of the microtubules (see 23).

B. Binding of ^{3}H-Vinblastine to Steady-State Microtubules

Investigation of the binding of ^{3}H-vinblastine to steady-state bovine brain microtubules has revealed the presence of at least two affinity classes of binding sites (23). A high affinity class of sites on the microtubules exhibiting very low binding stoichiometry was evident in the vinblastine concentration range between 7×10^{-7} M and $2-3 \times 10^{-6}$ M. A second class of sites having an apparent lower affinity and considerably higher capacity was also evident (data not shown). Binding in the high vinblastine concentration range occurred in association with peeling of protofilament strands at microtubule ends. Since some protofilament strands accompany the microtubules through the sucrose cushions used to isolate the microtubules after drug treatment, it has been difficult to determine whether the large numbers of sites evident at high drug concentration are due to the low affinity binding of vinblastine to tubulin along the surface of the microtubules, or to tubulin binding sites exposed on the lateral surfaces of peeled protofilament arrays.

Substoichiometric inhibition of microtubule assembly by vinblastine occurred in the vinblastine concentration range

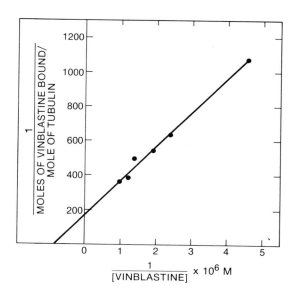

Figure 2.   Double-reciprocal plot: binding of ^{3}H-vinblastine in the high affinity concentration range to steady-state microtubules. All binding experiments were carried out at 30°C.

which titrated the high affinity binding sites, and it seems reasonable to hypothesize that the binding of vinblastine to this class of sites on the microtubules is responsible for the ability of the drug to poison microtubule assembly. The results of a typical binding stoichiometry experiment carried out in the high affinity concentration range is shown in Figure 2. The data indicate that there is a single affinity class of saturable sites with a binding affinity of 1.1×10^{-6} M, and a very small number of sites on each microtubule (5.9 molecules of vinblastine bound per 1000 molecules of tubulin in the microtubules). The stoichiometry data recalculated on the basis of mean microtubule length yield a value of 12 molecules of vinblastine per microtubule at saturating vinblastine concentration (this affinity range).

Similar binding affinities and stoichiometries were obtained in nine other similar experiments (Table I). The mean affinity for all ten experiments was 1.9×10^{-6} M, and the mean maximum stoichiometry was approximately 16-17 molecules of vinblastine bound per microtubule. The very low maximum stoichiometry obtained indicates that the high affinity sites are not distributed along the surface of the microtubules, but rather, are probably located at one or both ends of the microtubules.

A concentration of 1.38×10^{-7} M vinblastine, when added to the steady-state microtubules, inhibited the rate of net tubulin addition to the assembly ends by 50% (Figure 1). Using data from each of the vinblastine binding experiments summarized in Table I, we calculated the number of vinblastine molecules bound per microtubule at this added drug concentration for each of the ten individual experiments. A mean value for all experiments of 1.16 ($\pm$ 0.27) molecules of vinblastine per microtubule was obtained. Therefore, approximately one molecule of vinblastine is bound per microtubule when the assembly rate is inhibited by 50% (see Discussion Section).

It is important to emphasize that the steady-state microtubules used in this analysis depolymerized very slowly upon addition of a vinblastine concentration sufficient to block tubulin addition completely. Thus, adequate quantities of drug-treated polymer could be obtained for accurate determination of binding stoichiometries. Further, once bound to the microtubules, ^{3}H-vinblastine did not dissociate detectably from the microtubules during centrifugation through sucrose stabilizing buffer. Finally, the pellet of drug-treated polymer consisted only of microtubules, and did not contain any peeled protofilament arrays or other aggregated forms of microtubule protein (data not shown).

Vinblastine binding to the microtubules in the low concentration range was rapid, reaching saturation within 15 minutes. Further, the binding was rapidly and fully exchangeable.

Table I. Binding of ^{3}H-Vinblastine to Bovine Brain Microtubules at Steady-State[a]

Experiment number	Binding constant K_d (x10^{-6}M)	Mean microtubule length (μm)	Maximum molecules vinblastine bound/ microtubule	Molecules vinblastine bound/microtubule at half-maximal inhibition of assembly[b]
1	2.2	1.3	21	1.95
2	2.5	1.1	25	1.18
3	2.0	1.8	15	0.90
4	2.9	1.7	17	0.85
5	1.7	1.6	11	0.90
6	2.0	1.2	11	0.80
7	0.7	1.1	9	1.47
8	1.1	1.2	12	1.25
9	0.4	1.4	22	0.82
10	2.2	1.3	25	1.47
Mean values	1.9		16.8 ($\pm$4.3)	1.16 ($\pm$0.27)

[a]All experiments were similar to the one shown in Figure 2. Values in parentheses represent variance at the 95% confidence level (Student's t test).
[b]Calculated from the linear regression line of the binding data, using a value of 1.38 x 10^{-7} for the concentration of added vinblastine which inhibits the addition of tubulin to the assembly ends of steady-state microtubules by 50%.

This latter result contrasts with results of exchange experiments carried out with ^{3}H-colchicine bound at the net assembly ends of steady-state microtubules, in which case the rate of exchange was slow (29). The rapid reversibility of vinblastine binding to the high affinity sites makes it difficult to determine whether the labeled drug is bound at one or both ends of the microtubule.

Colchicine inhibits the assembly of steady-state microtubules by binding first to soluble tubulin, and then colchicine-tubulin complexes add to the net assembly ends of the microtubules and slow the rate of net tubulin addition (29,31). In order to investigate whether vinblastine itself or vinblastine-tubulin complexes were responsible for blocking microtubule assembly, we prepared three-layer sucrose gradients, with 20% sucrose in the top layers, 35% in the middle layers, and 50% in the bottom layers (all in stabilizing buffer). Middle layers contained ^{3}H-vinblastine at different concentrations in one set of tubes, and microtubule protein approximately at the critical concentration for assembly was included along with the labeled vinblastine in a second set of tubes. Microtubules in a steady-state suspension were layered on top of the gradients, then centrifuged through the three layers at 30°C. Upon entering the middle layers, the microtubules were exposed to labeled vinblastine, either in the presence or absence of microtubule protein. The results indicated that the microtubules bound the ^{3}H-vinblastine in the middle layers which contained no soluble microtubule protein to the same maximum extent as middle layers which contained the soluble microtubule protein (in both cases, 14 vinblastine molecules were bound per microtubule at saturating vinblastine concentration). The results suggest that vinblastine can bind directly to microtubules, presumably at their ends, without first being complexed with tubulin (discussed further in 23).

IV. DISCUSSION

The most likely location of the approximately 16-17 high affinity vinblastine binding sites on steady-state microtubules is at one or both ends of the microtubules. The rates of vinblastine exchange at these sites were too rapid to measure with the methods used in this study, so it was not possible to determine what proportion of the sites are located at the disassembly ends. At least half of the sites must be at the assembly ends (see below), so a reasonable working hypothesis is that half (approximately 8) sites are located at the assembly ends and the other half are located at the disassembly ends.

However, it is equally likely that an uneven distribution of sites exists due to possible unique tubulin geometries at the microtubule ends and the polar orientation of the tubulin molecule in the microtubule surface lattice.

The quantity of vinblastine bound per microtubule under conditions of half-maximal inhibition of assembly rate was approximately 1 molecule of vinblastine per microtubule. If an equal distribution of high affinity sites exists at the two microtubule ends, then at 50% inhibition of assembly, approximately 0.5 molecules of vinblastine would be bound at each microtubule end, or one vinblastine molecule bound at the net assembly end for every two microtubules. In this situation, the binding of a single vinblastine molecule to the net assembly end of a microtubule would be able to reduce the rate of tubulin addition to near zero. If all of the high affinity binding sites were located at the net assembly ends, then the binding of only two vinblastine molecules per microtubule would be sufficient to block assembly.

We believe that the binding of vinblastine to the microtubules we have titrated in the high affinity range does not reflect any copolymer formation between vinblastine or vinblastine-tubulin complexes and free tubulin, since the bound vinblastine is rapidly and completely exchangeable. If vinblastine were able to form copolymers, much higher stoichiometries would be observed.

Colchicine inhibits the rate of tubulin loss at the assembly ends of steady-state microtubules, as well as the gain, thus producing a kinetic cap at this microtubule end (29). We have also investigated whether vinblastine can induce an assembly end kinetic cap. We found that microtubules pulse-labeled at their assembly ends with ^{3}H-guanine nucleotide and treated with assembly-inhibiting concentrations of vinblastine lost their assembly end label (and, therefore, lost tubulin from the assembly ends) at a rate that was indistinguishable from that of non-drug-treated microtubules upon 3-fold dilution of the microtubule suspension (data not shown). We conclude that if vinblastine is able to induce a kinetic cap, the cap is substantially weaker than the one induced by colchicine.

The observation that vinblastine can bind rapidly and directly to the high affinity sites on the microtubules may be highly significant. The difficulty in understanding the mechanism of substoichiometric poisoning of microtubule assembly by colchicine arose because of evidence which indicated that the drug bound first to soluble tubulin, and it was the complex which most effectively blocked assembly. Since the ratio of free tubulin to colchicine-tubulin complex was high (e.g., 40:1, 31), it was difficult to imagine how the drug would work other than by kinetically slowing the rate of tubulin loss as well as the rate of gain at the microtubule assembly end. An assembly end kinetic cap is not required with vinblastine, if it interacts directly with the assembly ends of the microtubules as the data suggests.

It is likely that the mechanism of inhibition of microtubule assembly by vinblastine is simpler than that of colchicine. The kinetically rapid binding of vinblastine directly to a limited number of high affinity binding sites at the net assembly end of a microtubule could effectively block further tubulin addition regardless of the ratio of free tubulin to tubulin-vinblastine complex, so long as sufficient free drug were available to ensure that the binding sites at the microtubule assembly ends were constantly occupied by drug.

If the above analysis is correct, then one must ask how only one or two vinblastine molecules bound to the growing end of a microtubule prevent(s) further polymerization. One possibility is that polymerization occurs normally through a helical, step-wise addition of tubulin, with the addition of an incoming tubulin molecule at each growth (start) point dependent upon the proper three-dimensional placement of the previous tubulin molecule. Vinblastine is a large molecule (molecular weight of the free base, 814), and the binding of vinblastine to tubulin at just one or two growth points must be capable of producing sufficient misalignment of incoming tubulin molecules by steric hindrance to prevent further assembly of the polymer.

ACKNOWLEDGMENTS

We wish to thank the Eli Lilly and Company, Indianapolis, Indiana, for their generous gifts of vinblastine. We thank Dr. Richard Himes and Mr. Vincent Lee for valuable suggestions and discussions.

REFERENCES

1. George, P., Journey, L.J., and Goldstein, M.N., J. Nat. Cancer Inst. 35:355 (1965).
2. Wilson, L., and Friedkin, M., Biochemistry 6:3126 (1967).
3. Bensch, K.G., and Malawista, S.E., Nature (London) 218: 1176 (1968).

4. Owellen, R.J., Owens, A.H., and Donigian, D.W., Biochem. Biophys. Res. Commun. 47: 685 (1972).
5. Wilson, L., Anderson, K., and Chin, D., In "Cold Spring Harbor Conferences on Cell Proliferation" (R. Goldman, T. Pollard, and J. Rosenbaum, eds.) vol. 3, p. 1051, Cold Spring Harbor Laboratory, New York (1976).
6. Himes, R.H., Kersey, R.N., Heller-Bettinger, I., and Samson, F.E., Cancer Res. 36: 3798 (1976).
7. Gerzon, K., In "Anticancer Agents Based On Natural Product Models" (J. Douros and M. Cassidy, eds.), p. 271. Academic Press, New York (1980).
8. Schochet, S.S., Jr., Lambert, P.W., and Earle, K.M., J. Neuropathol. Exp. Neurol. 27:645 (1968).
9. Bensch, K.G., and Malawista, S.E., J. Cell Biol., 40: 95 (1969).
10. Strahs, K.R., and Sato, H., Exp. Cell Res. 80:10 (1973).
11. Fugiwara, K., and Tilney, L.G., Ann. N.Y. Acad. Sci. 253: 27 (1975).
12. Wilson, L., Morse, A., and Bryan, J., J. Mol. Biol. 121: 255 (1978).
13. Na, G.C., and Timasheff, S.N., Biochemistry 19:1347 (1980).
14. Na, G.C., and Timasheff, S.N., Biochemistry 19:1355 (1980).
15. Wilson, L., Bryan, J., Ruby, A., and Mazia, D., Proc. Nat. Acad. Sci. (USA) 66:807 (1970).
16. Olmsted, J.B., Carlson, K., Klebe, R., Ruddle, F., and Rosenbaum, J., Proc. Nat. Acad. Sci. (USA) 65:129 (1970).
17. Marantz, R., Ventilla, M., and Shelanski, M., Science 165:498 (1969).
18. Mori, H., and Kurokawa, M., Cell Structure and Function 4:163 (1979)
19. Wilson, L., Creswell, K.M., and Chin, D., Biochemistry 14: 5586 (1975).
20. Lee, J.C., Harrison, D., and Timasheff, S.N., J. Biol. Chem. 250:9276 (1975).
21. Bhattacharyya, B., and Wolff, J., Proc. Nat. Acad. Sci. (USA) 73: 2375 (1976).
22. Owellen, R.J., Hartke, C.A., Dickerson, R.M., and Hains, F.O., Cancer Res. 36:1499 (1976).
23. Wilson, L., Jordan, M.A., Morse, A., and Margolis, R.L., J. Mol. Biol. (in press).
24. Pierson, G.B., Burton, P.R., and Himes, R.H., J. Cell Biol. 76:223 (1978).
25. Warfield, R.K.N., and Bouck, G.B., Science 186:1219 (1974).
26. Asnes, C.F., and Wilson, L., Anal. Biochem. 98:64 (1979).

27. Amos, L.A., and Klug, A., J. Cell Sci., 14:523 (1974).
28. Tilney, L.G., Bryan, J., Bush, D., Fujiwara, K., Moos-
     eker, M.S., Murphy, D.B., and Snyder, D.H., J. Cell Biol.
     59:267 (1973).
29. Margolis, R.L., Rauch, C.T., and Wilson, L., Biochemistry
     19: 5550 (1980).
30. Margolis, R.L., and Wilson, L., Nature (London):293
     (1981).
31. Margolis, R.L., and Wilson, L., Proc. Nat. Acad. Sci.
     (USA) 74: 3466 (1977).

CHAPTER 8

QUANTITATIVE ANALYSIS OF ASSOCIATION
OF
CALMODULIN WITH TUBULIN

Hiromichi Kumagai
Eisuke Nishida
Hikoichi Sakai

Department of Biophysics and Biochemistry
Faculty of Science
University of Tokyo
Tokyo

INTRODUCTION

Since microtubule (MT) reconstitution *in vitro* was described using brain extracts (1), many works on the regulatory mechanism of MT assembly have been carried out. As one of the regulators of MT assembly, Ca^{2+} ion has been known to be an inhibitor of MT assembly (2,3). Nishida and Sakai (4) previously suggested the presence of Ca^{2+}-sensitizing factors of MT assembly in brain extracts. Recently Nishida *et al.* (5) and Marcum *et al.* (6) demonstrated that calmodulin, which has been known as a Ca^{2+}-dependent activator of cyclic nucleotide phosphodiesterase (7,8), modulates the extent of MT assembly in a Ca^{2+}-dependent fashion. We also found that calmodulin binds to tubulin in the presence of Ca^{2+} (9-11). This paper shows the evidence on the basis of the quantitative analysis of the binding between calmodulin and tubulin.

[1]*This study was supported by a grant-in-aid for scientific research from the Ministry of Education, Science and Culture, Japan (No.444074).*

I. QUALITATIVE EVIDENCE OF THE BINDING BETWEEN CALMODULIN
AND TUBULIN

   The binding of porcine brain calmodulin to tubulin has
been demonstrated using gel filtration and affinity chromatography (10,11).   A mixture of calmodulin and MT proteins
was subjected to a gel filtration column chromatography (Toyo
Soda high speed liquid chromatography, HLC803, 3000SW column)
in the presence or absence of Ca^{2+} ions (Fig. 1).
Calmodulin activity was found only in free calmodulin fraction in the absence of Ca^{2+} (Fig. 1A).   In contrast, a part
of calmodulin activity was found to shift into tubulin fraction only in the presence of Ca^{2+} (Fig. 1B).   This result
suggested that calmodulin associates with tubulin in a Ca^{2+}-
dependent manner.   It was also found that calmodulin was
adsorbed to tubulin-Sepharose 4B affinity column in the presence of Ca^{2+} and the adsorbed calmodulin was eluted by the
addition of EGTA (data not shown).
   We show here the effect of calmodulin on vinblastine
binding to tubulin as one of the evidences of their interaction.   Vinblastine binding was studied by measuring the

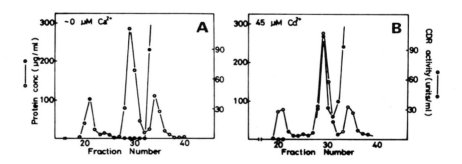

*Figure 1.   Binding of calmodulin to tubulin in the presence
of Ca^{2+}.*
*(A) A mixture of 2.7 mg/ml MT proteins and 0.8 mg/ml calmod-
ulin in a buffer solution (10 mM MES, pH 6.8, 0.5 mM $MgCl_2$,
200 mM NaCl, 1.2 mM EGTA) was subjected to a gel filtration
using a high speed liquid chromatography system (Toyo Soda
HLC803, double 3000SW columns in series).*
*(B) Same as A except that both the sample and the column
buffer contained 1.2 mM $CaCl_2$ (45 μM free Ca^{2+}).*
*Protein (O) was determined by the method of Lowry et al.
(16).   Calmodulin activity (●) was assayed based on activation of cAMP phosphodiesterase (10).*

intrinsic fluorescence of tubulin (Fig. 2).   When tubulin
solution was excited at 305 nm, the emission spectrum caused
by tryptophan residues gave a maximum at 337 nm.   Addition
of 9 μM vinblastine to the tubulin solution (0.4 mg/ml, 10 mM
MES, pH 6.8, 0.5 mM $MgCl_2$, 50 mM KCl and 1.2 mM EGTA) caused
the fluorescence intensity to decrease either in the presence
or the absence of Ca^{2+} (Fig. 2A).   To observe the effect of
calmodulin in the intrinsic fluorescence of tubulin, a mixture
of calmodulin (0.18 mg/ml) and tubulin (0.4 mg/ml) was excited
in the presence or absence of Ca^{2+} (Fig. 2B).   As already
reported by several workers, calmodulin does not contain
tryptophan residues (12).   Therefore, one can observe the
emission deriving only from tubulin by the excitation at 305
nm.   Both in the presence and absence of Ca^{2+}, we obtained
the same emission spectrum of a mixture of calmodulin and
tubulin.   However, the addition of vinblastine had different
effects on the emission spectrum depending on the presence or
absence of Ca^{2+}.   In the absence of Ca^{2+}, vinblastine caused
a decrease in fluorescence intensity.   In contrast, in

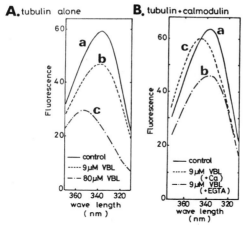

Figure 2.   Effects of Ca^{2+}-calmodulin on vinblastine binding
to tubulin.
(A) Emission spectrum of tubulin (0.4 mg/ml) in a medium containing 20 mM MES, pH 6.8, 0.5 mM $MgCl_2$, 50 mM KCl and 1.2 mM
EGTA or 0.09 mM $CaCl_2$ (a).   (b) or (c), same as (a) except
that 9 μM vinblastine (b) or 80 μM vinblastine (c) was contained. Ca^{2+} ions did not affect the spectrum.
(B) Emission spectrum of a mixture of tubulin and calmodulin
(0.18 mg/ml)(a).   The same spectrum was obtained either in
the presence or absence of Ca^{2+}.   (b), plus 9 μM vinblastine
in the absence of Ca^{2+}.   (c), plus 9μM vinblastine in the
presence of Ca^{2+}

the presence of Ca^{2+}, vinblastine induced a shift of the wave length giving a maximum emission from 337 nm to 350 nm, which is completely reversible depending on the Ca^{2+} concentration. These results also indicate that calmodulin associates with tubulin in a Ca^{2+}-dependent manner. Furthermore, confirmation of the binding was made by sedimentation analysis of the mixture of tubulin and calmodulin (data not shown).

II. QUANTITATIVE ANALYSIS OF THE BINDING BETWEEN CALMODULIN AND TUBULIN

   To analyze quantitatively the binding between calmodulin and tubulin, we carried out two series of analytical experiments. One is the frontal analysis using tubulin-Sepharose 4B. The other is the equilibrium gel filtration analysis developed by Hummel and Dreyer (13).

A. *Frontal Analysis*

   This method was originally developed by Kasai and Ishii (14) for the analysis of association of trypsin with trypsin inhibitors. The procedure of frontal analysis using tubulin-Sepharose 4B has been described elsewhere (11). Tubulin purified from porcine brain MT proteins by phosphocellulose column chromatography is first coupled to cyanogen bromide-activated Sepharose 4B. A known quantity of calmodulin, $(calmodulin)_o$, is applied continuously to tubulin-Sepharose 4B. Under the condition in which the interaction between calmodulin and tubulin is totally suppressed, calmodulin is eluted in a fraction designated as V_o, which represents the flow through volume of the column. In contrast, under the condition in which the interaction occurs, calmodulin is eluted after a delay, the elution volume of which is designated as V. This delay represents the amount of calmodulin adsorbed to the column. At equilibrium, free calmodulin concentration is $(calmodulin)_o$ and the amount of calmodulin adsorbed to the column is $(V-V_o)(calmodulin)_o$. Then the dissociation constant, K_d, can be calculated from the following equation:

$$K_d = (tubulin)_t/(V-V_o) - (calmodulin)_o,$$

where $(tubulin)_t$ represents the total amount of tubulin coupled to Sepharose 4B. Reforming the equation, we obtained

$$1/(calmodulin)_o(V-V_o) = K_d/(tubulin)_t(calmodulin)_o$$
$$+ 1/(tubulin)_t.$$

By the plot of $1/(calmodulin)_o(V-V_o)$ versus $1/(calmodulin)_o$, K_d value can be calculated from the intercept on the abscissa.   Figure 3 shows a linear relationship between them under the medium condition containing 55 mM KCl at 4°C, revealing a dissociation constant of 2.1 μM for the binding reaction between calmodulin and tubulin.

B. *Equilibrium Gel Filtration Analysis*

A Sephadex G-100 column was chosen for the equilibrium gel filtration analysis, which was developed by Hummel and Dreyer (13).   A mixture of calmodulin and tubulin was applied to this column pre-equilibrated with a buffer solution containing about 30 μg/ml calmodulin, followed by measurements of the amount of protein and calmodulin activity (Fig. 4). The appearance of a peak and a trough in the elution profile of calmodulin activity demonstrated that calmodulin binds to tubulin in the presence of Ca^{2+}.   In contrast, neither a peak nor a trough appeared in the absence of Ca^{2+}. This result strongly supports the idea that calmodulin binds to tubulin only in the presence of Ca^{2+}.   We determined the

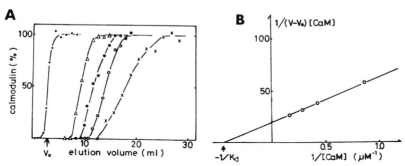

Figure 3.   *Frontal analysis of calmodulin on tubulin-Sepharose 4B.*
(A) Calmodulin was applied continuously to a tubulin-Sepharose 4B column (1 x 2.5 cm) in the absence of Ca^{2+} to measure the value $V_o(\bullet)$, which was determined to be 2.8 ml.   To measure the value V, 100(△), 60(◍), 40(○) or 20(✗) μg/ml calmodulin was loaded on the column in the presence of Ca^{2+}. Temperature was 4°C.   Buffer solution contained 10 mM MES, pH 6.8, 0.5 mM $MgCl_2$, 55 mM KCl and 0.5 mM $CaCl_2$.
(B) On the basis of the equation described in the text, $1/(calmodulin)_o(V-V_o)$ was plotted against $1/(calmodulin)_o$.

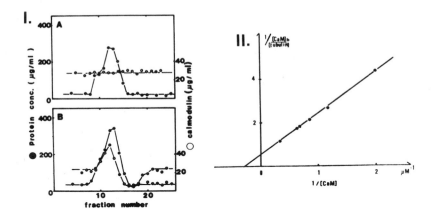

Figure 4.   *Equilibrium gel filtration column chromatography*
*   on Sephadex G-100.*
(I) A mixture of calmodulin (30µg/ml) and tubulin (3.6 mg/
ml) was loaded on a Sephadex G-100 column (0.6 x 42 cm) pre-
equilibrated with a medium (10 mM MES, pH 6.8, 0.5 mM MgCl₂
200 mM NaCl) containing the same concentration of calmodulin
in the absence (A) or presence (B) of Ca²⁺.
(II) The integrated amount of calmodulin in the peak or
trough areas was plotted against the reciprocal of the re-
spective free calmodulin concentration at which the column
was pre-equilibrated.

amount of calmodulin bound to tubulin by quantifying the peak
or the trough areas under conditions in which various amounts
of calmodulin were applied to the column together with a constant amount of tubulin.   By plotting the amount of calmodulin-tubulin complex obtained from the trough or the peak areas
of calmodulin activity against free calmodulin concentration,
we could calculate the K_d value from the intercept on the
abscissa.   This analysis also gave the stoichiometry of the
binding from the intercept on the ordinate.   Figure 4-II shows
that two moles calmodulin bind to one mole tubulin with a
dissociation constant of 3.5 µM under the medium condition
containing 200 mM NaCl at 4°C.
     We also determined the apparent molecular weight of cal-
modulin-tubulin complex by equilibrium gel filtration using
Sephadex G-200 column.   The measurement resulted in an apparent molecular weight of 150-180,000 for calmodulin-tubulin
complex as compared with 110,000 for tubulin.
     Table I shows the summary of the quantitative analysis of

Table I.   Summarized quantitative data on the binding
between calmodulin and tubulin.

(1) Hummel-Dreyer method

condition	K_d	stoichiometry (calmodulin/tubulin)
4°C, 200 mM NaCl	3.5 µM	1.9

(2) Frontal analysis

(i) Ionic strength (4°C)

KCl concentration	K_d
200 mM	4.0 µM
55 mM	2.1 µM

(ii) Temperature (55 mM KCl)

Temperature	K_d
4°C	2.1 µM
26°C	3.2 µM
35°C	5.0 µM

(1) The dissociation constant and stoichiometry of the bind-
ing between calmodulin and tubulin were obtained by the meth-
od described in the text.
(2) Effects of temperature and KCl concentration on the dis-
sociation constant were determined by the same method as
described in Figure 3 using tubulin-Sepharose 4B.

the binding between calmodulin and tubulin, in which the
effects of temperature and ionic strength on the binding are
also given as values measured by frontal analysis.

III. CALMODULIN-INDUCED INHIBITION OF TUBULIN POLYMERIZATION

Quantitative demonstration of the binding between calmodulin and tubulin suggests that calmodulin affects polymerization of purified tubulin in a Ca^{2+}-dependent manner.
Purified tubulin was polymerized in the presence of 8 %

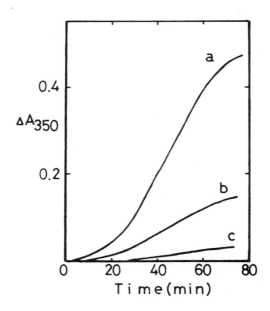

Figure 5.   *Effect of Ca^{2+}-calmodulin on tubulin polymeriz-*
             ation.
Purified tubulin (1.6 mg/ml) in a solution of 55 mM MES, pH
6.8, 11 mM $MgCl_2$, 0.9 mM GTP and 1.1 mM EGTA was polymerized
in the presence of 8 % DMSO and 0.6 mM $CaCl_2$ (2 μM free Ca^{2+}).
Tubulin polymerization was initiated by elevating temperature
and monitored by turbidimetry.   (a), control.   (b), plus
1 mg/ml calmodulin.   (c), plus 2 mg/ml calmodulin.

dimethylsulfoxide and magnesium (15).   Figure 5 shows that
calmodulin inhibited the polymerization in a concentration-
dependent fashion.   In contrast, in the absence of Ca^{2+},
calmodulin did not show a significant effect on tubulin polymerization (data not shown).   The same results were obtained
when glycerol and high concentration of magnesium were used
for MT assembly from purified tubulin alone.

IV. CONCLUSION

      Calmodulin inhibits MT assembly in a Ca^{2+}-dependent man-
ner.   That calmodulin binds to tubulin only in the presence
of Ca^{2+} is demonstrated by gel filtration, affinity chroma-

tography, tubulin's intrinsic fluorescence and sedimentation analysis.   Quantitative analyses of the binding demonstrate that two moles of calmodulin bind to one mole of tubulin with a dissociation constant of 2-5 μM.   Polymerization of tubulin dimer was inhibited by calmodulin in a concentration-dependent manner in the presence of Ca^{2+}.

REFERENCES

1. Weisenberg,R.C. (1972) *Science* *117*,1104-1105.
2. Haga,T., Abe,T., and Kurokawa,M. (1974) *FEBS Lett. 39*, 291-295.
3. Schliwa,M. (1976) *J.Cell Biol. 70*,527-540.
4. Nishida,E., and Sakai,H. (1977) *J.Biochem. 82,* 303-306.
5. Nishida,E., Kumagai,H., Ohtsuki,I., and Sakai,H. (1979) *J.Biochem. 85*,1257-1266.
6. Marcum,J.M., Dedman,J.R., Brinkley,B.R. and Means,A.R. (1978) *Proc.Natl.Acad.Sci.U.S.A. 75,* 3771-3775.
7. Cheung,W.Y. (1970) *Biochem.Biophys.Res.Commun. 38,* 533-538.
8. Kakiuchi,S., and Yamazaki,R. (1970) *Biochem.Biophys.Res. Commun. 41,* 1104-1110.
9. Kumagai,H., and Nishida,E. (1979) *J.Biochem. 85,* 1267-1274.
10. Kumagai,H., and Nishida,E. (1980) *Biomedical Res. 1,* 223-229.
11. Kumagai,H., Nishida,E., and Sakai,H. (1982) *J.Biochem. in press.*
12. Watterson,D.M., Sharief,F., and Vanaman,T.C. (1980) *J. Biol.Chem. 255,* 962-975.
13. Hummel,J.P., and Dreyer,W.J. (1962) *Biochim.Biophys.Acta 63*,532-534.
14. Kasai,K., and Ishii,S. (1978) *J.Biochem. 84,* 1061-1069.
15. Himes,R.H., Burton,P.R., and Gaito,J.M. (1977) *J.Biol. Chem. 252,* 6222-6228.
16. Lowry,O.H., Rosebrough,N.J., Farr,A.L., and Randall,R.J. (1951) *J.Biol.Chem. 193,* 265-275.

CHAPTER 9

Ca^{2+}- AND CALMODULIN-DEPENDENT FLIP-FLOP MECHANISM IN THE REGULATION OF MICROTUBULE ASSEMBLY-DISASSEMBLY.

Shiro Kakiuchi

Department of Neurochemistry
Institute of Higher Nervous Activity
Osaka University Medical School
Nakanoshima, Kita-ku, Osaka 530, Japan

I. INTRODUCTION

Weisenberg (1) was the first to demonstrate microtubule assembly *in vitro* by chelating Ca^{2+} in the medium by EGTA. Since then the concentration of Ca^{2+} has been thought to be physiological regulator governing microtubule assembly-disassembly (2,3). An attractive hypothesis is that this effect of Ca^{2+} may be mediated by calmodulin, a ubiquitous Ca^{2+}-dependent regulator in the animal and plant kingdoms. In support of this notion, Welsh *et al* (4) observed a characteristic localization of calmodulin in the chromosome-to-pole region of the mitotic apparatus visualized by immunofluorescence and Marcum *et al* (5) and Nishida *et al* (6) subsequently found that calmodulin both inhibits and reverses microtubule assembly *in vitro* in the presence of μM concentrations of Ca^{2+}. These results, strongly suggesting the implication of calmodulin in the Ca^{2+}-dependent microtubule disassembly, prompted us to investigate its mechanism. The experiments presented here in a review form reveal that tau (τ) factor in the microtubule is the target of the calmodulin action, *i.e.* the Ca^{2+}-dependent binding of calmodulin to tau factor caused disassembly of microtubules (7,8).

Biological Functions of Microtubules
and Related Structures

83

II. TAU FACTOR IS A CALMODULIN-BINDING PROTEIN

Microtubules, purified by cycles of assembly and disassembly *in vitro*, are composed of tubulin and several microtubule-associated proteins (MAPs). HMW-MAPs of $\sim$ 300,000 M_r (9-11) and a family of four closely related lower M_r (55,000 $\sim$ 62,000) proteins collectively termed tau (τ) factor (12) have been characterized recently for these non-tubulin accessory proteins. Therefore, we isolated and separated from each other these three protein species by column chromatographies (Fig. 1) and searched these proteins for the calmodulin-interacting activity.

Fig. 1. *SDS-Polyacrylamide gel electrophoresis of the microtubule proteins (7) : (a) microtubules, purified by 3 assembly-disassembly cycles (3XMT) ; (b) crude MAPs fraction containing both HMW-MAPs and tau factor ; (c) HMW-MAPs ; (d) tau factor ; (e) tubulin dimer.*

Microtubules, purified by 3 assembly-disassembly cycles (3XMT), was obtained from bovine brain by the temperature-dependent polymerization-depolymerization cycles as in (13). For crude MAPs, 3XMT pellets were homogenized with 100 mM MES buffer (pH 6.8), 1 mM EGTA, 1 mM $MgCl_2$, 1 M NaCl and 2 mM dithiothreitol. The mixture was kept in ice for 30 min and then immersed in a boiling water both for 15 min. It was then quickly chilled down and centrifuged at 200000 x g for 20 min. The supernatant fluid, containing crude MAPs, was saved and its solution medium was changed to 20 mM MES (pH 6.8), 80 mM KCl, 1 mM 2-mercaptoethanol and 1 mM $MgCl_2$ (medium A) plus 0.1 mM GTP and 1 mM EGTA by gel filtration. For HMW-MAPs and tau factor, the crude MAPs solution was applied to a column (1.2 x 44.5 cm) of Sephadex

G-200 equilibrated with medium A. Both protein species were eluted from the column with medium A.
HMW-MAPs and tau factor were further purified by
separate gel filtrations using Sepharose 4B and
Sephadex G-200, respectively. Tubulin was purified
from the 3XMT by a phosphocellulose column chromatography as in (14). PC-tubulin thus obtained was
further purified by a gel filtration chromatography
using Sephadex G-200. A fraction with M_r 120000 was
collected for the tubulin dimer.

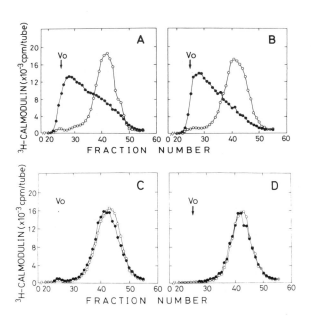

Fig. 2. Formation
of a protein com-
plex between tau
factor and [^{3}H]
calmodulin (7).
Mixtures of micro-
tubule proteins
and [^{3}H]calmodulin
were chromato-
graphed on a
Sephadex G-100
column in the
presence of either
Ca^{2+} (●) or EGTA
(O).   (A) Crude
MAPs fraction
containing both
HMW-MAPs and tau
factor; (B) tau
factor; (C) HMW-
MAPs; (D) PC-
tubulin.

The tubulin, HMW-MAPs and tau factor thus isolated were examined for their ability to bind to [^{3}H]
calmodulin. This was done by two different means,
i.e. a gel filtration column chromatography using
Sephadex G-100 (Fig. 2) and an affinity column
chromatography using calmodulin-Sepharose (Fig. 3).
In Fig. 2, mixtures of proteins and [^{3}H]calmodulin
were chromatographed on a gel column in the presence
and absence (+EGTA) of Ca^{2+} in the elution medium.
In the presence of Ca^{2+}, a shift of the [^{3}H]cal-
modulin peak toward the higher M_r region occurred
only with tau factor (Fig. 2B) but not with HMW-MAPs
(Fig. 2C) or PC-tubulin (Fig. 2D). The result indi-

cates the Ca^{2+}-dependent formation of a protein complex between tau factor and calmodulin.

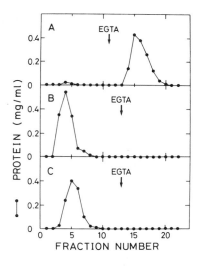

Fig. 3. Binding of tau factor to a calmodulin-Sepharose column (7). Either tau factor (A), HMW-MAPs (B), or tubulin dimer (C) was applied to a column of calmodulin-Sepharose. The column was first eluted with a medium containing Ca^{2+} and then, at an arrow indicated in the figure, with a medium containing EGTA.

A consistent result was obtained with the calmodulin-Sepharose column (see below). Only tau factor (Fig. 3A) but not HMW-MAPs (Fig. 3B) or PC-tubulin[1] (Fig. 3C) was retained in the affinity column in the presence of Ca^{2+} and then eluted from the column with EGTA. In combination with the result from the gel filtration chromatography, it is concluded that tau factor is a calmodulin-binding protein.

---

[1]*In the presence of 6 M urea, tubulin bound to the calmodulin-Sepharose to some extent. The reason for this is unclear at present.*

III. RECONSTITUTION OF Ca^{2+}-SENSITIVE MICROTUBULE ASSEMBLY SYSTEM WITH TUBULIN, TAU FACTOR AND CALMODULIN

When brain tubulin is freed of MAPs by means of chromatographic procedures, the tubulin becomes totally incompetent for self-assembly under standard polymerization conditions. Addition of either HMW-MAPs (14-18) or τ (12, 19-21) restored the capacity of tubulin to form microtubules. Although a considerable controversy has arisen as to the identity of the "true" polymerization-promoting activity, Weber and his colleagues (22,23) and Saudoval and Vandekerckhove (24) have shown that both HMW-MAPs and τ promote individually tubulin polymerization *in vitro*.

In the present study (7,8) we were able to reconstitute the Ca^{2+}-sensitive tubulin polymerization system with the purified tubulin, τ factor, and calmodulin as shown in Figs. 4 and 5. Assembly of PC-tubulin occurred in the presence of τ factor (Fig. 4a).

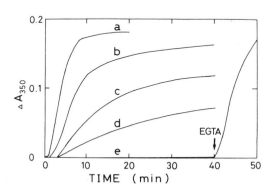

Fig. 4. Reconstitution of microtubule assembly with tubulin and tau factor and the inhibition of assembly by the addition of calmodulin(8). Calmodulin concentration in µg/ml : (a) 0; (b) 63; (c) 188; (d) 375 ; (e) 750. All tubes contained Ca^{2+}. At an arrow indicated on curve (e), EGTA was added to decrease free Ca^{2+} to 1 x 10^{-7}M.

However, this assembly was inhibited by the addition
of calmodulin in a dose-dependent fashion (Fig. 4b,
c, d and e). This effect of calmodulin was reversed
by the addition of EGTA (Fig. 4e), indicating that
the process is reversible depending upon the concentration of Ca^{2+}. In Fig. 5, the inhibitory effect
of calmodulin for the tubulin assembly was titrated
against the concentration of Ca^{2+}. From this figure,
3 µM was obtained for the concentration of Ca^{2+} required for half maximum inhibition.

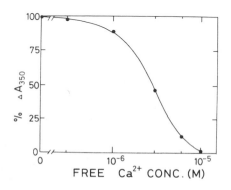

*Fig. 5. Degree of
the assembly of
tubulin as a func-
tion of Ca^{2+} concen-
tration (8). The
value with EGTA
(control) was taken
as 100 %.*

IV. CONCLUSION

Fig. 6 is the diagramatic presentation of the
flip-flop mechanism in the regulation of microtubule
assembly-disassembly. While tau factor promotes the
tubulin to assemble, Ca^{2+}-dependent binding of calmodulin to tau factor prevents the tau factor from
interacting with tubulin thus inhibiting or revers-
ing tubulin assembly. At decreased Ca^{2+} levels,
calmodulin-tau complex dissociate, making tau factor
available for the interaction of tau-tubulin, leading to assembly-formation. In this mechanism, the
concentration of Ca^{2+} acts as a flip-flop switch.
Recently, we have found another flip-flop switch
mechanism for the interaction of calmodulin and actin filaments. In this case, a 150000-M_r protein
(caldesmon) from chicken gizzard smooth muscle is
the calmodulin-binding protein (25,26). A comparison of both systems, *i.e.*, calmodulin-caldesmon-actin
and calmodulin-tau factor-tubulin, is summarized in
Fig. 6. Thus, the flip-flop switch mechanism as
shown here constitutes a type of principle regulating the cytoskeletal system by calmodulin.

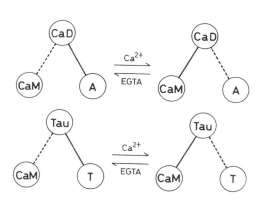

Fig. 6. Diagrammatic presentation of the flip-flop mechanism working in the calmodulin-caldesmon-actin system and the calmodulin-tau factor-tubulin system (8). CaM, calmodulin; CaD, caldesmon ; A, actin, T. tubulin.

ACKNOWLEDGEMENT
        I thank Miss Ayako Yoshikawa for typing this article.

REFERENCES

1. Weisenberg, R.C.,*Science 177* , 1104-1105 (1972).
2. Haga, T., Abe, T. and Kurokawa, M., *FEBS Lett. 39*, 291-295 (1974)
3. Nishida, E. and Sakai, H.,*J. Biochem. 82*, 303-306 (1977).
4. Welsh, M.J., Dedman, J.R., Brinkley, B.R. and Means, A.R.,*Proc. Natl. Acad. Sci. USA 75*, 1867-1871 (1978).
5. Marcum, J.M., Dedman, J.R., Brinkley, B.R. and Means, A.R.,*Proc. Natl. Acad. Sci. USA 75*, 3771-3775 (1978).
6. Nishida, E., Kumagai, H., Ohtsuki, I. and Sakai, H., *J. Biochem. 85,* 1257-1266 (1979).
7. Sobue, K., Fujita, M., Muramoto, Y. and Kakiuchi, S., *FEBS Lett. 132,* 137-140 (1981).
8. Kakiuchi, S. and Sobue, K., *FEBS Lett. 132,* 141-148 (1981).
9. Borisy, G.G., Olmsted, J.B., Marcum, J.M. and Allen, C.,*Fed. Proc. 33*, 167-174 (1974).
10. Dentler, W.L., Granett. S., Witman, G.B. and Rosenbaum, J.L.,*Proc. Natl. Acad. Sci. USA 71*, 1710-1714 (1974).
11. Keates, R.A. and Hall, R.H., *Nature 257*, 418-420 (1975).

12. Weingarten, M.D., Lockwood, A.H., Hwo, S.Y. and Kirschner, M.W., *Proc. Natl. Acad. Sci. USA 72*, 1858-1862 (1975).
13. Berkowitz, S.A., Katagiri, J., Binder, H.K. and Williams, R.C. Jr., *Biochemistry 16*, 5610-5617 (1977).
14. Sloboda, R.D., Dentler, W.L. and Rosenbaum, J.L., *Biochemistry 15*, 4487-4505 (1976).
15. Murphy, D.B. and Borisy, G.G., *Proc. Natl. Acad. Sci. USA 72*, 2696-2700 (1975).
16. Dentler, W.L., Grannet, S. and Rosenbaum, J.L., *J. Cell Biol. 65*, 237-241 (1975).
17. Murphy, D.B., Valley, R.B. and Borisy, G.G., *Biochemistry 16*, 2598-2605 (1977).
18. Murphy, D.B., Johnson, K.A. and Borisy, G.G., *J. Mol. Biol. 117*, 33-52 (1977).
19. Cleveland, D.W., Hwo, S.Y. and Kirschner, M.W., *J. Mol. Biol. 116*, 207-225 (1977).
20. Fellous, A., Francon, J., Lennon, A.M. and Nunez, J., *Eur. J. Biochem. 78*, 167-174 (1977).
21. Francon, J., Fellous, A., Lennon, A.M. and Nunez, J., *Eur. J. Biochem. 85*, 45-53 (1978).
22. Herzog, W. and Weber, K., *Eur. J. Biochem. 92*, 1-8 (1978).
23. Sandoval, I.V. and Weber, K., *J. Biol. chem. 255*, 8952-8954 (1980).
24. Sandoval, I.V. and Vandekerckhove, J.S., *J. Biol. Chem. 256*, 8795-8800 (1981).
25. Sobue, K., Muramoto, Y., Fujita, M. and Kakiuchi, S., *Biochem. International 2*, 469-476 (1981).
26. Sobue, K., Muramoto, Y., Fujita, M. and Kakiuchi, S., *Proc. Natl. Acad. Sci. USA 78*, 5652-5655 (1981).

CHAPTER 10

INTRA-AXOPODIAL PARTICLE MOVEMENT
AND AXOPODIAL SURFACE MOTILITY
IN *Echinosphaerium akamae*

Toshinobu Suzaki[1]

Zoological Laboratory
Faculty of Science
Hiroshima University
Hiroshima

Yoshinobu Shigenaka

Department of Information and Behavior Science
Faculty of Integrated Arts and Sciences
Hiroshima University
Hiroshima

I. INTRODUCTION

Axopodia of a heliozoan *Echinosphaerium* are multifunctional cytoplasmic protrusions which work through their shortening and successive elongation at cell division (1), cell fusion (2), cell locomotion (1 , 3), cell adhesion to substratum (1 , 3), and also food capturing (4). In addition to them, two types of motilities have been reported on heliozoan axopodia, which do not involve any transformation of the axopodia themselves: One of them is a unidirectional surface motility (5), and the other is a bidirectional intra-axopodial movement of the so-called electron-dense granules (6, 7), which fulfills some of the criteria for saltatory movement proposed by Rebhun (8). Although little is known about the mechanisms of these motilities, the fact that both

[1]*Present address: Department of Developmental Biology, RSBS, The Australian National University, Canberra City, Australia*

Biological Functions of Microtubules
and Related Structures

91

of the motilities are associated closely with the axopodial
surface membrane has raised a possibility that both of them
might be mediated by similar mechanisms (9, 5). In addition,
Edds (6) has clearly shown that the intra-axopodial motility
does not necessarily require the axonemal microtubules,
suggesting that some kind of microtubule-independent system,
probably acto-myosin system, might be involved.

In the present study, we have examined (i) the ultrastructural basis for the motility of dense granules by
employing an improved fixation procedure for the heliozoan
axopodia (4), and (ii) the effects of various reagents on the
motilities, in a special reference to the comparison between
these two types of motilities.

II. MATERIALS AND METHODS

A large heliozoan *Echinosphaerium akamae* (10) used in this
study was originally collected in Yoshinaga-cho, Wake-gun,
Okayama-ken, Japan, and has been cultured in our laboratory as
previously described (4). For a light microscopy, a Nikon
Apophoto microscope equipped with Nomarski differential
interference optics, and a highly sensitive polarization
microscope (Nikon Rectifier) were employed. Particle movement
was measured directly from the microscope using an ocular
micrometer and a stop watch. In this study, we could make
detailed observations on the motility of intra-axopodial
particles, because they are much bigger in *E*. *akamae* (0.4 to
0.6 μm in diameter) than those in other heliozoan species such
as *E*. *nucleofilum* so that we could trace them easily under the
microscope. Axopodial surface motility was monitored by using
polystyrene microspheres (0.3 μm in diameter) as described by
Bloodgood (5). For an electron microscopy, materials were
prepared according to Suzaki *et al*. (4) by using a glutaraldehyde fixative containing 0.25 mg/ml ruthenium red (Kataya-
ma Chemical, Osaka).

III. RESULTS

A. *Intra-axopodial Particle Movement*

A lot of axopodia are radiating from the spherical cell
body of *E*. *akamae* as shown in Fig. 1. Inside each axopodium,
there can be seen the so-called axoneme, which is composed of
a bundle of microtubules as cytoskeletal elements. They can

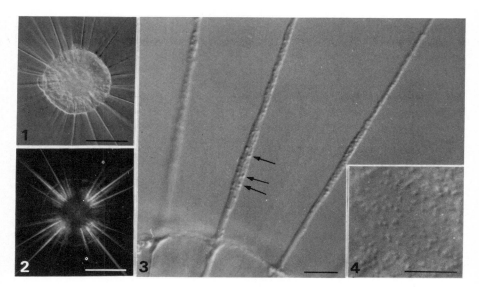

FIGURE 1. Light micrograph of Echinosphaerium. X 120.
Bar: 100 µm. FIGURE 2. Polarization micrograph of the same
organism as shown in Fig. 1. X 120. Bar: 100 µm. FIGURE 3.
Light micrograph at proximal region of axopodia. A lot of
particles are located inside the axopodia as indicated by
arrows. X 970. Bar: 10 µm. FIGURE 4. Light micrograph of
cell body surface, showing the presence of the particles.
X 1,550. Bar: 10 µm.

be easily detected by using a polarization microscope (Fig. 2).
Fig. 3 is an enlarged micrograph of the proximal region of
axopodia, representing many light-refractile particles inside
the axopodia (Fig. 3, arrows). These particles, called
electron-dense granules, are always moving up and down along
the axopodial axis in the so-called saltatory manner. The
highest velocity of such a particle movement measured so far
was about 1.5 µm/sec. The dense granule is a kind of extrusion granule, whose contents are expelled from the axopodial
surface as the cell captures prey organisms. Dense granules
are also present just beneath the cell body surface (Fig. 4).
Also in this region, saltatory movement of the granules could
be observed. They could not be observed at all inside the
cell body itself.
    Motile activity of the dense granules was investigated at
several points from the base to the distal region of axopodia,
separately (Fig. 5). At each point, the moving distance
within 10 seconds was measured. The motile activity at cell
body surface was relatively low as compared with those in

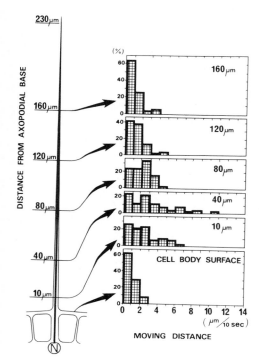

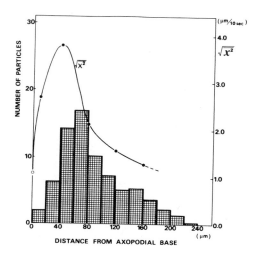

FIGURE 5. Histograms
of moving distance of
dense granules at each
axopodial level and cell
body surface. In each
graph, abscissa indi-
cates the moving dis-
tance within 10 seconds,
and ordinate the frequency.

FIGURE 6. Motile
activity and distribution
of dense granules at each
axopodial level. RMS dis-
tances of dense granules
were plotted with closed
circles. Open circle re-
presents the value obtained from cell body surface.

basal or middle regions of axopodia. At the distal region, the motility was also found to be relatively low. Although it is not shown here in these histograms, any difference was not observed between the granules moving toward the tip and those toward the base in the manner of movement. Based on the results shown in Fig. 5, root of the mean value of the square of x (RMS distance) was chosen as an index of motility, where the x corresponds to the moving distance of a dense granule within 10 seconds. As is clearly shown in Fig. 6, the RMS distance does not give a uniform value, but varies as a function of distance from the axopodial base. The motility became maximum at about 40 μm level from the axopodial base. In Fig. 6, distribution of the dense granules at each axopodial level from the base is also presented. This histogram of the particle density is quite similar in shape to the curve of the motility. Each axopodium contained 60 to 90 granules, and about 250 axopodia are radiating from the cell body. Therefore, total number of the dense granules within axopodia is about 20,000. Granule density at cell surface area was calculated to be 20 to 40 per 100 μm^2. Since average value of cell body diameter was about 160 μm, 20,000 to 30,000 granules are estimated to be located at the cell body surface of a single organism.

Inside the axopodia, there are only two types of filamentous structures, i. e., the microtubules and the X-body as shown in Fig. 7 and 8. The microtubular axoneme is located at the center of the axopodium as a cytoskeleton (Fig. 7). The X-body, which has been considered to be indispensable for a rapid axopodial contraction, is running in parallel with the axoneme throughout the axopodial length. The electron-dense granules are always observed to be attached closely to the axopodial surface membrane with a distance of only 5 to 10 nm. They are sometimes observed to be attached with the microtubules (Fig. 7), or to the X-body (Fig. 8). No microfilaments could be detected inside the axopodia. Fig. 9 represents a transverse view of a single dense granule located just beneath the cell body surface membrane. In this region, there can be observed no filamentous structures such as microtubules, microfilaments, or X-bodies at all. The dense granule, which is enclosed with a single membrane, is closely apposed only to the inner surface of the cell body surface membrane. Distance between the two membrane was just same as observed inside the axopodia, i. e., 5 to 10 nm. Any special structures such as connecting bridges could not be detected between the two membranes.

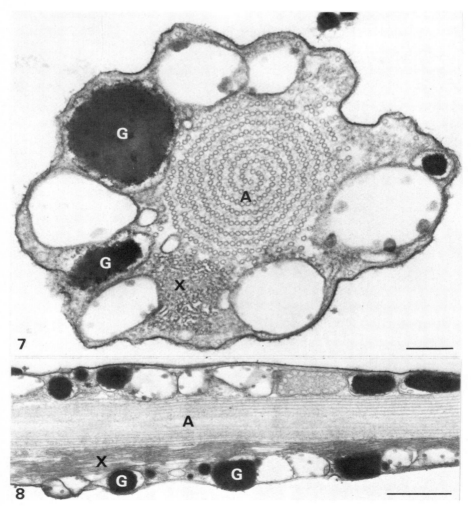

FIGURE 7. Electron micrograph of a cross section of an axopodium, representing dense granules (G), an axoneme (A), and an X-body (X). X 68,000. Bar: 200 nm.
FIGURE 8. A longitudinal section of an axopodium. A: Axoneme, X: X-body, G: Dense granules. X 19,000. Bar: 1 μm.

B. Axopodial Surface Motility

In addition to the motility of intra-axopodial particles, another type of motility, the axopodial surface motility, has been reported in *E. nucleofilum* (5). The same motility could be observed in *E. akamae*. As already known in *E. nucleofilum*,

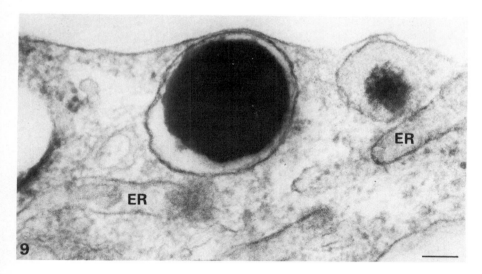

FIGURE 9. Transverse view of a dense granule at cell body surface area. The dense granule is closely apposed to the inner surface of the cell body surface membrane. ER: Endoplasmic reticulum. X 107,000. Bar: 100 nm.

the extracellular marker particles such as polystyrene microspheres attached to the axopodial surface exhibit unidirectional movement toward the axopodial tip. Besides the polystyrene beads, carbon particles, carmin particles, squid ink, and some kind of bacteria were also found to be transported along the axopodial surface. Once such a marker particle attached to the axopodium, it begins to move centrifugally without any stopping until it reaches the tip. Velocity of the marker particle was quite constant in the range of 1.6 to 1.9 μm/sec. In this respect, the axopodial surface motility seems to be quite different from the intra-axopodial one.

C. Effects of Various Agents on the Motilities

In order to elucidate the mechanisms of these intra- and extra-axopodial motilities, inhibitory effects of various reagents were examined. As shown in Table 1, the intra-axopodial particle movement was inhibited by treatment with energy-inhibitors, protein-synthesis inhibitor, diamide, chloral hydrate and also vanadate. Anti-mitotic drugs such as colchicine and vinblastine did not inhibit the motility at all, while the axopodia became shortened by these treatments because of disassembly of the axonemal microtubules. Moreover, cytochalasin B did not affect the intra-axopodial motility.

TABLE I. Inhibitory Effects of Various Agents on the
          Motilities

Treatments	Inhibition of Intra-axopodial Particle Movement	Inhibition of Surface Motility	Axopodial Resorption
NaN$_3$ (0.5 mM, 2 hr)	+	−	+
DNP (0.1 mM, 2 hr)	+	−	+
KCN (0.2 mM, 2 hr)	+ +	−	+
Cycloheximide (0.03-3 mM, 2 hr)	+	−	+
Colchicine (20 mM, 30 min)	−	+	+
Vinblastine (0.1 mM, 30 min)	−	−	+
Cytochalasin B (0.1 mM, 2 hr)	−	−	−
PCMB (0.1 mM, 1 hr)	−	−	+
Diamide (0.1 mM, 30 min)	+ +	−	−
Chloral hydrate (5 mM, 10 min)	+ +	−	−
Vanadate (1 mM, 30 min)	+ +	−	+
Ouabain (5 mM, 1 hr)	−	−	+
Low temperature (2°C, 2 hr)	+	−	+

In case of extra-axopodial surface motility, on the other
hand, most of the reagents other than colchicine did not
inhibit the motility as shown in the right column of Table I.
     It was found that the dense granules became to be detouch-
ed from the axopodial surface by treatment with chloral
hydrate or vanadate, while other structures including microtubules and X-body were not affected. Figs. 10 and 11 are
the electron micrographs of an axopodium fixed after complete
cessation of intra-axopodial motility by treatment with
chloral hydrate, representing that the distance between the
membrane of dense granules and axopodial surface membrane
became widened remarkably to be about 20 to 30 nm. Fig. 12
is an electron micrograph taken from the control axopodium and
printed at the same magnification as Fig. 11.

IV. DISCUSSION

A. *Intra-axopodial Motility*

In this study, we employed an improved fixation method by
using ruthenium red in glutaraldehyde fixative. This fixation
procedure enabled us to preserve a quite labile contractile

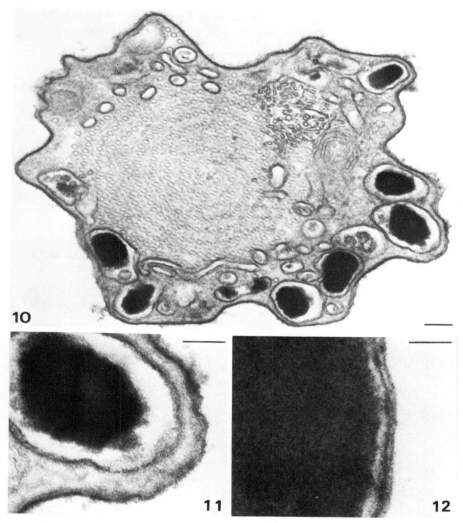

FIGURE 10. Electron micrograph of a cross section of an
axopodium after the treatment with chloral hydrate, representing that the granules are detouched from the inner surface of
the axopodial surface membrane. X 84,000. Bar: 100 nm.

FIGURE 11. An enlarged figure of Fig. 10. X 250,000.
Bar: 50 nm.

FIGURE 12. A transverse section of a dense granule inside
the control axopodium, showing a close proximity between the
membrane of dense granule and the axopodial surface membrane.
X 250,000. Bar: 50 nm.

structure called X-body which was located inside the axopodia
(4). Both microtubules and microfilaments were also
preserved much better than by the conventional fixation method
recommended by Roth *et al.* (11). As shown in the present
study, there could not be detected any filamentous structures
such as microtubules and microfilaments just beneath the cell
body surface, where dense granules also exhibit saltatory
movements. Microtubules are located only inside the axopodia
including their proximal regions, and microfilaments could be
observed only deep inside the cell body and also in food
vacuole-forming pseudopodium (4). Edds (12) has Shown that
the actin-containing microfilaments are present in *Echinosphaerium* cells. He has supposed, therefore, that the
particle movement might be mediated by microfilaments or acto-
myosin system, while they could not be detected by the
electron microscopy probably because of some problem in
fixation technique. It seems to be unlikely, however, that
the microfilaments had been present inside the axopodia but
they were disrupted by fixation, because a lot of microfilaments were observed constantly in other cell regions such
as in the food vacuole-forming pseudopodia. The fact that
cytochalasin B did not affect the particle movement also
supports the possibility that the acto-myosin system is not
involved in such a motility. In addition, it was also demonstrated in the present study that the dense granules became
detouched from the cell body surface membrane by the treatment with chloral hydrate or vanadate. It is considered from
these results that a membrane-membrane interaction between the
dense granules and the cell surface membrane might be indispensable for inducing the movement of these granules.

It was found that the intra-axopodial particles move much
more actively in the basal region of axopodia as compared
with the cell body surface area. Inside the axopodia, the
dense granules are always apposed closely to the inner surface
of axopodial membrane just in the same manner as in the cell
body surface area. Sometimes, they were observed to be
closely attached to the microtubules or the X-body. Although
neither of these filamentous structures could be considered
as essential elements for the particle movement, such a close
proximity seems to implicate the possibility that the microtubules and/or the X-bodies might be involved in acceleration
of the particle motility.

The present experiments have shown that the particle
movememt was inhibited by the treatments with several kinds
of reagents including energy- and protein synthesis-inhibitors,
and vanadate which is a specific inhibitor of dynein ATPase.
These results suggest that the membrane-membrane interaction
between the membrane of each dense granule and the axopodial
surface membrane might be generated by some kind of energy-

dependent enzymatic process, in which ATPase activity of
dynein might be involved.

B. Extra-axopodial Surface Motility

Experiments on the effects of various reagents on the
axopodial surface motility revealed that this type of motility
was energy-independent and not inhibited by the treatment with
most of the reagents tested. The inhibitory effect of
colchicine cannot be regarded as a result of disassembly of
microtubules, because the treatment with vinblastine did not
affect the motility. These results, therefore, give us a
strong impression that this event might be non-biological in
nature.

Similar surface motilities have also been reported on
Chlamydomonas flagella (13,14), cilia of sea urchin embryo
(15), and centrohelidian heliozoa, *Heterophrys* (16). In contrast to the surface motility of *Echinosphaerium,* all of these
motilities are not unidirectional, but they were characterized
to be saltatory. Effects of various reagents and environmen-
tal conditions were also examined on *Chlamydomonas* flagellar
surface motility(14), which appeared to be quite different
from those on the axopodial surface motility of *Echinosphaerium*. In case of *Chlamydomonas* flagella, the surface motility
was inhibited with several kinds of reagents such as sodium
citrate, cycloheximide, chloral hydrate, and also by cold
treatment. Although the mechanism of axopodial surface
motility remains to be elucidated in future, it was revealed
in this study that the mechanism of such a motility had to be
fairly unique, being different from that of intra-axopodial
particle movement and also the surface motility of *Chlamydomonas* flagella.

The axopodial surface motility seems to play quite important roles on the gliding cell movement and conveyance of
various materials along the axopodia. When the cell is held
with two plates, the cell begins to exhibit the so-called
gliding movement (3). From our preliminary experiments,
the velocity of gliding cell movement was found to be almost
identical to that of the marker particles along the axopodial
surface, and this type of motility might be ascribed to the
interaction between the outer surface of axopodial membrane
and substrates. In addition to the gliding movement, there
are two other fundamental cell functions being associated
with the axopodial surface motility. One of them is concerned
with the conveyance of excretory materials which have been
expelled from the cell body surface by exocytosis, and the
other is related to the removal of adhered and useless
materials from the axopodial tip as demonstrated earlier by

Kitching (17, 18).

V. SUMMARY

Two types of membrane-associated motilities of a large heliozoan *Echinosphaerium akamae* were studied to elucidate the mechanisms and the relationship between them. The characteristics of these motilities, the intra-axopodial particle movement and the axopodial surface motility, were compared and found to be quite different from each other. Several reagents such as energy-inhibitors, cycloheximide, chloral hydrate, diamide, and vanadate, and also cold treatment inhibited the intra-axopodial motility, while the surface motility was inhibited only by the colchicine treatment. Morphological evidences were also presented, indicating that the intra-axopodial motility was mediated by an interaction between the membrane of particles and the inner surface of axopodial membrane. Mechanism of axopodial surface motility is regarded to be quite different from that of the intra-axopodial motility. It was strongly suggested that neither acto-myosin nor dynein-microtubule system might be involved in both events.

ACKNOWLEDGMENTS

The present study was supported partially by a research fund to T. Suzaki from the Kazato's Research Promotion Committee.

REFERENCES

1. Suzaki, T., Shigenaka, Y., and Takeda, Y., *Cell Struct. Funct. 3*, 209 (1978).
2. Shigenaka, Y., and Kaneda, M., *Annot. Zool. Japon. 52*, 28 (1979).
3. Watters, C., *J. Cell Sci. 3*, 231 (1968).
4. Suzaki, T., Shigenaka, Y., Watanabe, S., and Toyohara, A., *J. Cell Sci. 42*, 61 (1980).

5. Bloodgood, R. A., *Cell Biol. Int. Rep. 2*, 171 (1978).
6. Edds, K. T., *J. Cell Biol. 66*, 145 (1975).
7. Fitzharris, T. P., Bloodgood, R. A., and McIntosh, J.R., *J. Mechanochem. Cell Motility 1*, 117(1972).
8. Rebhun, L. I., *Int. Rev. Cytol. 32*, 93 (1972).
9. Bardelę, C. F., *Z. Naturforsch. 31*, 190 (1976).
10. Shigenaka, Y., Watanabe, K. and Suzaki, T.,*Annot. Zool. Japon. 53*, 103 (1980).
11. Roth, L. E. , Pihlaja, D. J., and Shigenaka, Y., *J. Ultrastruct. Res. 30*, 7 (1970).
12. Edds, K. T., *J. Cell Biol. 66*, 156 (1975).
13. Bloodgood, R. A., *J. Cell Biol. 75*, 983 (1977).
14. Bloodgood, R. A., Leffler, E. M., and Bojczuk, A. T., *J. Cell Biol. 82*, 664 (1979).
15. Bloodgood, R. A., *J. Exp. Zool. 213*, 293 (1980).
16. Troyer, D., *Nature 254*, 696 (1975).
17. Kitching, J. A., *J. Exp. Biol. 37*, 407 (1960).
18. Kitching, J. A., *J. Exp. Biol. 39*, 359 (1962).

CHAPTER 11

RAPID CONTRACTION OF THE MICROTUBULE-CONTAINING AXOPODIA
IN A LARGE HELIOZOAN *ECHINOSPHAERIUM*

Yoshinobu Shigenaka
Kazuhide Yano
Reiko Yogosawa

Department of Information and Behavior Science
Faculty of Integrated Arts and Sciences
Hiroshima University
Hiroshima

Toshinobu Suzaki[1]

Zoological Laboratory
Faculty of Science
Hiroshima University
Hiroshima

I. INTRODUCTION

From the cell surface, a heliozoan *Echinosphaerium* extends a number of needle-like axopodia, each of which contains several hundreds of microtubules as the cytoskeletal elements inside (1). These axopodia, moreover, are known to play some fundamental cell functions, e. g., cell locomotion (2, 3, 4), cell-to-cell recognition (3, 5), cell fusion (2, 3, 6, 7), and cell-to-substratum adhesion and the subsequent cell division (2). All of these cell functions, as reported already by us, are controllable by themselves and executed by dis- and re-assembly of the cytoskeletal elements, microtu-

[1]*Present address: Department of Developmental Biology, RSBS, The Australian National University, Canberra City, Australia*

105

bules, which are located inside the actively functioning axopodia.

Quite recently, on the other hand, the rapid axopodial
contraction was found to occur in *Echinosphaerium* during food
capture and ingestion process (5), suggesting that the axopodia might contain some special contractile elements. However, no further investigation has been made on this phenomenon. In the present study, therefore, we aimed to detect or
identify the contractile elements and also, if possible, to
elucidate the mechanism of rapid axopodial contraction.

II. MATERIALS AND METHODS

Living samples of *Echinosphaerium akamae* (8) were originally collected from a pond in Yoshinaga-cho, Wake-gun, Okayama-ken, and cultured at $22 \pm 1°C$ with small protozoans as
food sources in 0.01% Knop solution. Prior to every experiment, some heliozoan cells were kept in the standardized salt
solution (0.2 mM $CaCl_2$, 2 mM KCl, 0.05 mM $MgCl_2$, 1 mM Tris-HCl
at pH 7.3) for more than 1 hr to make them adapt to the solution. All of chemical reagents were dissolved directly into
this solution (control medium) to examine their effects on the
heliozoan cells as shown in Table I.

In the electrical stimulation, platinum wires (0.5 mm ∅)
were employed as electrodes and set at an interval of 5 mm in
a glass chamber. A heliozoan cell was put just in the middle
of the two electrodes and stimulated with a rectangular pulse

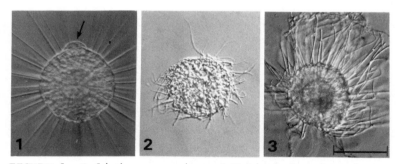

FIGURE 1. A living organism of Echinosphaerium akamae,
showing a number of radiating axopodia and a normally-func-
tioning contractile vacuole (arrow). FIGURE 2. An organism
fixed with the RR-free GA fixative for 30 min, showing fairly
coagulated protoplasm and conspicuous degradation of axopodia.
FIGURE 3. Another organism fixed with the RR-containing GA
fixative for 30 min, showing the well-preserved cellular components. Compare with Fig. 2. Bar=100 μm.

(0.1 to 2.0 V, 1 sec) by using an electrical stimulater, Ni-hon Khoden, MSE-3. Five minutes were chosen as the intervals of stimulation so that the axopodia might recover from the stimulation effect.

For electron microscopy, we tried to improve the fixation method (1, 9) especially for axopodia because the axopodial contraction should be analyzed in detail at the fine structur-al level. As the results, it was found that there were two critical points in glutaraldehyde (GA) fixative; (i) addition of ruthenium red (RR) at 0.25 mg/ml to the GA fixative, and (ii) replacement of phosphate buffer by cacodylate buffer to prevent the formation of precipitation. The results are il-lustrated in Figs. 1 to 3. By using this method, the organisms were fixed with the RR-containing GA fixative for some option-al time periods, followed by post-fixation with the OsO_4 fixa-tive recommended by Shigenaka et al. (9). Ultrathin sections were made from the samples embedded in Spurr's resin and proc-essed by staining and observation by the method described al-ready elsewhere (5).

III. RESULTS AND DISCUSSION

A. *Induction and Inhibition of Axopodial Contraction*

By our group (5), the prey organism attached to a helio-zoan axopodium was found to be conveyed toward the cell sur-face by means of 'axopodial contraction' or 'axopodial flow'. However, the mechanism of these phenomena remained uncertain. At first, therefore, the present study aimed to clarify what kind of mechanism was functioning on the first event, *i. e.*, axopodial contraction.

When analyzed on 16 mm cine films, the axopodial contract-ion induced with being touched by prey organisms was recogniz-ed to be accomplished through a whole length of axopodia (200 to 300 μm) within 40 msec, demonstrating that the velocity of contraction was more than 5 mm/sec. As the next step, we tried to find other factors resulting in such a rapid con-traction of axopodia. As the results, the following agents, in addition to the prey organisms, were found to induce the axopodial contraction; (i) carbon particles, (ii) carmine par-ticles coated with albumen, (iii) anion exchange resin parti-cles, and (iv) poly-L-lysine coated glass particles. On the other hand, cation exchange resin particles and uncoated car-mine or glass particles never resulted in the contraction. Also when the axopodium was touched with the tip of a glass or metal needle, such a contraction could not be induced at all

as reported also in *Actinophrys sol* (10, 11). From these results, it is considered that the rapid axopodial contraction
might occur, at its earliest step, in collaboration with activation of chemoreceptors and/or surface negative charge on
the axopodial membranes.

From these results, it was presumed that the axopodial
contraction might be correlated with certain excitation of the
axopodial membranes. In consequence, an electrical stimulation (1.0 V/5 mm, 1 sec, rectangular pulse) was applied extra-
cellularly to the heliozoan cell, revealing that the stimulation induced the axopodial contraction as well as in the above-
mentioned agents. From further detailed experiments, moreover, it was recognized that the contraction occurred only at
the cathode or at the anode when an electric current was
switched on or off, respectively. This is considered to agree
with the well-known law of polar excitation. These results
strongly suggest that such an electrical stimulation causes
depolarization of the axopodial membranes, through which Ca^{2+}
influx might be enhanced to induce the axopodial contraction
and also the breakdown of axonemal microtubules simultaneously.

TABLE I. *Effects of Divalent Cations and Energy*
Inhibitor on Axopodial Contraction

Test Solution	Induction of Axopodial Contraction with	
	Carbon Particle	Electrical Stimulation
Control medium (CM)[a]	+	+
Ca-free CM	−	−
0.01 mM MnCl$_2$ in CM	±	±
0.5 mM MnCl$_2$ in CM	−	±
1.0 mM MnCl$_2$ in CM	−	−
0.5 mM NaN$_3$ in CM[b]	−	−

[a]CM: 0.2 mM CaCl$_2$, 0.2 mM KCl, 0.05 mM MgCl$_2$, 1 mM Tris-
    HCl (pH 7.3)
[b]Examined after 60-min treatment

Table I reveals the effects of divalent cations and energy
inhibitor on the axopodial contraction which was induced with
carbon particles or electrical stimulation (1.0 V/5 mm, 1

sec). In the electrical stimulation, the effects of reagents were examined only at the cathodal contraction, that is, only when the electricity was switched on, because the anodal contraction might be affected directly by the cathodal contraction which occurred just 1 sec before. In Ca^{2+}-free medium, the axopodial contraction was not induced with carbon particles or electrical stimulation. On the other hand, Mn^{2+} was found to inhibit the contraction completely at 0.5 mM in carbon particles and at 1 mM in electrical stimulation, showing that 0.01 mM and 0.5 mM were critical concentrations in Mn^{2+}, respectively. In this experiment, the Mn^{2+} ions are presumed to have acted as inhibitors of Ca^{2+} channels as suggested by Hagiwara & Byerly (12). From these results, therefore, it is considered that the axopodial contraction occurs with Ca^{2+} influx through the axopodial membrane, depending on the extracellular Ca^{2+} concentrations.

In 0.5 mM NaN_3, on the other hand, the axopodial contraction was inhibited after 60-min treatment when the contractile vacuole started to become inactive but the axopodia did not shorten in length (Table I). This result suggests that the axopodial contraction is also energy-dependent, although some further and detailed experiments are necessarily required.

B. *Ultrastructural Changes of the Possible Contractile Elements of Axopodia*

When a control organism was fixed with the RR-containing GA fixative, the so-called X-body (5, 6), in addition to the axonemal microtubules, was preserved best of all the fixatives we have tried before. Figures 4 and 5 illustrate that the X-body is composed of a number of tubular filaments which vary 10 to 20 nm in diameter, being located just beneath the axopodial membrane and beside the axonemal microtubules. As shown by an arrow in Fig. 6, moreover, the tubular filaments composing the X-body were found to be constricted at some places to make nodal or moniliform structures (ca. 20 X 30 nm each).

On the other hand, it is known by us (5) that the axopodial beads are usually formed as the prey organism is being ingested by the axopodial flow as shown in Fig. 7. When the beads were examined at fine structural level, the X-body was recognized to be considerably granulated and accumulated inside the beads (Fig. 8). Similar granulation of the X-body, furthermore, was found to be induced also by the treatment with 10 mM $CaCl_2$ for 10 min (Fig. 9). All of these results suggest that the axopodial beading might result directly from the granulation of X-body which was initiated with Ca^{2+} influx

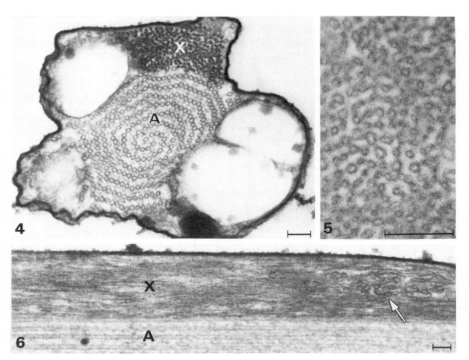

FIGURE 4. A cross section of the heliozoan axopodium,
showing the well preserved X-body (X) and the axonemal microtubules (A). FIGURE 5. An enlarged electron micrograph of
the tubular elements (10 to 20 nm in diameter) composing the
X-body. FIGURE 6. A longitudinal section of the axopodium,
showing that the tubular elements of X-body (X) are granulated at an arrow and running parallel to the axoneme (A). Bar=
100 nm in all figures.

and occurred at many levels through the axopodium.
     When the RR-free GA fixative was employed, moreover, it
was found that the prolonged GA fixation caused also the similar granulation of X-body at various degrees (Figs. 10 and
11), suggesting that the RR might inhibit Ca^{2+} influx through
the axopodial membrane especially by means of masking the membrane surface including the receptor sites. Furthermore, the
granulated X-bodies were checked in the staining property; the
control organism was fixed with the RR-free fixative, stained
en bloc with 1% uranyl magnesium acetate (UMA) just before the
subsequent dehydration, and observed without a further electron staining. As shown in Fig. 12, the UMA did not stain the
axonemal microtubules themselves but the X-bodies as well as
in the axopodial membranes and other membranous components.

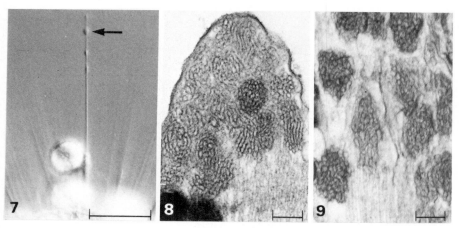

FIGURE 7. A light micrograph showing a newly formed bead (arrow) of the axopodium fixed when the prey organism is being ingested by an axopodial flow. FIGURE 8. An oblique section of the bead, showing that the X-bodies were granulated everywhere. FIGURE 9. The granulated X-bodies appeared after the treatment with 10 mM $CaCl_2$ for 10 min. Bar=50 µm in Fig. 7 and 100 µm in Figs. 8 and 9.

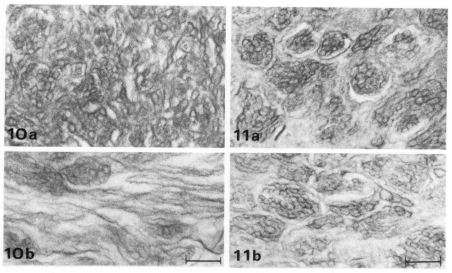

FIGURES 10 and 11. Electron micrographs of the granulating X-bodies resulted from the prolonged GA fixation (RR-free GA fixative, 30 min). Cross and longitudinal sections of them are indicated by a and b, respectively. Bar=100 nm in all figures.

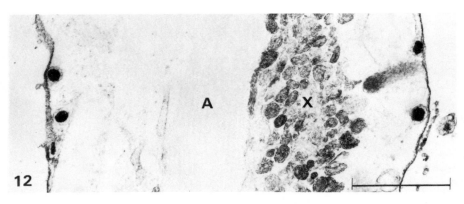

FIGURE 12. An oblique section through an axopodium of
the organism which was fixed with RR-free GA fixative and
stained en bloc with 1% UMA for 30 min. Except for the axonemal microtubules (A), X-bodies (X) and other membranous
components were found to be stained conspicuously with UMA.
Bar=1 μm.

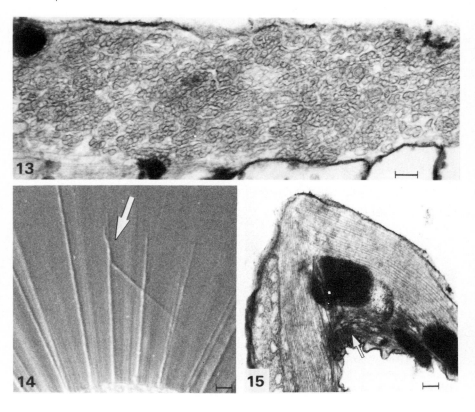

FIGURE 13.   An electron micrograph showing the granulat-
ion of X-bodies, which was induced by an electrical stimulat-
ion (1.0 V/5 mm, 1 sec).   FIGURE 14.   A light micrograph of
the folded axopodium (arrow) resulted from the same stimulat-
ion as in Fig. 13.   FIGURE 15.   An electron micrograph of the
electrically stimulated axopodium, in which X-bodies are lo-
cated inside the folded region.   Bars=100 nm, 10 µm and 1 µm
in Figs. 13 to 15, respectively.

This result suggests that the X-body may not be composed of
microtubule proteins, tubulins.

Finally, the morphological changes were examined when the
axopodial contraction was induced with an electrical stimulation (1.0 V/5 mm, 1 sec).   As shown in Fig. 13, the contracted
axopodium was found to be featured by a considerable granulation of the X-body as well as in food ingestion, Ca^{2+} ion
treatment, and prolonged GA fixation.   When the electrical
stimulation was applied, moreover, we often found the peculiar
phenomenon, *i. e.,* folding of the axopodium as shown by an
arrow in Fig. 14.   A section through the folded axopodium
(Fig. 15) revealed that the axonemal microtubules were folded
or broken and the X-body was granulated and located only at
the inner side of the folded point.   Moreover, the so-called
actin-like filaments were not detected inside the contracted
axopodium.   This suggests that the granulation of X-body might
be related closely to the folding as well as to the contraction of axopodia.

Based on the present data, it was summarized that the tubular X-body might be transformed into the granular one (Fig.
16) by Ca^{2+} influx through the axopodial membranes, and thereby, that the X-body is indispensable for the rapid contraction
and the folding of axopodia.

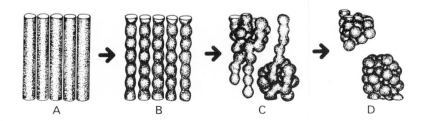

A   B   C   D

FIGURE 16.   A schematic presentation showing the possible
process of speedy granulation of tubular X-body (A through D).
Refer to the text in detail.

IV. SUMMARY

In the present study, the rapid axopodial contraction in
Echinosphaerium akamae was investigated, especially so that
the mechanism of contraction might be clarified. At first,
the axopodial contraction was found to be induced with various
factors which might affect the chemoreceptors of axopodial
membranes, and also with the extra-cellularly applied electrical stimulation. From further experiments on Mn^{2+} treatment,
it was considered that Ca^{2+} influx might be involved in such
a rapid contraction of axopodia. Next, the electron microscopy of contracted axopodia revealed that the axopodial contraction was featured by two big events as follows; (i) folding of the axonemal microtubules instead of the complete disassembly of them, and (ii) rapid transformation of the X-body
from tubular elements to granular ones, resulting in rapid
axopodial contraction or folding. Especially, the second
event might have resulted from Ca^{2+} influx through the axopodial membranes.

REFERENCES

1. Roth, L. E., Pihlaja, D. J., & Shigenaka, Y., *J. Ultra-
   Struc. Res. 30*, 7 (1970).
2. Suzaki, T., Shigenaka, Y., & Takeda, Y., *Cell Struct.
   Funct. 3*, 209 (1978).
3. Toyohara, A., Maruoka, T., & Shigenaka, Y., *Bull. Biol.
   Soc. Hiroshima Univ. 43*, 35 (1977).
4. Watters, C., *J. Cell Sci. 3*, 231 (1968).
5. Suzaki, T., Shigenaka, Y., Watanabe, S., & Toyohara, A.,
   J. Cell Sci. 42, 61 (1980).
6. Shigenaka, Y., & Kaneda, M., *Annot. Zool. Japon. 52*, 28
   (1979).
7. Shigenaka, Y., Ogura, T., & Maruoka, T., *Zool. Mag. 85*,
   65 (1976).
8. Shigenaka, Y., Watanabe, K., & Suzaki, T., *Annot. Zool.
   Japon. 53*, 103 (1980).
9. Shigenaka, Y., Roth, L. E., & Pihlaja, D. J., *J. Cell Sci.
   8*, 127 (1971).
10. Kitching, J. A., *J. Exp. Biol. 37*, 407 (1960).
11. Ockleford, C. D., & Tucker, J. B., *J. Ultrastruct. Res.
    44*, 369 (1973).
12. Hagiwara, S., & Byerly, L., *Fed. Proc. 40*, 2220 (1981).

CHAPTER 12

CALMODULIN IN THE CILIA OF *TETRAHYMENA*

Yoshio Watanabe and *Kazuo Ohnishi*

Institute of Biological Sciences
The University of Tsukuba
Ibaraki, Japan

Ciliary reversal is one of several well-known calcium-dependent biological phenomena found in nature and ubiquitous calmodulin is considered as a strong candidate for the target molecule of calcium in this phenomenon. Occurrence of calmodulin in cilia or flagella has been suggested by much indirect evidence [1-6]. However, Van Eldik *et al*. have recently reported that a protein isolated from *Chlamydomonas* flagella by phenothiazine-Sepharose affinity chromatography did not activate brain cyclic nucleotide phosphodiesterase at all, in spite of the fact that the protein possessed several characteristic properties previously thought to be unique features of calmodulin [7]. The flagella protein they isolated was a protein similar to but not identical with calmodulin. Therefore, they indicated that for the identification of a protein included in cilia or flagella as calmodulin, several criteria should be scrutinized using a test protein purified to homogeneity. There have been no reports based on such strict criteria which have conclusively demonstrated that calmodulin is included in cilia or flagella. We therefore have tried to purify a *Tetrahymena* ciliary protein previously thought to be a calmodulin and to ascertain whether the purified ciliary protein may be identified as a calmodulin.

I. DIRECT EVIDENCE FOR THE OCCURRENCE OF CALMODULIN IN
    TETRAHYMENA CILIA

We have purified a ciliary protein by applying a method for isolating calmodulin from *Tetrahymena* whole cells [8] to

Biological Functions of Microtubules
and Related Structures

115

the isolated cilia of *Tetrahymena*. The ciliary protein purified to homogeneity co-migrated with *Tetrahymena* whole cell
calmodulin in SDS-polyacrylamide gel electrophoresis (Fig. 1),
and showed a calcium-dependent mobility change in alkali-glycerol gel electrophoresis, as did whole cell calmodulin [9].
In addition, the ciliary protein reacted with an anti-*Tetrahymena* calmodulin [5] to give rise to a single precipitin line
which was completely confluent with the line formed between
the *Tetrahymena* whole cell calmodulin (or cell body calmodulin) and the antiserum [9]. Fig. 2 shows the curves of dose-
dependent brain cAMP phosphodiesterase activation by the ciliary protein and calmodulin from the deciliated cell bodies.
The calcium-dependent activation curve of the ciliary protein
was identical with that of cell body calmodulin.

     Thus, we may conclude that the ciliary protein we iso-
lated is in fact *Tetrahymena* calmodulin and that therefore
calmodulin is definitely included in *Tetrahymena* cilia.

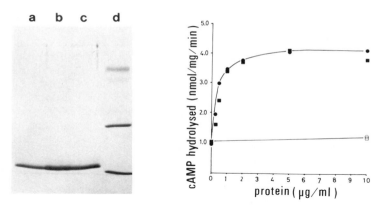

     *Fig. 1 (left). Co-migration of ciliary protein and whole
cell calmodulin in SDS-polyacrylamide gel electrophoresis.
(a) Ciliary protein; (b) a mixture of ciliary protein and
whole cell calmodulin; (c) whole cell calmodulin; (d) a mixture of molecular weight marker, ovalbumin, chymotrypsinogen A
and ribonuclease A. Electrophoresis was performed in 12.5%
polyacrylamide gel after Laemmli [10]. (Ohnishi, Suzuki and
Watanabe [9], reproduced by permission.)*

     *Fig. 2 (right). Activation of porcine brain cAMP phos-
phodiesterase by ciliary protein (circles) and cell body calmodulin (rectangles).   The reaction was performed in the
presence of 1 mM Ca^{2+} (filled symbols) or 1 mM EGTA (open
symbols).   (Ohnishi, Suzuki and Watanabe [9], reproduced by
permission.)*

II. DISTRIBUTIONS OF CALMODULIN AND ITS COUNTERPART WITHIN *TETRAHYMENA* CILIUM

When 8 M urea-extract of whole cilia was subjected to al-kali-glycerol gel electrophoresis in the presence of 2 mM EGTA, a fast-migrating calmodulin band was clearly seen; but in the presence of 2 mM Ca^{2+} the band disappeared completely (Fig. 3A). This suggests that calmodulin and its counterpart form a complex in a calcium-dependent manner. To confirm the formation, the gel lanes, shown in Fig. 3A, were further elec-trophoresed in a second dimension in the presence of 5 mM EGTA. In the secondary electrophoresis, calmodulin emerged from the top of the gel in which it had been initially elec-trophoresed in the presence of Ca^{2+}, while calmodulin emerged from only the fast-migrating band of the gel which had been initially electrophoresed in the presence of EGTA (Fig. 3B). The results indicate that calmodulin can form a calcium-dependent complex with a certain ciliary protein.

We then prepared ciliary subfractions according to a method similar to that of Gibbons [12] and electrophoresed them in the alkali-glycerol gel with or without Ca^{2+}. As shown in Fig. 4, the bulk of calmodulin was present in the membrane-matrix fraction and outer-doublet microtubule fraction.

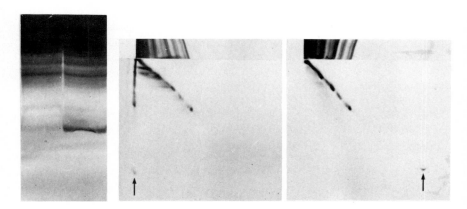

Fig. 3.   Two dimensional gel electrophoresis of whole cilia extract.   In A, 8 M urea-extract of cilia was electro-phoresed in alkali-glycerol gel [11] with 2 mM Ca^{2+} (a) and 2 mM EGTA (b).   The two lanes were loaded separately on the second dimensional gels and electrophoresed in the presence of 5 mM EGTA (B).   Arrows indicate calmodulin spots. (Ohnishi, Suzuki and Watanabe [9], reproduced by permission.)

Furthermore, the presence of the calmodulin counterpart was
strongly suggested in the outer-doublet microtubule fraction:
the fast-migrating calmodulin band seen in the presence of
EGTA disappeared in the presence of Ca^{2+} as shown in Fig. 4.

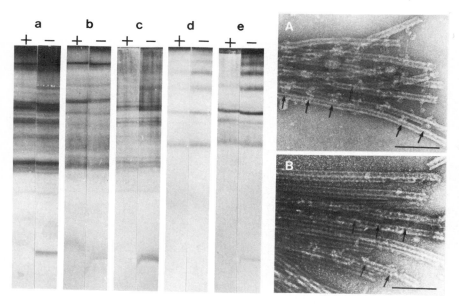

*Fig. 4 (left). Alkali-glycerol gel electrophoresis of
subciliary fractions. (a) Whole cilia; (b) membrane-matrix
fraction; (c) axoneme fraction; (d) crude dynein fraction;
(e) outer-doublet microtubule fraction. The subciliary
fractions were solubilized in equal volumes of 8 M urea and
electrophoresed in the presence of 2 mM Ca^{2+} (+) and 2 mM EGTA
(-). (Ohnishi, Suzuki and Watanabe [9], reproduced by permission.)*

*Fig. 5 (right). Localization of calmodulin within cilium
as revealed by indirect immunoelectron microscopy. Outer-
doublet microtubule fraction was washed with Tris-Mg^{2+}-NaCl
(2 mM $MgCl_2$, 145 mM NaCl, 1 mM mercaptoethanol, 30 mM Tris-HCl,
pH 8.3) including either 1 mM Ca^{2+} (A) or 1 mM EGTA (B). Each
sample was incubated with anti-calmodulin antibody and then
with ferritin-conjugated anti-rabbit IgG antibody. Appropriate
dilutions for the first and second antibodies were about 1/80
and 1/1000, respectively. Arrows indicate periodic structures
which are connecting neighbouring doublet microtubules with
each other. Bar represents 0.2 μm. (Ohnishi, Suzuki and
Watanabe [9], reproduced by permission.)*

On the other hand, the presence of the calmodulin counterpart
was not suggested in the membrane-matrix fraction and crude
dynein fraction. This might be due to an insufficiently small
quantity of the counterpart molecule in these fractions or to
the denaturation (loss of capacity for complex fromation) of
the counterpart molecule caused by 8 M-urea treatment.
However, as for the outer-doublet microtubule fraction, calcium-dependent complex formation between calmodulin and its
counterpart occurred in the electrophoretic system, similar to
that seen in the complex forming between calmodulin and troponin I or between troponin C and troponin I [11]. Gitelman and
Witman [2] and Blum *et al.* [13] also proposed the presence of
calmodulin in the outer-doublet microtubules of *Chlamydomonas*
flagella and *Tetrahymena* cilia, respectively. However, they
have not yet investigated the localization of calmodulin. In
the present study, we were able to show for the first time the
calmodulin localization within ciliary axonemes using immunoelectron microscopy (Fig. 5). We found that immunoferritin
particles were localized along the longitudinal axis of the
microtubule at regular intervals of 90 nm, which coincided
well with the spacing of radial spokes or interdoublet links.
In the outer-doublet microtubule preparation, no spoke head
was seen and ferritin particles were shown to be clustered on
the structures laterally linking a doublet microtubule to an
adjacent one.

Thus, it is most likely that calmodulin is localized on
the interdoublet links, and that the calmodulin counterpart is
contained within the interdoublet link.

III. EFFECT OF CALMODULIN-INHIBITORS ON CILIARY MOVEMENT

Aside from the specific localization of calmodulin in the
ciliary axonemes, we observed that a potent inhibitor of calmodulin function, trifluoperazine or chlorpromazine, caused
living *Tetrahymena* cells to shift from forward swimming to
backwark swimming at a final concentration of 40 μM [5]. This
urged us to investigate the effect of trifluoperazine or
chlorpromazine on calcium-dependent ciliary reversal in the
Triton model of *Paramecium* [14], since reactivation of the
Triton model of *Tetrahymena* is not as easily achieved. Before
the trial, we confirmed that calmodulins isolated from *Tetrahymena* and *Paramecium* were indistinguishable from each other
in terms of several characteristics such as their antigenic
properties [5], activation activities against brain phosphodiesterase and *Tetrahymena* guanylate cyclase [15] and in addition confirmed that trifluoperazine was able to potently inhibit the activation of these enzymes by *Tetrahymena* calmodu-

lin [5].

As shown in Table I, calcium-dependent ciliary reversal in the *Paramecium* Triton model was not affected by the addition of trifluoperazine (or chlorpromazine) even at a final concentration of 100 µM. This suggested that i) trifluoperazine might not inhibit all of the diverse functions of calmodulin; ii) trifluoperazine-binding sites of calmodulin might have been occupied by the association between calmodulin and a certain ciliary protein before the addition of trifluoperazine; or iii) a ciliary calcium-binding protein other than calmodulin might play a crucial role in the calcium-dependent ciliary reversal.

TABLE I. *Effects of Calmodulin Inhibitors on the Direction of Ciliary Movement of Triton Model of* Paramecium

Preincubation		Reactivation		Swimmingᵇ
Ca^{2+}	TFP(CPZ)ᵃ	Ca^{2+}	TFP(CPZ)	
		−	−	F
		+	−	B
		+	+	B
−	+	−	+	F
−	+	+	+	B
+	+	−	+	Fᶜ
+	+	+	+	Bᶜ
		− → +	−	F → B
		− → +	+	F → B
		+ → −	−	B → F
		+ → −	+	B → F

ᵃTFP(CPZ), *Trifluoperazine (chlorpromazine).*
ᵇF, *Forward swimming; B, backward swimming.*
ᶜ*Swimming velocity was markedly reduced.*

IV. EVIDENCE SUGGESTING THE OCCURRENCE OF ANOTHER CALCIUM-BINDING PROTEIN IN *TETRAHYMENA* CILIA

To detect any calcium-binding protein other than calmodulin, the isolated *Tetrahymena* cilia were sonicated and heat-treated (95°C for 30 min), and the soluble ciliary protein fraction which was recovered by TCA precipitation was subjected to alkali-glycerol gel electrophoresis with or without Ca^{2+}. As shown in Fig. 6a, a protein band showing a cal-

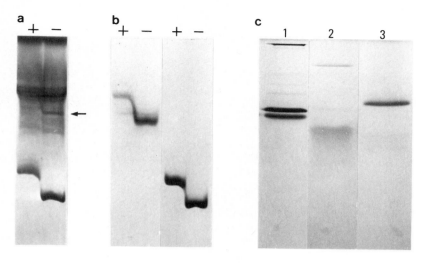

Fig. 6. Another calcium-binding protein present in
Tetrahymena *cilia. (a) Ciliary protein preparation was sub-*
jected to alkali-glycerol gel electrophoresis with $Ca^{2+}(+)$ *and*
EGTA (-). For the preparation of the ciliary protein, see the
text. An arrow indicates a putative calcium-binding protein.
(b) Calmodulin (right) and another calcium-binding protein
purified partially (left) were subjected to alkali-glycerol
gel electrophoresis with Ca^{2+} *(+) and EGTA (-). (c) These pro-*
teins were electrophoresed in SDS-polyacrylamide gel.
1, Marker proteins (myosin light chains, ribonuclease A and
cytochrome C); 2, another calcium-binding protein; 3, cal-
modulin.

cium-dependent mobility change (arrow) was detected in addition to the calmodulin band. The protein did not appear to be
an artifact caused by the heat treatment or TCA treatment,
since the protein band was clearly seen when the native ciliary protein fraction was fractionated by ammonium sulfate fractionation (60 - 100% saturation) and subjected to alkali gel
electrophoresis. This putative calcium-binding protein could
be partially purified by cutting out the band containing this
protein from the preparative electrophoretic gel. The purified protein showed a calcium-dependent mobility change as
had calmodulin (Fig. 6b), but its mobility in the alkali gel
electrophoresis was estimated as 51% while the migration distance of calmodulin was taken as 100. The apparent molecular
weight of this protein was much smaller (8,000 - 10,000) than
that of calmodulin (15,000) as revealed by SDS-polyacrylamide
gel electrophoresis (Fig. 6c). We shall hereafter refer to
the small ciliary calcium-binding protein from *Tetrahymena* as
scTCBP for convenience.

We had previously considered the possibility that the
scTCBP might be a degraded fragment of *Tetrahymena* calmodulin.
However, that does not seem to be the case, because i) although a degraded fragment appeared when calmodulin was digested in a supernatant (10,000 g) of *Tetrahymena* which included protease *in situ,* the fragment was quite different from
scTCBP and migrated much faster than calmodulin in the alkali-
glycerol gel; ii) when reactivity of scTCBP to anti-*Tetrahymena* calmodulin was tested in an Ouchterlony agar plate, no
precipitin line was formed between them; and iii) when *Tetrahymena* calmodulin and scTCBP were treated with cyanogen
bromide, calmodulin was easily cleaved into several small
fragments and virtually no intact molecule remained, whereas
scTCBP was not cleaved under the same conditions.

Thus, we consider that *Tetrahymena* cilia contain two
types of calcium-binding proteins, calmodulin and scTCBP.
Biochemical studies to date on scTCBP are not yet sufficient,
but we are tempted to speculate that scTCBP has several points
of similarity to the vitamin D-induced calcium-binding proteins from mammalian intestine [16].

V.   CONCLUDING REMARKS

We isolated calmodulin from *Tetrahymena* cilia to obtain
direct evidence for the occurrence of calmodulin in the organelle.   The ciliary calmodulin was identical with calmodulins
from deciliated cell bodies and whole cells in terms of molecular size, calcium-dependent change in electrophoretic
mobility, antigenicity and activation of brain cyclic nucleotide phosphodiesterase.   Within the cilium, calmodulin was
mainly present in the membrane plus matrix fraction and outer-
doublet microtubule fraction.   In the latter case, calmodulin
was localized on the interdoublet links.

We also detected another small calcium-binding protein in
Tetrahymena cilia and purified it.

The biological roles of calmodulin and the small ciliary
calcium-binding protein in ciliary reversal remain to be
elucidated, but the evidence presented here may be useful in
the future for explaining how these two types of calcium-
binding proteins are involved in the control of ciliary
movement.

REFERENCES

1. Jamieson, G. A. Jr., Vanaman, T. C., and Blum, J. J.,
    Proc. Natl. Acad. Sci. U.S.A. 76, 6471 (1979).
2. Gitelman, S. E., and Witman, G. B., *J. Cell Biol.* 98,
    764 (1980).
3. Satir, B. H., Garofalo, R. S., Gilligan, D. M., and
    Maihle, N. J., *Ann. NY Acad. Sci.* 356, 83 (1980).
4. Maihle, N. J., Dedman, J. R., Means, A. R., Chafouleas,
    J. G., and Satir, B. H., *J. Cell Biol.* 89, 695 (1981).
5. Suzuki, Y., Ohnishi, K., and Watanabe, Y., *Exptl. Cell
    Res.* In press.
6. Watanabe, Y., and Nozawa, Y., *In* "Calcium and Cell
    Function" Vol. II (W. Y. Cheung, ed.), p. 297. Academic
    Press, New York, (1981).
7. Van Eldik, L. J., Piperno, G., and Watterson, M., *Proc.
    Natl. Acad. Sci. U.S.A.* 77, 4779 (1980).
8. Suzuki, Y., Nagao, S., Abe, K., Hirabayashi, T., and
    Watanabe, Y., *J. Biochem.* 89, 333 (1981).
9. Ohnishi, K., Suzuki, Y., and Watanabe, Y., *Exptl. Cell
    Res.* In press.
10. Laemmli, U. K. , *Nature* 227, 680 (1970).
11. Suzuki, Y., Hirabayashi, T., and Watanabe, Y., *Biochem.
    Biophys. Res. Commun.* 90, 253 (1979).
12. Gibbons, I. R., *Proc. Natl. Acad. Sci. U.S.A.* 50, 1002
    (1963).
13. Blum, J. J., Hayes, A., Jamieson, G. A. Jr., and Vanaman,
    T. C., *J. Cell Biol.* 87, 386 (1980).
14. Naitoh, Y., and Kaneko, H., *Science* 176, 523 (1972).
15. Kudo, S., Ohnishi, K., Muto, Y., Watanabe, Y., and
    Nozawa, Y., *Biochem. Int.* 3, 255 (1981).
16. Wasserman, R. H., Fullmer, C. S., and Taylor, A. N., *In*
    "Vitamin D" (D. E. M. Lawson, ed.), p. 133. Academic
    Press, New York, (1978).

CHAPTER 13

MOLECULAR COMPOSITION AND STRUCTURE OF DYNEIN ARMS[1]

Yoko Yano
Hideo Mohri

Department of Biology
College of General Education
University of Tokyo, Tokyo

Chikashi Toyoshima
Takeyuki Wakabayashi

Department of Physics
Faculty of Science
University of Tokyo, Tokyo

More than a decade has past since the name "tubulin" was proposed for the main constituent of microtubules, and the idea that a system consisting of tubulin and dynein is functioning in eucaryotic cells in addition to the actomyosin system emerged (1,2). In the movement of cilia and flagella, the sliding filament model has been generally accepted. There is abundant evidence that active sliding displacement between adjacent doublet microtubules occurs as a result of dynein-microtubule interaction (3-5). Dynein, which resides in two rows of arms along each doublet microtubule, forms cyclic cross-bridges with an adjacent doublet and converts the chemical energy of ATP to mechanical movement. However, our present knowledge about the dynein molecule is rather scarce both in biochemical and morphological aspects. To elucidate the cross-bridge mechanism, it is important to examine the

1. This work was supported in part by a Grant-in-Aid for Special Project Research and a Grant-in-Aid for Scientific Research (H. M.), a Grant-in-Aid for Special Project Research and a Grant-in-Aid for Scientific Research (T. W.) from the Ministry of Education, Science and Culture of Japan.

structure of the dynein arms.   In the present report we
describe the morphology and substructure of the dynein arms.

I. MORPHOLOGY OF DYNEIN ARMS BOUND TO DOUBLET MICROTUBULES

    The structure of the dynein arms has been studied using
negatively contrasted images of ciliary axonemes of *Tetrahymena* disintegrated in a Mg-ATP solution, such as are shown
in Fig. 1.   Analyzing similar images, Warner and Mitchell (6,
7) and Satir (8) claimed that the free dynein arms on the
doublet are extended and can be seen to be comprised of three
spherical subunits, and that these subunits resemble in size
and shape the 9.3 nm particles found in the high salt extract
from the axonemes (9).   They postulated that these particles
correspond to the monomeric form of dynein.   They also noted
that the dynein arms between adjacent doublets show somewhat

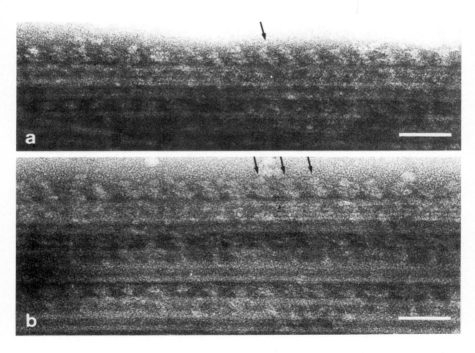

FIGURE 1.   Tetrahymena *ciliary axonemes disintegrated in
Mg-ATP.   Negatively stained with uranyl acetate (pH 5.2).   The
bars indicate 50 nm.   Arrows indicate the projections which
are apparently not composed of three subunits.   (See text for
details.)*

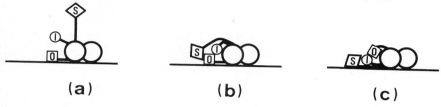

(a)   **(b)**   **(c)**

FIGURE 2. Schematic illustration of a cross-sectional view of a free doublet on a support film prepared for electron microscopy, when the preservation of the specimen is ideal (a), and not ideal (b), (c).

different features, termed a flattened configuration (8). According to these authors, such an apparent difference in shape would represent different stages in the cross-bridge cycle of the dynein arms.

As we analyze these images, it is important to note that the structure of the 9+2 system is very complicated and that two rows of dynein arms and a row of radial spokes are located on the A-tubule of the doublet. Furthermore, the images recorded on electron micrographs are only the projections of particular structures. Thus, a question arises as to whether the apparent difference in shape really represents a difference in conformation of the dynein cross-bridge. The possible situations which we would encounter are schematically illustrated in Fig. 2. If the preservation of the specimen is ideal, the doublet will lie on a support film in a manner shown in Fig. 2(a). In this case, since the staggering of the two rows of arms is smaller than the dimension of one arm along the tubule axis (10, 11), the two rows of arms are recorded, almost inevitably, superimposed, when the negative staining reagent contrasts all of these components. If the preservation is not ideal, not only the two rows of arms but also a row of radial spokes may be superimposed as shown in Fig. 2(b) and (c).

In micrographs such as those shown in Fig. 1, we cannot distinguish all of these components. Therefore, we cannot claim that particular projections observed here represent a single row of dynein arms, or individual arms. For the present, we use the term "projection" rather than "arm" to describe the structure projected from the doublets.

In Fig. 1(a), it is certain that some projections on the

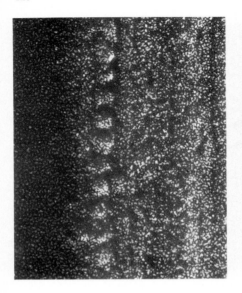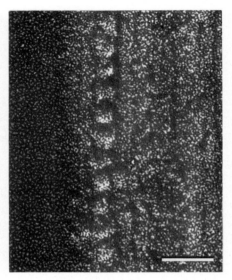

FIGURE 3. Stereo pair of a mechanically frayed axoneme of
Tetrahymena *cilia, glycerol dried and bidirectionally shadowed*
with platinum-carbon (12) at an elevation angle of 25°.
Tilting angles ± 7°. The bar indicates 50 nm.

uppermost doublet appear to be composed of three subunits.
However, the projection indicated by an arrow, appears to be
separated into only two *particles,* and we cannot identify the
projection as a single arm. In Fig. 1(b), many projections,
such as those indicated by arrows, appear to be composed of
two *particles* rather than three *subunits.* The length of the
globular part of the *particle* is 16.8 ± 1.1 nm, while that of
the projection observed in the space between the juxtaposed
doublets and that of the "extended" projections are 16.7 ±
1.5 nm and 25.3 ± 2.5 nm, respectively. The size and shape of
the *particle* are very similar to the former and the length of
the globular part is evidently smaller than the latter.

The original micrograph of Warner and Mitchell (Fig. 9a of
(6)) also shows these features: the projections, especially
the ones at the lower end of the row of free projections,
consist of two *particles* rather than three *subunits* and these
particles resemble the projections observed between the
juxtaposed doublets. Further, the projection located at the
end of the row of free projections is composed of only one
particle, and a thin rod which connects the globular part to
the doublet is recognizable.

All these observations suggest that the extended configuration of the projections can be interpreted as an overlap of

two rows of dynein arms.

Further precise information can be obtained either with the shadow casting technique which reveals the relief of the specimen, or by analyzing a reconstituted specimen which consists of a single row of arms and a doublet.

Fig. 3 presents a stereo pair of a mechanically frayed axoneme which was glycerol dried and bidirectionally shadowed with platinum-carbon (12). Stereo views allow us to inspect the three-dimensional structure of the specimen. A row of globular heads, which are not so elongated, and the dark region between the globular parts and doublet are observed at the left side of the axoneme. These images suggest the existence of a relatively thin rod or stalk, mentioned previously, connecting the globular part of the projection and the doublet. Such a thin rod was also recognized in the images of negatively stained specimens.

It is known that the dynein extracted from axonemes rebinds to exhaustively dialyzed doublet microtubules which have no projections, and reconstitutes the arm structure. If the amount of dynein is low enough to avoid overlapping, it should be possible to observe the morphology of individual dynein arms. We examined these specimens and found that the individual arms apparently consist of a single globular part, and that the proximal region connected to the doublet is slender. These features are consistent with the micrographs obtained by Takahashi and Tonomura (13). The measured dimensions of the globular part are in good agreement with those of the single particle in the row of free projections and those of the projections situated between the juxtaposed doublets: the length of the globular part is evidently shorter than that of the "extended projections". Hitherto all the data presented were derived from specimens of *Tetrahymena* cilia, but almost the same results were obtained concerning the dynein arms of sea urchin sperm flagella.

II. MORPHOLOGY OF 21S DYNEIN ISOLATED FROM AXONEMES

Electron microscopy of the particles of isolated dynein was performed. The 21S fraction (14) of dynein extracted from sea urchin (*Strongylocentrotus* and *Pseudocentrotus*) sperm flagellar axonemes was used, since 21S dynein is known to be derived from the outer arms of the axonemes, whereas the 30S dynein of *Tetrahymena* cilia may contain both outer and inner arms. Moreover, the 21S dynein can reconstitute the arm structure, and also restore the beat frequency of demembranated and reactivated sea urchin sperm without outer arms (15).

Fig.4 shows electron micrographs of a negatively stained

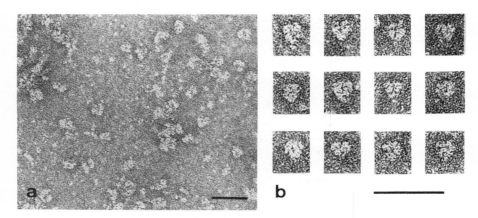

FIGURE 4. *21S dynein of sea urchin sperm flagella, negatively stained with 1% uranyl oxalate (pH 6.5). A large field of view (a) and the selected images (b). The bars indicate 50 nm.*

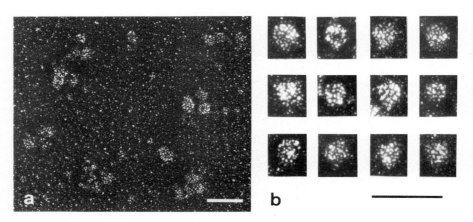

FIGURE 5. *21S dynein of sea urchin sperm flagella, glycerol dried and rotary shadowed with platinum-carbon (12) at an elevation angle of 15°. A large field of view (a) and the selected images (b). The bars indicate 50 nm.*

21S dynein fraction. The shape of these particles resembles a rounded triangle. The dimensions of these particles are 16.7 $\pm$ 1.1 nm in length and 14.2 $\pm$ 1.3 nm in width. These values are consistent with those of single *particles* observed in the row of free projections.

The 21S particles were also visualized using rotary shadowing technique. As shown in Fig. 5, similar morphology to that of the negatively stained 21S particles was observed.

From these observations, we consider that the dynein arm is not a linear polymer of three spherical subunits, but consists of a single globular particle and that a thin rod presumably connects the globular part and the doublet.

III. SUBSTRUCTURE OF 21S DYNEIN

As the next step, the substructure of the 21S particle was studied. The SDS gel electrophoresis pattern reveals that the 21S fraction contains several polypeptides, that is, two heavy chains termed A_α and A_β (16), three intermediate chains and several light chains (Fig. 6). The molar ratio of A_α and A_β is almost unity throughout the 21S fractions.

Various values have been reported for the molecular weight of the dynein heavy chain due to the lack of suitable high molecular weight markers. As markers, we used bovine serum albumin (BSA) oligomers cross-linked with glutaraldehyde (18). The mobility and the logarithm of molecular weight of BSA oligomers showed very good linearity in a 3% gel using the

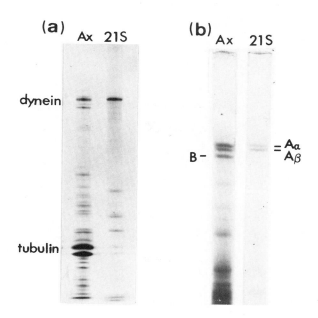

FIGURE 6.   SDS polyacrylamide gel electrophoretic pattern of whole axonemes and the 21S dynein fraction of sea urchin sperm flagella. (a) 4-15% acrylamide gradient gel (16). (b) Upper region of the 2.8% gel (17).

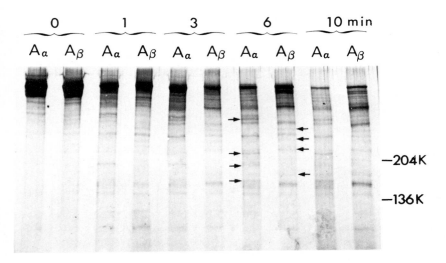

FIGURE 7. *Time course of the degradation of the heavy chains,* A_α *and* A_β*, by trypsin digestion at 37°C in the presence of SDS. After electrophoresis on a 6% gel (21), the gel was silver-stained (22).*

(a)

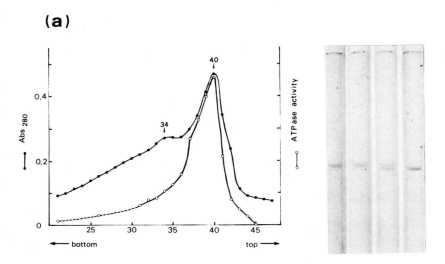

FIGURE 8. *Sedimentation profile of the Tris-EDTA dialyzed 21S fraction by 5-20% sucrose density gradient centrifugation (a). Absorbance at 280 nm (•) and ATPase activity in arbitrary unit (○). High molecular weight regions of 2.8% gel patterns (b) of fractions in (a).*

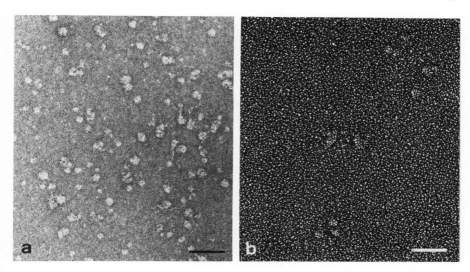

*FIGURE 9. Dissociated particles after dialysis of the 21S
fraction against a Tris-EDTA solution. Negatively stained
with 1% uranyl oxalate (pH 6.8) (a), and rotary shadowed with
platinum-carbon (b). The bars indicate 50 nm.*

phosphate buffer system (19), although heavy chains could not
be separated into two bands. A value of around 500K daltons
was obtained for the weight of an A-band heavy chain.
      In order to analyze the two heavy chains, the 21S dynein
fraction was reacted with a fluorescent reagent N-(7-
dimethylamino-4-methyl coumarinyl) maleimide (DACM) (20), and
then fractionated by SDS gel electrophoresis. A_α and A_β
chains were collected separately from the gel of DACM-labelled
specimens, and were subjected to chemical analysis.
      Fig. 7 shows a silver-stained gel pattern of the heavy
chains digested with trypsin in the presence of SDS. It
reveals that some bands are different, as indicated by arrows.
Preliminary results of amino acid analysis indicate that the
A_α and A_β chains are similar to each other, although the
cysteine content of the A_α chain is higher than that of the A_β
chain.
      The 21S particle can be dissociated into smaller particles
by dialyzing it against Tris-EDTA solution. These were partially fractionated by sucrose density gradient centrifugation. As illustrated in Fig. 8, the sedimentation profile
showed two peaks. The major component of fractions 34 and 40
were the A_α and A_β chains, respectively. In the SDS gel pattern of fraction 34, a faint A_β band was also observed,
whereas in that of fraction 40, only an A_β band was visible in

the high molecular weight region. The profile of ATPase activity coincided almost exactly with that of A_β chain in this
case.

   In order to check the affinity of these dissociated subunits for doublet microtubules, co-sedimentation experiments
were carried out. The results showed that the A_α subunit has
a higher affinity than the A_β subunit in the presence of divalent cations, especially calcium ion.

   All these biochemical data indicate that the A_β chain is
not a fragment of the A_α chain, and that there are two
distinct types of heavy chains in the 21S fraction.

   The dissociated products of the 21S fraction could also be
visualized in the electron microscope. Fig. 9 presents the
images of the dissociated 21S components negatively stained
(a) and rotary shadowed (b). They look considerably smaller
than the 21S particles, and the dimensions of the two subunits
are nearly the same.

   At present, we cannot propose a precise model of the sub-
molecular structure of the 21S particle. However, it is
encouraging that sometimes we can observe the substructure in
the images of the negatively stained 21S particles (23). We
expect that further image analysis will reveal the sub-
molecular architecture of the dynein arms.

ACKNOWLEDGMENTS

   The authors wish to thank Dr. I. Mabuchi for valuable
suggestions in biochemical analysis and Dr. M. Yamazaki for
the use of his amino acid analyzer. Thanks are also due to
Dr. M. M. Pratt for reading through the manuscript.

REFERENCES

1. Mohri, H., *Nature 217*, 1053 (1968).
2. Mohri, H., *Biochim. Biophys. Acta 456*, 85 (1976).
3. Satir, P., *J. Cell Biol. 39*, 77 (1968).
4. Summers, K. E., and Gibbons, I. R., *Proc. Natl. Acad.
    Sci. U.S.A. 68*, 3092 (1971).
5. Gibbons, B. H., and Gibbons, I. R., *J. Cell Biol. 54*, 75
    (1972).
6. Warner, F. D., and Mitchell, D. R., *J. Cell Biol. 76*, 261
    (1978).
7. Warner, F. D., and Mitchell, D. R., *Intern. Rev. Cytol.
    66*, 1 (1980).
8. Satir, P., *in* "The Spermatozoon" (D. W. Fawcett and J. M.

Bedford, eds.), p. 81. Urban and Schwarzenberg,
Baltimore, (1979).

9. Warner, F. D., Mitchell, D. R., and Perkins, C. R., *J.
    Mol. Biol.* *114*, 367 (1977).

10. Warner, F. D., *J. Cell Sci.* *20*, 101 (1976).

11. Amos, L. A., Linck, R. W., and Klug, A., *in* "Cell
    Motility" (R. D. Goldman, T. D. Pollard, and J. L.
    Rosenbaum, eds.), Book C, p. 847. Cold Spring Harbor
    Laboratory, Cold Spring Harbor, (1976).

12. Tyler, J. M., and Branton, D., *J. Ultratstruct. Res.* *71*,
    95 (1980).

13. Takahashi, M., and Tonomura, Y., *J. Biochem.* *84*, 1339
    (1978).

14. Gibbons, I. R., and Fronk, E., *J. Biol. Chem.* *254*, 187
    (1979).

15. Gibbons, B. H., and Gibbons, I. R., *J. Biol. Chem.* *254*,
    197 (1979).

16. Bell, C. W., Fronk, E., and Gibbons, I. R., *J. Supramol.
    Struct.* *11*, 311 (1979).

17. Mabuchi, I., and Shimizu, T., *J. Biochem.* *76*, 991 (1974).

18. Payne, J. W., *Biochem. J.* *135*, 867 (1973).

19. Weber, K., and Osborn, M., *J. Biol. Chem.* *244*, 4406
    (1969).

20. Yamamoto, K., Sekine, T., and Kanaoka, Y., *Anal. Biochem.*
    79, 83 (1977).

21. Laemmli, U. K., *Nature* *227*, 680 (1970).

22. Oakley, B. R., Kirsch, D. R., and Morris, N. R., *Anal.
    Biochem.* *105*, 361 (1980).

23. Yano, Y., and Miki-Noumura, T., *Biomed. Res.* *2*, 73
    (1981).

CHAPTER 14

DYNEIN AND ITS ROLE IN CELL MOTILITY

Christopher W. Bell

Pacific Biomedical Research Center
University of Hawaii
Honolulu, Hawaii
USA

I. INTRODUCTION

This paper combines a short discussion of the status of dynein regarding general cellular motility with a brief report on the work carried out in this laboratory on the detailed structural and functional characterization of latent activity dynein-1 (LAD-1) from sea urchin sperm flagella. I believe this combined approached (which was editorially inspired) may effectively point out the difficulties involved in identifying the participation of dynein in cellular processes other than flagellar and ciliary motility, and aid in the clarification of these difficulties by suggesting more stringent structural *and functional* criteria for the identification of dynein than have, hitherto, been generally used. In the remainder of this introduction, I wish to develop the idea of dynein as a molecule that is functionally dependent on other structures (specifically microtubules) and, more importantly, conceptually dependent to a large extent on other, better defined systems.

A. *Dynein and Microtubules*

Dynein is inevitably thought of in conjunction with microtubules. The functional and structural link is made

This work was supported by grant HD 10002 from the National Institute of Child Health and Human Development.

explicit by the large body of data relating to the action
of dynein arms in driving the motility of cilia and eukaryotic
flagella (1-4). A more conceptual link between dynein and
microtubules is found in the general assumption that dynein-
like proteins isolated from sources other than cilia and
flagella are intimately involved in the functioning of
microtubule-based systems, even when the evidence for this
relationship is, at best, indirect (5-7). In this we see
the beginnings of a conceptual paradigm, that wherever
dynein is found in the eukaryotic cell, its function is
likely to involve motile processes mediated by microtubules.
The converse, however, does not appear to be true, and
microtubules are generally recognized in their own right
as structurally and conceptually distinct entities with no
absolute dependence on dynein (8).

B. *Dynein and Myosin*

The paradigm alluded to above is also mirrored to a
certain extent in studies on actin and myosin, with actin
filaments (microfilaments) taking the place of microtubules
and myosin taking the place of dynein. Indeed, since the
demonstration of the sliding tubule mechanism of ciliary
and flagellar motility (9,10), and even to a certain extent
before this, the relatively well understood actomyosin
system, in both its muscle and non-muscle forms, has been
a fruitful source of ideas for the investigation of the
more recently defined dynein-microtubule system. In particular, myosin has tended to assume the status of a role
model for dynein, influencing choice of experimental methods
(e.g., limited proteolytic digestion (11,12)) and ideas on
the structure and function of the molecule (e.g., kinetic
schemes for cyclic dynein-ATP-tubulin interactions based
on those for myosin-ATP-actin interactions (13,14)). However,
despite the basic similarity represented by the sliding
filament mechanism, all dyneins so far studied are structurally and enzymatically very different from myosin (15,16),
and the dynein-microtubule system is organizationally
quite distinct from the actomyosin system (17,18).
The use of muscle research as a conceptual foundation
for the investigation of ciliary and flagellar motility
was almost inevitable. The danger is that too great a
dependence for ideas on the example of the actomyosin
system may cause the wrong questions to be asked about the
dynein-microtubule system, and may inhibit the recognition
in it of new and different features.

C. Dynein and Dynein

One of the major reasons for the frequent appeal to the results of myosin research is that myosin molecules, with a few well-defined exceptions (e.g., *Acanthamoeba* myosin I (19)) appear to be physically and enzymatically similar over a wide variety of species and cell types, although differences do appear at more detailed levels of comparison (20). Dynein research, however, is afflicted by a diversity and multiplicity within its subject material which has hindered the recognition of the fundamental properties of the dynein molecule. This diversity shows up at a relatively gross level in the recognition in cilia and flagella of two dynein arms, the inner and outer arms, that appear structurally different in the electron microscope (21). Physical heterogeneity is displayed in most cilia and flagella in the existence of two extractable forms of dynein, one sedimenting at 18-30S, the other sedimenting at 10-14S (22-26). It is possible that one of these forms may represent the outer and the other form the inner dynein arm, but the observation in some cases of a reversible conversion of the fast sedimenting form of dynein into a slow sedimenting form obscures this possibility (25,27,28), and recourse has been made to the method of sodium dodecyl sulfate polyacrylamide gel electrophoresis (SDS-PAGE) to discover the basis of these sedimentation differences. Where this has been done, it has been found that the polypeptide compositions of the fast forms of dynein (and their derivatives) are distinct from the slow forms of dynein (23-25,29). This high resolution technique has also given the clearest demonstration of the structural heterogeneity of dynein between species, within species and even within a single organelle. The use of SDS-PAGE rapidly established the possession of very high molecular weight polypeptide chains as a diagnostic for dynein (30,31). However, as Fig. 1 shows, these bands are manifold, and subject to redefinition as electrophoretic resolution increases. The values reported for the molecular weights of these polypeptide chains have varied widely between about 300,000 and 600,000, but it is becoming increasingly apparent that these differences are largely extrinsic, arising primarily from the choice of electrophoretic buffer system and from the lack of adequate calibration polypeptides (15). Regardless of this, it may be legitimately questioned whether the identification of all of these high molecular weight polypeptides with dynein is justified. The answer to this question hinges on the non-trivial problem of defining what dynein is and does. In the case of cilia and flagella,

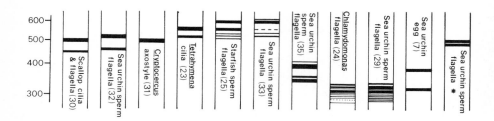

Fig. 1. High molecular weight polypeptide compositions of dyneins and dynein-like proteins from various sources. Apparent molecular weight scale (in thousands) at left. Numbers in brackets are references (*see - Yano et al., these proceedings). Adapted, with additions, from Mabuchi (33).

where the obvious morphological markers of the inner and outer arms are present, the answer appears to be a qualified yes. In the most extreme case of dynein polypeptide multiplicity so far studied, 9 out of 10 high molecular weight components of *Chlamydomonas* flagella have been implicated in arm structure (34). Four of these have been shown to be associated, in a non-overlapping manner, with two distinct ATPases located in the outer arm (24). Similar results, described in more detail later, are available for some of the high molecular weight polypeptides from sea urchin sperm flagella. The recent identification of distinct populations of intermediate and low molecular weight polypeptides specifically associated with different high molecular weight dynein polypeptide ensembles has opened up new levels of complexity and diversity in the study of dynein (24,29).

Even when differences that appear to arise from technique are eliminated, it is abundantly clear that dynein displays considerable inter- and intraspecific heterogeneity at levels of resolution where its counterpart, myosin, appears essentially homogeneous. The question of definition--what is dynein?--is therefore both valid and important. A useful and accurate definition should avoid parochial bias, but should nevertheless be derived from detailed biochemical and structural knowledge of a protein or proteins that are universally recognized as dynein. The only structures that are so recognized are the arms of ciliary and flagellar axonemes, of which dynein-1, comprising most or all of the outer arm in the flagellar axoneme of sea urchin sperm, is an example.

II.   DYNEIN-1:   A WELL DEFINED DYNEIN

The recent work carried out in this laboratory on the structure and function of dynein-1 from sperm flagellar axonemes of the Hawaiian sea urchin *Tripneustes gratilla* makes this one of the best characterized dyneins (3,12,27,28). Selected results from this work are briefly summarized below.

A.   *Results*

The outer arms of the sperm flagellar axonemes of *Tripneustes* can be solubilized in an apparently specific manner by incubation of the axonemes for about 15 min at $0^{\circ}C$ in a solution containing 0.6 M NaCl and 4 mM Mg^{2+} at pH 7 (abbrev. 0.6 M $NaCl/Mg^{2+}$) (27) (Fig. 2A).   The 0.6 M $NaCl/Mg^{2+}$ extract possesses a low latent Mg^{2+}-ATPase activity associated with a 21S particle, LAD-1, that is the major component of the extract.   This ATPase activity can be activated by a variety of procedures, the most effective being incubation with 0.1% Triton X-100 which causes a stable, 10-fold activation (27).   In this condition both the major protein component and the ATPase activity sediment at 10-14S.   The ATPase activity of LAD-1 is also activated up to ten-fold after rebinding to salt-extracted flagellar axonemes (27,35).   The rebinding of the 21S particle results in the specific reappearance of the outer dynein arms in the extracted axonemes.   Furthermore, the rebinding of LAD-1 to the KCl-extracted axonemes of reactivated sea urchin sperm causes their beat frequency, which, after KCl-extraction, drops from ~30 Hz to ~15 Hz, to increase to about 25 Hz (3) (using LAD-1 prepared from osmotically rather than Triton-demembranated sperm, the beat frequency increases as high as 29 Hz (28)).   LAD-1 whose ATPase activity has in any way been previously activated neither rebinds to nor induces beat frequency increase in KCl-extracted reactivated sperm.   These observations establish that the latent 21S form of dynein 1 represents most or all of the outer arm in sperm flagellar axonemes, and that this latent form probably retains many of the natural functional capabilities of the *in situ* outer arm.

Electrophoresis shows that the latent 21S dynein-1 contains nine different polypeptide chains, with apparent molecular weights ranging from about 330,000 to 14,000 (29). Sucrose density gradient sedimentation followed by electrophoresis shows that, after activation by 0.1% Triton X-100, the 21S dynein-1 particle breaks down to at least two particles overlapping in sedimentation velocity at about 10-14S (Fig. 2B). The slower particle contains the A_α heavy chain and the faster the A_β heavy chain and intermediate chains (28). ATPase activity is high and peaks evenly *between* the peaks of the two particles, suggesting that both possess approximately equal ATPase activity. A more effective separation of these particles is achieved by dialysis of the 21S dynein-1 against a low ionic strength imidazole/EDTA solution (pH 7) followed by sucrose density gradient sedimentation in the same conditions. In this case the A_β heavy chain and intermediate chains all sediment at about 9S and the A_α heavy chain aggregates in a smeared "peak" at 12 to 30S (28) (Fig. 2C). If the A_β/intermediate chain (A_β/IC) fraction and A_α chain fractions from such gradients are taken separately and resedimented in the original 0.6 M NaCl/Mg^{2+} conditions, *no 21S particles are reformed* (28). In the A_α fraction, the A_α chain has disaggregated and now sediments at about 12S along with a peak of ATPase activity (Fig. 2D). In the A_β/IC fraction, the A_β chain and intermediate chain 1 now cosediment at about 14S, while intermediate chains 2 and 3 also cosediment but show a wider distribution, peaking at about 9S and 17S; ATPase activity cosediments with A_β chain and intermediate chain 1 (Fig. 2E). Neither the A_α nor the A_β chain associated ATPase activities display the degree of latency observed in LAD-1. If the A_α chain fraction and A_β/IC fraction are mixed before resedimentation in 0.6 M NaCl/Mg^{2+}, up to 60-70% of the protein now sediments as a 21S particle with the same heavy and intermediate chain polypeptide composition as the original 21S LAD-1 (28); the remaining protein sediments at 9-14S (Fig. 2F). The fact that both the A_α and A_β/IC fractions are required for the reformation of 21S particles is the first direct evidence that LAD-1 is a *single species of particle* containing both the A_α and A_β chains in equal quantity, and not merely a mixture of fortuitously cosedimenting particles containing the A_α and A_β chains separately.

The reformed 21S particles regain a substantial degree of latency of ATPase activity, but they are not fully functional, being unable to restore beat frequency to outer arm depleted reactivated sperm. However, they appear partially competent to bind to doublet tubules since they

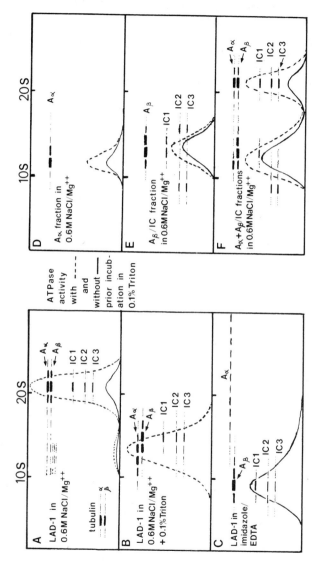

Fig. 2. Diagram showing typical distributions of heavy and intermediate chains (IC) of LAD-1 and its subfractions, under various conditions, in fractionated sucrose density gradients. Typical ATPase activities are superimposed (activities are only qualitatively comparable between the various gradients). Approximate sedimentation coefficient scale shown at top. Appearances correspond to sedimentation in 5–20% w/v sucrose gradients at 35,000 rpm and 4°C for 15 h in a Beckman SW 41 rotor followed by fractionation of gradients and analysis of fractions by SDS-PAGE. See text for further details.

temporarily block the ATP-dependent restoration of beat frequency by subsequently added intact LAD-1 (28). The Aβ/IC fraction also appears able to block the rebinding of LAD-1, although the very rapid removal of this blockage on the addition of ATP suggests it is weaker in this ability than the reformed 21S particle. The dramatic effect of ATP in this case suggests that the Aβ/IC fraction may only bind to the ATP-sensitive dynein arm binding sites on the B-tubules of the axoneme (28). The A_α fraction appears to have no effect on the restoration of beat frequency by LAD-1 to outer arm depleted sperm. However, it appears that the A_α chain may be involved in the binding of the dynein arm to the A-tubule binding site. This is suggested by the ability of the LAD-1 polypeptides to be extracted simply by the addition of ATP (in the *absence* of 0.6 M NaCl) after moderate tryptic digestion of axonemes (12). Such extraction indicates that the A-tubule binding site of the dynein arm is weakened or destroyed by trypsin, whereas the ATP-sensitive B-tubule binding site remains intact. Gel electrophoresis of these extracted LAD-1 polypeptides (which, at such moderate levels of digestion, remain associated in a 21S particle) shows that the Aβ chain is unaffected, but that the A_α chain has been cleaved to a single fragment of about 80% of its original apparent molecular weight (12). This selective digestion suggests that the A_α chain is involved in the dynein arm/A-tubule interaction that is attacked by trypsin.

B. *Interpretation*

The 21S dynein-1 that constitutes the outer arm of sperm flagellar axonemes from *Tripneustes* can be relatively easily separated into three different subunits. The Aβ/IC1 subunit, contains the Aβ heavy chain and intermediate chain 1 in apparently equal molar ratio; the IC2/3 subunit contains intermediate chains 2 and 3, again in equal molar ratio; lastly, the A_α subunit contains only the A_α heavy chain. The distribution of the light chains among these subunits has not yet been determined. The subunits also differ in other properties. Two, the A_α suburit and the Aβ/IC1 subunit, possess ATPase activity but the parameters of these activities differ (28). Most notably, the ATPase activity of the A_α subunit exhibits a lability in the absence of Mg^{2+} which is not displayed by that of the Aβ/IC1 subunit. Various types of evidence suggest that A_α subunit is involved with the stable interaction of the dynein-1 molecule with the A-tubule, while the Aβ/IC1 subunit participates in the ATP-sensitive B-tubule interaction. The role of the IC2/3 subunit is unclear at present, although preliminary data suggest that it can bind to neither the A_α subunit nor the A_β/IC1 subunit alone, but that both must be present simultaneously for binding to occur (28). Although further detailed study of the interactions and properties of the subunits, coupled, ideally, with electron microscopic evidence, is obviously required, a topological model of the 21S dynein-1 molecule can be derived from these data (12) (Fig. 3).

Since LAD-1 from *Tripneustes* is the only dynein that has been analyzed at this level, the generality of the organization depicted below remains to be discovered. However, the demonstration, in a 21S outer arm dynein from the sea urchin *Pseudocentrotus,* of two separate ATPases and a polypeptide composition very similar to that of *Tripneustes* (see Yano *et al.,* in these proceedings) indicates that this organization of dynein molecules may be at least common to the echinoderms. A more general validity for this model is suggested by the work on *Chlamydomonas* flagellar dynein which shows two different ATPases, one sedimenting at 18S, the other at 12S, associated with the outer arm (24,34), although it has not yet been demonstrated that these ATPases initially coexist in the same arm structure.

It is obvious from the results given earlier that considerable functional significance resides in the possession of ATPase latency by *Tripneustes* dynein-1. However, the relevance of this property to other systems is not yet clear since, although the latency of other dyneins is usually much less marked, data are generally not available

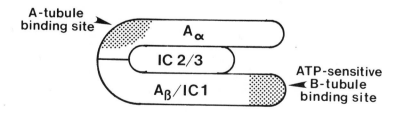

Fig. 3. Topological model of LAD-1 molecule showing relative arrangement of distinct subunits. Only those doublet tubule binding sites for which there is evidence are shown: others may exist. See text for further details.

to determine whether these isolated dyneins could be considered
functional. Nevertheless, there are reasons, further
discussed below, for believing that the possession of a
latent ATPase activity may be an important criterion for
identifying a *functional* dynein.

III. DYNEIN AND MOTILITY

    As dynein is studied at higher resolution and in func-
tionally active states, it becomes possible to define its
properties, and hence its role, with fewer references to
supposedly analogous systems. With a list of these properties
it should be possible to decide whether various proteins
found outside of cilia and flagella are in fact cytoplasmic
forms of dynein. The following list of properties might
be considered:

    1. In keeping with known ciliary and flagellar dyneins,
a putative dynein might be expected to sediment in the
range 18-30S and/or 10-14S, and to contain large polypeptide
chains with molecular weights in the region of 300,000-
500,000. The general applicability of the more detailed
levels of structural organization described earlier for
LAD-1 is too uncertain to be included here.
    2. The protein must possess ATPase activity. Most data
suggest that this activity will require the presence of
divalent cations (15,33).
    3. Hill, Jencks and others have argued (36,37) that all
cyclic coupled vectorial systems that employ ATP to perform
work (this includes such processes as ion transport and
muscle contraction) must follow a set of rules that prevent
ATP hydrolysis from taking place without work being done.
This can be, and is observed to be, achieved by having the
rate of the ATPase be very low in the absence of the "second
substrate" (i.e., actin, in the case of muscle). The preparation of LAD-1 suggests that the action of dynein is probably
also governed by such rules. This leads to a third property:
the protein should be isolatable in a state possessing a low
latent ATPase activity, which may be activated by the addition
of microtubules (or by other means).
    4. Dynein in axonemes possesses two dissimilar micro-
tubule binding sites. That binding to the A-tubule is relatively stable, but the B-tubule binding site is ATP-sensitive
(38,39). Haimo *et al.* have shown that these sites in
Chlamydomonas dynein are satisfied in a similar manner by
in vitro polymerized, MAP-free cytoplasmic microtubules, and

that ATP dissociates the microtubules from the B-tubule binding site of the dynein (40). A further property, there- fore, is that the protein should possess two dissimilar, spatially separated microtubule (or tubulin (7)) binding sites, one of which should be ATP sensitive.

5. There is convincing evidence that, in cilia and fla- gella, the dynein arms, rather than any other accessory axo- nemal structures, are the primary transducers of free-energy into motility (1-4). In other words, the dynein displays *motile functionality,* causing the sliding of doublet micro- tubules relative to one another. A further requisite of any putative dynein, therefore, is that it should demonstrate a motile function in some form of microtubule-based system.

The emphasis in this list of properties is on function, since that is the final measure against which any dynein-like protein must be judged. There has been considerable argument both for and against the involvement of dynein in general cellular motility. The evidence in use is basically of two types. First, there are *in situ* experiments, in which attempts are made to disturb the behavior of whole cells or cell organelles (e.g., mitotic spindles) by the addition of some modifying agent specific for dynein (41). The problem with these experiments is that it is very difficult to rule out *non-specific* effects. Vanadate, for instance, although highly inhibitory for dynein (42,43), is known to inhibit several other transphosphorylating enzymes (44); furthermore, in whole cells it is possible that vanadate could be con- verted to non-inhibitory forms (45). Similar arguments (46) compromise the *in situ* use of even the apparently excep- tionally dynein-specific inhibitor 9-(*erythro*-2-hydroxy-3- nonyl)adenine (47). Similarly, although the inhibition of chromosome movement in isolated mitotic apparatus by an antiserum directed against flagellar dynein is strongly suggestive (48), a more conclusive demonstration of the participation of dynein must await the preparation of mono- clonal antibodies against dynein.

The second type of evidence is based on the isolation of dynein-like proteins from various cytoplasmic sources (5-7, 49). The problem here is that the dynein-like proteins have been shown to satisfy only the structural requirements (prop- erty 1) and the ATPase requirement (property 2). The identi- fying functional properties 3, 4 and 5 (which are vital, because of the structural variability noted in *bona fide* dyneins from cilia and flagella) have not been demonstrated with the isolated proteins.

The argument about the participation of dynein in cellu- lar motility outside of cilia and flagella must therefore be

considered moot at present, owing to lack of sufficiently
clear data. If it is wished to apply the reductionist desig-
nation dynein to a hypothetical component of a cellular
motile system, then recourse must be made to reductionist
methods, and a protein should be isolated that displays the
properties of dynein as listed above. The development of
reactivatable "models" of cytoplasmic motile systems, such
as the particle translocation systems of fish scale melanophores (reported in these proceedings by Rosenbaum) or the
mitotic apparatus (50), is an invaluable step, providing for
the first time the possibility of unambiguously testing the
functional competence of dynein and "dynein-like" proteins
in stimulating various forms of cellular motility.

ACKNOWLEDGMENTS

    The work described in section II was carried out in col-
laboration with Drs. Wen-Jing Y. Tang, Winfield S. Sale,
Barbara H. Gibbons and Ian R. Gibbons, and I thank them and
Mr. John A. Evans for many stimulating discussions.

REFERENCES

1. Summers, K. E. and Gibbons, I. R., *J. Cell Biol. 58,*
    618 (1973).
2. Gibbons, B. H. and Gibbons, I. R., *J. Cell Sci. 13,* 337
    (1973).
3. Gibbons, B. H. and Gibbons, I. R., *J. Biol. Chem. 254,*
    197 (1979).
4. Yano, Y. and Miki-Noumura, T., *J. Cell Sci. 48,* 223
    (1981).
5. Weisenberg, R. and Taylor, E. W., *Exptl. Cell Res. 53,*
    372 (1968).
6. Gaskin, F., Kramer, S. B., Cantor, C. R., Adelstein, R.
    and Shelanski, M. L., *FEBS Lett. 40,* 281 (1974).
7. Pratt, M. M., *Develop. Biol. 74,* 364 (1980).
8. Dustin, P., "Microtubules", Springer Verlag, Berlin
    (1978).
9. Satir, P., *J. Cell Biol. 39,* 77 (1968).
10. Summers, K. E. and Gibbons, I. R., *Proc. Natl. Acad. Sci.
    U.S.A. 68,* 3092 (1971).
11. Ogawa, K., *Biochim. Biophys. Acta 293,* 514 (1973).
12. Bell, C. W. and Gibbons, I. R., *J. Biol. Chem.* in press
    (1982).

13. Blum, J. J. and Hines, M., *Q. Rev. Biophys.* *12*, 103 (1979).
14. Sale, W. S. and Gibbons, I. R., *J. Cell Biol.* *82*, 291 (1979).
15. Bell, C. W., Fraser, C., Sale, W. S., Tang, W.-J. Y. and Gibbons, I. R., *in* "Methods in Cell Biology" Vol. 24A (L. Wilson, ed.) p. 373. Academic Press, New York (1982).
16. Taylor, E. W., *CRC Crit. Rev. Biochem.* *6*, 103 (1979).
17. Gibbons, I. R., *J. Cell Biol.* *91*, 107s (1981).
18. Huxley, H. E., *in* "The Structure and Function of Muscle" Vol. 1, part 1 (G. H. Bourne, ed.) p. 301. Academic Press, New York (1972).
19. Pollard, T. D. and Korn, E. D., *J. Biol. Chem.* *248*, 4682 (1973).
20. Pollard, T. D., *J. Cell Biol.* *91*, 156s (1981).
21. Allen, R. D., *J. Cell Biol.* *37*, 825 (1968).
22. Gibbons, I. R. and Rowe, A. J., *Science* *149*, 424 (1965).
23. Mabuchi, I. and Shimizu, T., *J. Biochem. (Tokyo)* *76*, 991 (1974).
24. Piperno, G. and Luck, D. J. L. *J. Biol. Chem.* *254*, 3084 (1979).
25. Mabuchi, I., Shimizu, T. and Mabuchi, Y., *Arch. Biochem. Biophys.* *176*, 564 (1976).
26. Mohri, H., Hasegawa, S., Yamamoto, M. and Murakami, S., *Sci. Pap. Coll. Gen. Educ., Univ. Tokyo* *19*, 195 (1969).
27. Gibbons, I. R. and Fronk, E., *J. Biol. Chem.* *254*, 187 (1979).
28. Tang, W.-J. Y., Bell, C. W., Sale, W. S. and Gibbons, I. R., *J. Biol. Chem.* in press (1982).
29. Bell, C. W., Fronk, E. and Gibbons, I. R., *J. Supramol. Struct.* *11*, 311 (1979).
30. Linck, R. W., *J. Cell Sci.* *12*, 951 (1973).
31. Mooseker, M. S. and Tilney, L. G., *J. Cell Biol.* *56*, 13 (1973).
32. Kincaid, H. L., Gibbons, B. H. and Gibbons, I. R., *J. Supramol. Struct.* *1*, 461 (1973).
33. Mabuchi, I., *Horizons Biochem. Biophys.* *5*, 1 (1978).
34. Huang, B., Piperno, G. and Luck, D. J. L., *J. Biol. Chem.* *254*, 3091 (1979).
35. Gibbons, I. R., Fronk, E., Gibbons, B. H. and Ogawa, K., *in* "Cell Motility" Book C (R. Goldman, T. Pollard and J. Rosenbaum, eds.) p. 915. Cold Spring Harbor, New York (1976).
36. Hill, T. L., "Free Energy Transduction in Biology" Academic Press, New York (1977).
37. Jencks, W. P., *in* "Advances in Enzymology" Vol. 51 (A. Meister, ed.). Wiley, New York (1980).

38. Takahashi, M. and Tonomura, Y., *J. Biochem. (Tokyo) 84,* 1339 (1978).
39. Mitchell, D. R. and Warner, F. D., *J. Cell Biol.* 87, 84 (1980).
40. Haimo, L. T., Telzer, B. R. and Rosenbaum, J. L., *Proc. Natl. Acad. Sci. U.S.A. 76,* 5759 (1979).
41. Cande, W. Z. and Wolniak, S. M., *J. Cell Biol.* 79, 573 (1978).
42. Kobayashi, T., Martensen, T., Nath, J. and Flavin, M., *Biophys. Biochem. Res. Commun. 81,* 1313 (1978).
43. Gibbons, I. R., Cosson, M. P., Evans, J. A., Gibbons, B. H., Houck, B., Martinson, K. H., Sale, W. S., Tang, W.-J. Y., *Proc. Natl. Acad. Sci. U.S.A. 75,* 2220 (1978).
44. Simons, T. J. B., *Nature (London) 281,* 337 (1979).
45. Buckley, I. and Stewart, M., *J. Cell Biol.* 91, 323a (1981).
46. Frieden, C., Kurz, L. C. and Gilbert, H. R., *Biochemistry 19,* 5303 (1980).
47. Penningroth, S. M., Cheung, A., Bouchard, P., Gagnon, C. and Bardin, C. W., *J. Cell Biol.* 87, 34a (1980).
48. Sakai, H., Mabuchi, I., Shimoda, S., Kuriyama, R., Ogawa, K. and Mohri, H., *Dev. Growth Differ. 18,* 211 (1976).
49. Pratt, M. M., Otter, T., Salmon, E. D., *J. Cell Biol.* 86, 738 (1980).
50. Cande, W. Z., *J. Cell Biol.* 87, 326 (1980).

CHAPTER 15

CYCLIC AMP AND INITIATION OF FLAGELLAR
MOVEMENT IN RAINBOW TROUT SPERMATOZOA

Makoto Okuno

Department of Biology
University of Tokyo
Tokyo, Japan

Masaaki Morisawa

Ocean Research Institute
University of Tokyo
Tokyo, Japan

Although teleosts are very popular sources for studying
gametes, relatively few descriptions of sperm motility have
been accumulated. Encouraged by the findings by Schlenk and
Kahmann (1) that potassium suppresses sperm motility of
rainbow trout, Morisawa *et al.* (2, 3, 4) have shown that
potassium ions, which are contained at high concentration in
seminal plasma, suppress sperm motility in the sperm duct,
and the decrease in potassium concentration surrounding the
spawned spermatozoa is the external factor which initiates
sperm motility in the salmonidae generally.
Recently we have shown that cAMP is the intracellular
factor which triggers the initiation of flagellar movement
resulting in sperm motility, by using the rainbow trout
spermatozoa demembranated with Triton X-100 (5). In the
present paper, we describe more detailed story concerning the
roles of potassium and cyclic nucleotides on the initiation
mechanism of flagellar movement in rainbow trout spermatozoa.

MATERIALS AND METHODS

The spermatozoa were collected directly, by inserting a

pipette into the sperm duct of the mature rainbow trout, *Salmo gairdneri,* and stocked on ice without dilution for several hours during the experiments.

In order to observe sperm motility, the semen was immersed in 100 volumes of solution containing 100 mM NaCl on a glass slide in the presence or absence of various concentrations of KCl, RbCl, LiCl, inhibitors of respiration (NaN$_3$, KCN and antimycin A) or an uncoupler (CCCP; carbonyl cyanide m-chloro-phenylhydrazone). The solutions were buffered with 10 mM HEPES (N-2-hydroxyethylpiperazine-N'-2-ethanesulfonic acid) and NaOH at pH 7.7.

The rate of respiration of the intact spermatozoa was determined by a conventional platinum electrode (Rank Brothers Co. Ltd.). The incubation vessel with electrode was filled with solutions containing 100 mM NaCl and 10 mM HEPES, pH 7.7 in the presence or absence of KCl, inhibitors of respiration or the uncoupler. The recording of oxygen consumption started immediately after adding 20 μl of semen to the vessel and the decrease of oxygen in the solution was measured for 2 minutes.

Demembranation and reactivation of spermatozoa were carried out as follows: 10 μl of semen were mixed with 0.2 ml of the extraction solution containing 0.15 M KCl, 0.5 mM MgCl$_2$, 10 mM CaCl$_2$, 0.5 mM EDTA, 2 mM Tris-HCl, 1 mM DTT (dithiothreitol) and 0.04 % (v/v) Triton X-100, pH 8.0, and gently shaken for 30 seconds on ice. A drop of the demembranated sperm suspension (usually 10 μl) was suspended in 1 ml of the reactivation solution containing 0.15 M KCl, 2 mM MgCl$_2$, 0.5 mM CaCl$_2$, 2 mM EGTA, 20 mM Tris-HCl, 1 mM DTT, 2 % (w/v) PEG (polyethyleneglycol) and additional supplements, such as ATP, cAMP etc., pH 8.0, at room temperature. The sperm suspension was placed on a glass slide and observed by dark field microscopy. Movement of spermatozoa was recorded on 35 mm films as illustrated previously (5) or by video camera and VTR. The velocity of spermatozoa swimming straight forward was measured on the video display.

For electron microscopy, spermatozoa were demembranated under the same conditions with the extraction solution, and then suspended in a solution which was essentially the same as the reactivation solution except that Tris-HCl was replaced by NaH$_2$PO$_4$-Na$_2$HPO$_4$ buffer, and PEG and DTT were omitted. The sperm suspension was centrifuged at 3000 rpm and the pellet was fixed with 2 % glutaraldehyde followed by 1 % osmic acid.

Antimycin A, CCCP, trypsin, phosphodiesterase, hexokinase and all of nucleotides except ADP were purchased from Sigma Chem. Co.. ADP was from Oriental Yeast Co. Ltd.. Water was deionized and glass-distilled.

All of the experiments except the demembranation step, which was made on ice, were carried out at room temperature, 20° $\pm$ 2° C.

RESULTS

Effects of Potassium on Sperm Motility and Respiration

When the semen was diluted in the solution containing 100 mM NaCl and 10 mM HEPES at pH 7.7, the spermatozoa began to swim actively and motility continued for approximately 20 sec. The velocity, determined by video recording, was 260 µm/sec. When the semen was diluted in a solution containing 100 mM NaCl and various concentrations of potassium, spermatozoa swam actively only at potassium concentrations below 1.5 mM, and the swimming period decreased with increasing potassium concentration. Sperm motility was completely suppressed in the presence of 3 mM potassium (Fig. 1).

As shown in Table 1, spermatozoa swimming actively in NaCl-HEPES solution exhibited an oxygen consumption of 0.3 ml O_2/ml semen/hr. When the spermatozoa were diluted in 100 mM NaCl solution containing 10 mM KCN, 10 mM NaN_3 or 50 µM antimycin A, the rate of respiration decreased remarkably and

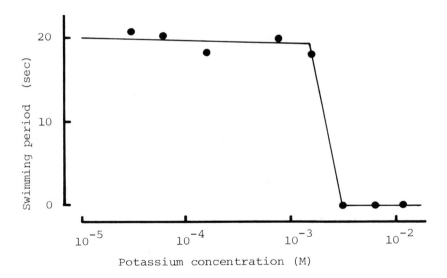

Fig. 1.   Effect of potassium concentration on the motility of the intact spermatozoa of the rainbow trout.   The swimming period of the spermatozoa is measured when semen is diluted in a solution containing 100 mM NaCl and 10 mM HEPES, pH 7.7, and represented as an index of motility.   The number of swimming spermatozoa is almost 100 % at 1.5 mM or less KCl.   No sperm movement is observed in the presence of 3 mM or more KCl.

Table 1.   *Effects of Inhibitors on Oxygen Consumption of
Intact Rainbow Trout Spermatozoa*

Inhibitors*	Oxygen consumption $\pm$ S. D. ($\mu l/ml$ semen/hr)
None	278 $\pm$ 19
10 mM KCl	267 $\pm$ 13
10 mM KCN	16 $\pm$ 13
10 mM NaN$_3$	20 $\pm$ 7
50 µM Antimycin A	14 $\pm$ 7
5 µM CCCP	361 $\pm$ 29

*Inhibitors are added in the solution containing 100 mM
NaCl and 10 mM HEPES, pH 7.7.

the spermatozoa became immotile. CCCP is a well known
uncoupler of oxidative phosphorylation in mitochondria which
causes an inhibition of ATP synthesis. Five µM CCCP also
inhibited rainbow trout sperm motility, however, respiration
was not inhibited and, in fact, was enhanced. The rate of
oxygen consumption of immotile spermatozoa, in the solution
containing 10 mM KCl, was approximately the same as that of
motile spermatozoa in NaCl-HEPES solution.

Demembranated Spermatozoa

When the flagella demembranated with Triton X-100 were
observed by dark field microscopy, the light intensity along
the flagellum was somewhat uneven. This was probably due to
small fragments derived from the plasma membrane of spermatozoa which attached to the flagellar axoneme during the
treatment with Triton X-100 and remained in the reactivation
solution. Electron microscopy revealed that the plasma
membrane of the axoneme and mitochondria were completely
removed, however, amorphous vesicles were often observed close
to the axoneme (data not shown).

Reactivation of the Demembranated Spermatozoa

When spermatozoa were demembranated with Triton X-100 and
suspended in the reactivation solution containing MgATP^{2-} at
a concentration between 0.1 and 1.0 mM most of the spermatozoa

remained quiescent. The shape of flagella was straight or
slightly bent indicating that they were in a relaxed state.
As described in the following paragraphs, the demembranated
flagella were able to produce bending waves at a lower
concentration of cAMP when the concentration of $MgATP^{2-}$ was
decreased. When the concentration of $MgATP^{2-}$ was decreased
down to 0.05 mM, some spermatozoa (less than 20 %) vibrated
their heads irregularly without progressive movement. The
vibration, however, stopped when the spermatozoa were demembranated and reactivated in the presence of phosphodiesterase.
Thus, the head vibration, produced by bending of the flagella
in the short proximal region, seemed to be associated with the

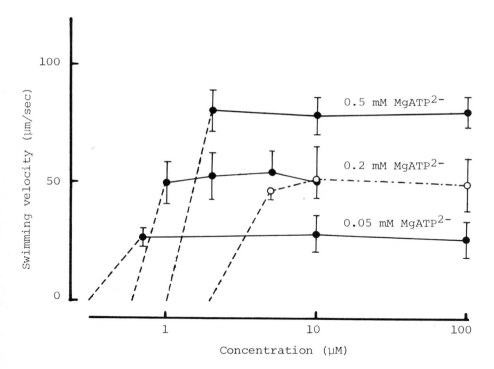

Fig. 2.   Motility of the demembranated spermatozoa at
various concentrations of cAMP (●) and cGMP (○). $MgATP^{2-}$
concentration is maintained at 0.05 mM, 0.2 mM or 0.5 mM.
Each broken line starts on the abscissa, indicating the cAMP
concentration which induces more than half of the spermatozoa
to begin head vibration. Each broken line ends with a circle,
indicating the concentration of cAMP or cGMP at which more
than half of spermatozoa show forward progressive movement.
Vertical bars represent standard deviations.

presence of a low concentration of cAMP, which probably
remained in the axoneme after the demembranation.

When cAMP at a concentration below 0.3 μM was added to the
reactivation solution containing 0.05 mM MgATP^{2-}, some
spermatozoa exhibited head vibration or bending of the short
proximal region of the flagella. An increase in the cAMP
concentration caused an increase in the number of head
vibrating spermatozoa. More than half of the spermatozoa
vibrated their heads at 0.3 μM cAMP. When the cAMP concentration was increased up to 0.7 μM, more than half of
spermatozoa began to swim forward progressively at a velocity
of 27 μm/sec (Fig. 2).

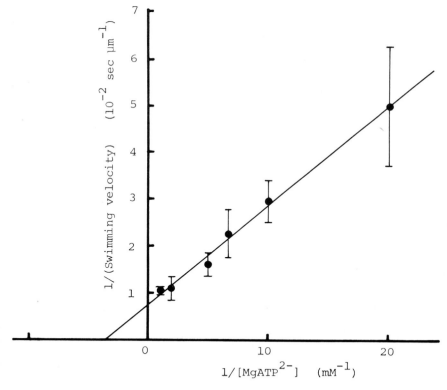

Fig. 3. *Double reciprocal plots of the swimming velocity
of the demembranated spermatozoa at different MgATP^{2-}
concentrations. More than 10 spermatozoa, which swim straight
forward, are measured at each condition. The best fit of the
data to a straight line is obtained by the least squares
method. Vertical bars represent standard deviations.*

When the MgATP^{2-} concentration was 0.2 mM, the cAMP concentration required to attain the same degree of reactivation as in the presence of 0.05 mM MgATP^{2-} increased. More than half of the spermatozoa vibrated heads and only a few percent of them moved progressively with a random path at various velocities in the presence of 0.6 µM cAMP. The number of swimming spermatozoa increased remarkably up to 50 % at 1 µM cAMP, nevertheless a considerable portion of spermatozoa swam with small circular or helical paths. The swimming path became straight in the presence of 5 µM or more cAMP. The velocity of the swimming spermatozoa attained a plateau (55 µm/sec) at 1 µM cAMP and maintained the same level at higher concentration of cAMP.

When the MgATP^{2-} concentration was increased up to 0.5 mM, the required cAMP concentration, for 50 % progressive movement of spermatozoa, became 2 µM. The swimming velocity of spermatozoa was 80 µm/sec and the velocity was again independent of a wide range of cAMP concentrations.

As shown by open circles, cGMP also induced sperm movement in the same manner as cAMP, however, a five times higher concentration was required in the presence of 0.2 mM MgATP^{2-}. The swimming velocity of the spermatozoa reactivated in the presence of cGMP was almost the same as that attained in the presence of cAMP, and was independent of cGMP concentration.

Fig. 3 shows the typical pattern of double reciprocal plots for the swimming velocity at different MgATP^{2-} concentrations in the presence of 2 µM cAMP. Almost 100 % reactivation was obtained at all of the concentrations of MgATP^{2-} in the experimental range. The line obtained by the least squares method gave a Km value of 0.3 mM, which is approximately equivalent to that of sea urchin spermatozoa (6). The maximum velocity was 130 µm/sec, which is approximately half of the swimming velocity of the intact spermatozoa of the rainbow trout.

The effects of nucleotides on sperm motility are summarized in Table 2. When the demembranated spermatozoa were suspended in a solution containing 20 µM cAMP and 0.2 or 1.0 mM nucleotide, they were reactivated only in the presence of ATP. When the sperm motility was examined in the presence of 0.2 mM nucleotide in addition to 0.2 mM MgATP^{2-}, only cAMP and cGMP were able to initiate progressive movement of spermatozoa.

Our previous report (5) described the requirement for a high concentration of cAMP (0.5 mM) for reactivation. In the present reactivation condition, on the other hand, only micro molar concentrations of cAMP were necessary for 100 % reactivation. This difference in dose dependency was due to a difference in the extraction step. Lowering the temperature of the extraction and/or decreasing the concentration of

Table 2.   Effects of Nucleotides on Flagellar Movement of the Demembranated Spermatozoa of the Rainbow Trout

Nucleotide	In the presence of 0.2 mM MgATP^{2-} and 0.2 mM nucleotide	In the presence of 20 μM cAMP and nucleotide	
	Flagellar movement observed		
		0.2 mM	1 mM
AMP	−	−	−
ADP	−	−*	−*
ATP	−	+	+
GMP	−	−	−
GDP	−	−	−
GTP	−	−	−
ITP	−	−	−
CTP	−	−	−
UTP	−	−	−
cAMP	+	−	−
cGMP	+	−	−
none	−	−	−

The solution was preincubated with 10 mM glucose and 10 μg/ml hexokinase for 1 hr at room temperature (7).

Triton X-100 in the extraction solution increased the cAMP sensitivity of spermatozoa.

Trypsin Treatment of Demembranated Spermatozoa

When the demembranated spermatozoa were suspended in re-activation solution containing 0.2 mM MgATP^{2-} and 1 to 2 μg/ml trypsin in the absence of cAMP, spermatozoa were initially quiescent for several seconds. Within 15 seconds, some of spermatozoa began to vibrate their heads and then start to swim. Some spermatozoa began to swim abruptly without head vibration. Finally, approximately half of spermatozoa exhibited progressive movement, and the rest of spermatozoa remained quiescent during the incubation. All of axonemes disintegrated into individual outer doublet microtubules within four minutes whether the spermatozoa were reactivated to move or not.

DISCUSSION

Recently, we have shown that the potassium concentration in the seminal plasma of salmonid fishes such as rainbow trout, chum salmon and masu salmon is considerably higher than that in the blood plasma (3, 4). Since several mM of potassium completely inhibits sperm motility in rainbow trout (Fig. 1) as well as other salmonid species, potassium ions are a common external factor to suppress sperm motility in the sperm duct of salmonid fishes (2, 4).

We wish to ask which area of the spermatozoon, head, midpiece or flagellum, is the target site for potassium in inhibiting the sperm motility. In our preliminary observation, the motility of the intact spermatozoa was not inhibited by the inhibitors of the glycolytic pathway, such as monoiodoacetic acid or NaF. However, the sperm motility was completely inhibited by inhibitors of the respiratory chain, such as NaN_3, KCN and antimycin A, and the uncoupler of oxidative phosphorylation, CCCP. Therefore, it seems likely that these systems rather than glycolysis are indispensable for supplying the energy source, ATP, to produce the flagellar movement. Exogenous potassium, a potent inhibitor of intact sperm motility, however, neither suppressed nor enhanced the oxygen consumption, indicating that the mechanism which regulates the initiation of flagellar movement with potassium is not involved in mitochondria, but may be found in the flagellar apparatus.

In our previous paper (5), we demonstrated that cAMP (0.5 mM) initiates flagellar movement in rainbow trout spermatozoa demembranated and reactivated in the presence of 0.15 M KCl at room temperature. Since this concentration of potassium is high enough to inhibit the motility of the intact spermatozoa, it is likely that potassium is not coupled to the axoneme directly but mediated by the plasma membrane of the flagellum.

We improved the sensitivity of cAMP in initiating flagellar movement. When the spermatozoa were demembranated on ice, they were reactivated in the presence of cAMP at a concentration of only a few μM. As shown in Fig. 2, the swimming velocity, an index of flagellar movement, depends on $MgATP^{2-}$ concentration, not on cAMP. However, the transition from immotile to motile spermatozoa requires cAMP, apparently suggesting that cAMP is the intracellular factor which triggers the initiation of flagellar movement. In last several years, some investigators have demonstrated that cAMP improves the motility of demembranated mammalian spermatozoa when they are reactivated in the presence of $MgATP^{2-}$ (8, 9, 10), although none of them have succeeded in demonstrating

a direct relationship between cAMP and the initiation of
flagellar movement, except in tunicate spermatozoa (11).

Recently, there has been much evidence that a protein
kinase mediates the effects of cAMP via the phosphorylation of
target proteins (12). This cAMP dependent protein phosphorylation has been observed in bull spermatozoa (13). Murofushi
et al. showed that the phosphorylation of proteins by a cAMP
dependent protein kinase is important in flagellar movement
of sea urchin spermatozoa (in these proceedings).

Our studies seems to add some indirect evidence that cAMP
dependent protein kinase is involved in the initiation mechanism of flagellar movement in rainbow trout spermatozoa. As
described previously, when the spermatozoa are released from
suppression by potassium, by suspension in a potassium-free
solution, and subsequently demembranated, the flagella become
motile in the absence of cAMP to a certain extent (5). This
result suggests that the reduction of potassium concentration
surrounding the intact spermatozoa induces an increase in
intraflagellar cAMP concentration, and endogenous cAMP
dependent phosphorylation occurrs in some protein[s] necessary
to initiate flagellar movement. The other indirect evidence
seems to be provided by the partial digestion experiment, in
which almost half of the demembranated spermatozoa become
motile during trypsin treatment in the absence of cAMP. This
result seems to imply that an enzymatic modification or
release of some component[s] of the axoneme plays a significant role in the initiation of flagellar movement. The cAMP
dependent protein kinase may participate in this modification

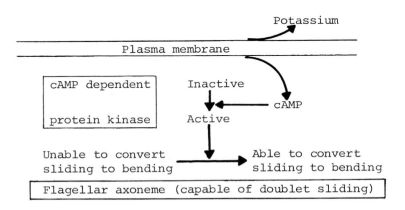

*Fig. 4. A schematic diagram showing the hypothesized
mechanism of potassium and cAMP dependent initiation of
flagellar movement.*

or release system in intact spermatozoa, while it can be replaced by trypsin in the demembranated spermatozoa.

From another point of view, the trypsin experiment seems to provide evidence that the flagellar axoneme is already capable of producing active sliding of outer doublet microtubules in the absence of cAMP at the stage, because disintegration of the axoneme into individual outer doublet microtubules can be observed during the trypsin treatment in the presence of MgATP^{2-} without generation of flagellar movement.

Considering the observations of the effects of potassium and cAMP on the initiation of flagellar movement, we would like to propose a hypothesis which is schematically diagrammed in Fig. 4. This model suggests that a decrease in external potassium ion concentration increases the intraflagellar cAMP concentration, which activates some component[s] of the axoneme. The component[s], which might be a cAMP dependent protein kinase, modifies the axoneme (already capable of outer doublet sliding) from the inactive state to the active state which allows for generating the flagellar bending wave.

ACKNOWLEDGMENT :

We thank Dr. M. M. Pratt for preparation of the manuscript and Mrs. S. Morisawa for electron microscopy. We also thank the directors and staffs of Oshino Branch of Yamanashi Fish Culture Center and Kumagaya Branch of Saitama Fisheries Experiment Station for supplying animals. This work has been supported in part by grant-in-aid from the Ministry of Education, Science and Culture of Japan and Ito-Foundation for the Advancement of Ichthyology to M. M.

REFERENCES

1. Schlenk, W. Jr. and Kahmann, H., *Biochem. Z. 295,* 283 (1938).
2. Morisawa, M. and Suzuki, K., *Science, 210,* 1145 (1980).
3. Morisawa, M., Hirano, T. and Suzuki, K., *Comp. Biochem. Physiol. 64A,* 325 (1979).
4. Morisawa, M., Suzuki, K. and Morisawa, S., in preparation.
5. Morisawa, M. and Okuno, M., *Nature,* in press.
6. Okuno, M. and Brokaw, C. J., *J. Cell Sci. 38,* 105 (1979).
7. Okuno, M., *J. Cell Biol. 85,* 712 (1980).
8. Lindemann, C. B., *Cell, 13,* 9 (1978).
9. Mohri, H. and Yanagimachi, R., *Exp. Cell Res. 127,* 191 (1980).

10. Tamblyn, T. M. and First, N. L., *Arch. Biochem. Biophys.* *181*, 208 (1977).
11. Brokaw, C. J., *Cell Motility*, in press.
12. Greengard, P., *Science*, *199*, 146 (1978).
13. Brandt, H. and Hoskins, D. D., *J. Biol. Chem. 255*, 982 (1980).

CHAPTER 16

INVOLVEMENT OF CYCLIC AMP-DEPENDENT PROTEIN KINASE AND A PROTEIN FACTOR IN THE REGULATION OF THE MOTILITY OF SEA URCHIN AND STARFISH SPERMATOZOA[1]

Hiromu Murofushi
Koichi Ishiguro
Hikoichi Sakai

Department of Biophysics and Biochemistry
Faculty of Science
University of Tokyo
Tokyo

I. INTRODUCTION

Since Summers and Gibbons reported ATP-induced sliding of outer fibers in trypsinized flagellar axonemes (1), it has been widely accepted that the movement of flagella and cilia derives from local sliding of outer doublets coupled with ATP hydrolysis by dynein arms. Triton models of the sea urchin spermatozoa developed by Gibbons and Gibbons (2) also favor studies of flagellar motility on a physiological as well as biochemical basis. Many a datum has accumulated on the mechanism of flagellar and ciliary movement using these techniques. But little is known on the control mechanism of flagellar or ciliary movement.

In the case of cilia, it has been shown that calcium ions play an important role in the regulation of the motility (3). At the same time, cyclic nucleotide has frequently been proposed as a regulator of mammalian sperm flagellar motility. It has been reported that cyclic nucleotide phosphodiesterase inhibitors increase the motility of mammalian spermatozoa

[1]*This study was supported by a grant-in-aid for scientific research from the Ministry of Education, Science and Culture, Japan (No. 444074)*

along with an accumulation of cyclic AMP in the cells (4).
Mohri and Yanagimachi (5) reported that Triton models of
immature mammalian sperm increase motility by the addition of
cyclic AMP.

In addition to these physiological data, it was reported
that spermatozoa from various sources contain enzymes related
to cyclic nucleotides (6-12).

These data lead to the hypothesis that the cyclic nucleotide system is involved in the regulatory mechanism of flagellar movement. In the course of our study on protein kinases
in cilia and flagella, we found that these organelles contained large amounts of cyclic nucleotide-dependent protein
kinases. We purified protein kinases from *Tetrahymena* cilia
and sea urchin spermatozoa, and determined the enzymic and
chemical natures of the enzymes (9,13,14).

The next step was to introduce protein kinase to the motile system, namely Triton models of sea urchin spermatozoa.
Recently we found that cyclic AMP-dependent protein kinase
and a protein factor isolated from sperm extract induce
motility reactivation of sea urchin sperm Triton models (15).
Here we present our data on the roles of the kinase and the
protein factor in the control of flagellar motility.

II. MATERIALS AND METHODS

Spermatozoa of the sea urchins, *Hemicentrotus pulcherrimus*,
Pseudocentrotus depressus and *Anthocidaris crassispina*, and
those of starfish, *Asterias amurensis*, were used as experimental materials.

Triton models of spermatozoa were prepared according to
the method of Gibbons and Gibbons (2) with slight modifications (15). Extraction of soluble materials from spermatozoa
of sea urchins or starfish and preparation of DEAE-adsorbed
fraction have been reported (15).

Protein kinase activity was measured according to the
method of Murofushi (13) using calf thymus histones as substrate. The procedure used to prepare cyclic AMP-dependent
protein kinase from sea urchin or starfish spermatozoa was
described elsewhere (15). Protein kinase inhibitor was prepared from rabbit skeletal muscle after Walsh *et al.* (16).

III. EFFECTS OF DEAE-ADSORBED FRACTION ON THE MOTILITY
OF TRITON MODELS

In the course of the preparation of Triton models, some

components are extracted by Triton X-100, leaving the axoneme apparently intact in terms of motility. Many studies have been reported on the motility of Triton models, but very little attention has been paid to the materials which are extracted by the detergent. Our purpose is to study the soluble materials affecting motility of the Triton models of spermatozoa.

Spermatozoa were extracted with a solution containing Nonidet P-40, and the extract was applied to a DEAE-cellulose column. After the unadsorbed materials were washed out completely, the adsorbed materials were eluted, followed by dialysis to obtain the DEAE-adsorbed fraction. The effects of this fraction on the motility of Triton models were studied.

Triton models prepared by our procedure gradually ceased to undulate. Cyclic AMP alone did not show any effects on the motility of Triton models. When the DEAE-adsorbed fraction was added to the models, they quickly stopped beating, but began to move again after further addition of cyclic AMP

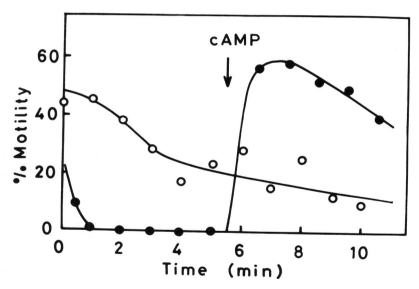

Fig. 1 Time course of changes in % motility of Triton models in the presence or absence of DEAE-adsorbed fraction. ○: control, without addition of DEAE-adsorbed fraction. ●: 10 µl of DEAE-adsorbed fraction containing 20 µg of protein was added to 20 µl of Triton model suspension. Cyclic AMP was added, at the point indicated by the arrow, to a final concentration of 10 µM. (Redrawn from Ishiguro et al. (15) by permission of J. Cell Biol.)

(Fig. 1). When the Triton models were added to the mixture of the DEAE-adsorbed fraction and cyclic AMP, a high level of % motility was maintained.

It should be mentioned here that the demembranated sperm flagella without head and middle piece sometimes seen in the preparation of Triton models were also reactivated to undulate along with the intact sperm models by cyclic AMP in the presence of the DEAE-adsorbed fraction. This suggests that the target of some components in the DEAE-adsorbed fraction is not the heads nor middle pieces but the flagellar axonemes.

Figure 2 shows the dose responses of the DEAE-adsorbed fraction on % motility of the Triton models in the presence of cyclic AMP. When the maximal levels of % motility versus the amount of DEAE-adsorbed fraction were plotted (b), a linearity was observed until saturation of the DEAE-adsorbed

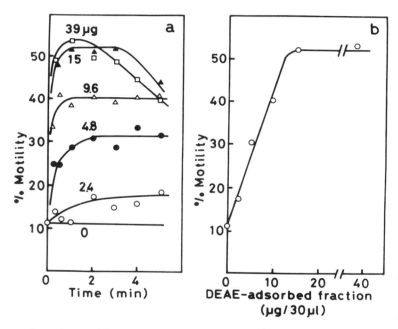

Fig. 2. Effect of the concentration of DEAE-adsorbed fraction on % motility of the Triton model in the presence of cyclic AMP. Time course of % motility after the addition of cyclic AMP (a). Figures on each curve indicate the amount of protein in DEAE-adsorbed fraction added to the Triton model suspension. Relationship between the maximal levels of % motility and amount of DEAE-adsorbed fraction (b). (Redrawn from Ishiguro et al. (15) by permission of J. Cell Biol.)

fraction. This suggests that the mode of reactivation caused by the DEAE-adsorbed fraction with the addition of cyclic AMP is stoichiometric rather than catalytic.

Percent motility was preferentially elevated by low concentration of cyclic AMP (Fig. 3). Higher concentration of cyclic GMP was necessary for the reactivation. 5'-AMP was totally inactive. It is evident that this reactivation is specific for cyclic AMP.

It should be mentioned that when the motility of Triton models was repressed in the presence of the DEAE-adsorbed fraction, further addition of 1 mM ATP never initiated beating of the Triton models (Fig. 3). The motionless states of the models were not due to exhaustion of ATP in the model suspension. In fact, the concentration of the Triton models in our

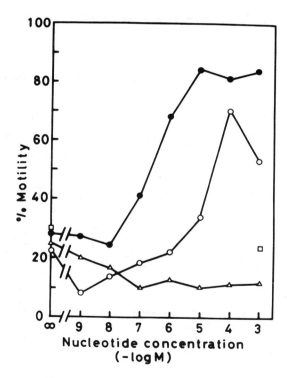

Fig. 3. Nucleotide specificity for the reactivation of
Triton models. Percent motility was determined in the
presence of 39 µg of DEAE-adsorbed fraction at 3 min after
the addition of various concentrations of nucleotide.
cyclic AMP, ● ; cyclic GMP, ○ ; 5'-AMP, △ ; ATP, □ .
(Redrawn from Ishiguro et al. (15) by permission of J. Cell
Biol.)

experiments was too low to hydrolyze most of the ATP in the assay system. Therefore, it is reasonable to consider that cyclic AMP and components in the DEAE-adsorbed fraction affected the motile apparatus directly not *via* ATP generating system.

It should also be mentioned that the effects of cyclic AMP and cyclic GMP on the motility reactivation parallel the effects of the cyclic nucleotides on the enzymic activity of the cyclic AMP-dependent protein kinase in spermatozoa. It is quite reasonable to consider that this cyclic AMP-mediated reactivation of the Triton model motility is mediated by cyclic AMP-dependent protein kinase.

IV. PARTICIPATION OF CYCLIC AMP-DEPENDENT PROTEIN KINASE IN THE REACTIVATION OF TRITON MODELS

In order to examine participation of cyclic AMP-dependent protein kinase in motility reactivation of sperm models, we employed a specific protein inhibitor of the enzyme prepared from rabbit skeletal muscle. It was reported that this

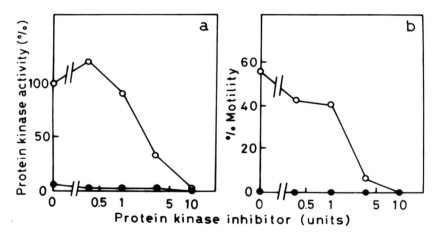

Fig. 4. Effects of protein kinase inhibitor on the activity of protein kinase (a) and the activity to reactivate Triton models (b) in the presence of DEAE-adsorbed fraction. Protein kinase activity was measured in the same condition as was used for motility determination except for the addition of histones for substrate and the omission of Triton models. ○, with cyclic AMP; ●, without cyclic AMP. (Redrawn from Ishiguro et al. (15) by permission of J. Cell Biol.)

R C

Fig. 5.   SDS-PAGE pattern of purified starfish protein kinase.   R, regulatory subunit.   C, catalytic subunit.

inhibitor blocks the activity of cyclic AMP-dependent protein kinases from a wide variety of sources.

When an increasing amount of the inhibitor was added to the DEAE-adsorbed fraction, the activity of this fraction to reactivate Triton models decreased (Fig. 4b) along with a concomitant decrease in the enzyme activity of the cyclic AMP-dependent protein kinase in the same DEAE-adsorbed fraction (a).   This good correlation strongly supports the supposition that this reactivation of Triton models is mediated by cyclic AMP-dependent protein kinase.

To obtain direct evidence on the role of cyclic AMP-dependent protein kinase in the Triton model reactivation, we used purified protein kinase prepared from sea urchin or

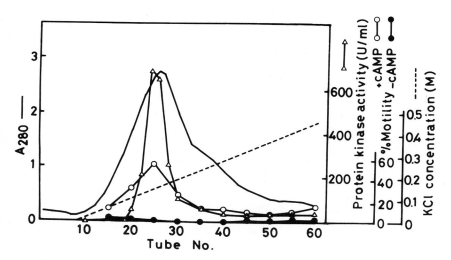

Fig. 6.   DEAE-cellulose column chromatography of sperm extract.   Extract of Hemicentrotus spermatozoa was applied to a DEAE-cellulose column pre-equilibrated with 20 mM Tris-HCl (pH 8.1) containing 0.1 mM DTT.   After unadsorbed materials were washed out, elution was carried out by a linear gradient of KCl from 0 to 0.5 M.   Activity of each fraction for enhancing % motility of the Triton models was measured in the presence of purified protein kinase exogenously added.

starfish spermatozoa.

Spermatozoa were extracted with a medium containing Nonidet P-40, and the sperm extract was chromatographed on DEAE-cellulose and hydroxyapatite, followed by high speed liquid chromatography using gel filtration columns.

Figure 5 shows the SDS-PAGE pattern of the purified starfish sperm protein kinase. The purified enzyme consists of two kinds of polypeptides having molecular weights of 42,000 and 38,000. Judging from the data obtained by separation of the two components using a cyclic AMP-immobilized column and reconstitution of the holoenzyme (data not shown), 42,000 and 38,000 components were determined to be the regulatory subunit and the catalytic subunit, respectively.

The ability of the enzyme fractions obtained in the course of the enzyme purification to reactivate Triton models was determined. The DEAE-cellulose fraction showed an ability to reactivate the models in a cyclic AMP-dependent manner. But the hydroxyapatite fraction and the high speed liquid chromatography fraction were unable to reactivate Triton models.

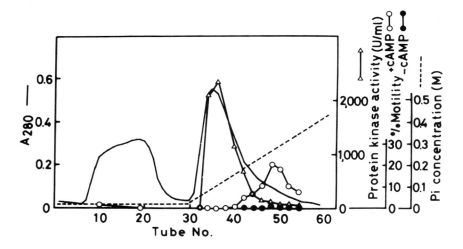

Fig. 7. Hydroxyapatite column chromatography of the DEAE-cellulose fraction. Active fractions in the DEAE-cellulose fraction to reactivate Triton models were applied to a hydroxyapatite column pre-equilibrated with 20 mM potassium phosphate (pH 7.0) containing 0.1 mM DTT. After the column was washed with the same buffer, adsorbed materials were eluted with a linear gradient of phosphate buffer from 20 mM to 0.5 M. Activity of each fraction to stimulate the motility of Triton models was measured in the presence of purified protein kinase.

From these data, it is clear that cyclic AMP-dependent protein kinase alone is incapable of reactivating Triton models. It can be considered that some factors other than the protein kinase were also necessary for the reactivation and that these factors had been separated from the protein kinase in the course of the enzyme purification.

V. ROLE OF A PROTEIN FACTOR IN THE REACTIVATION OF TRITON MODELS

Figure 6 shows the pattern of DEAE-cellulose column chromatography of the sea urchin sperm extract. A peak of cyclic AMP-dependent protein kinase was obtained. Each fraction was supplemented with purified protein kinase, and the activity to reactivate sperm model motility was determined in the absence or presence of cyclic AMP. A peak of the activity for cyclic

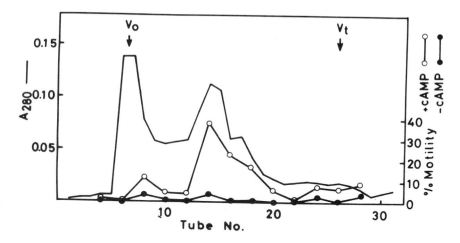

Fig. 8. High speed liquid chromatography of the hydroxy-apatite fraction. Active fractions in the hydroxyapatite chromatography to stimulate sperm model motility were combined and condensed by ammonium sulfate precipitation (70 % saturation) and subjected to gel filtration using an apparatus for high speed liquid chromatography (HLC-803 system equipped with two G 3000 SW columns in series, Toyo Soda Manufacturing Co.). The columns were pre-equilibrated with 10 mM Tris-HCl (pH 8.1)-0.25 M KCl-2 mM MgCl₂-0.1 mM DTT. Activity of each fraction for enhancing % motility of the sperm models was measured in the presence of purified protein kinase. Vo, void volume; Vt, column volume.

TABLE I. *Sensitivity of the Factor to Proteolytic or*
          Heat Treatment[a]

	% Motility	
Treatment	- *cyclic AMP*	+ *cyclic AMP*
None	14	49
Trypsin[b]	18	18
Heat[c]	16	15

[a]*The factor fraction (hydroxyapatite fraction, 1.2 μg protein) was treated and was mixed with protein kinase (hydroxyapatite fraction, 3.7 units of protein kinase activity), followed by the addition of Triton model suspension. Percent motility was measured in the absence or presence of 10 μM cyclic AMP.*

[b]*Trypsin was added to the factor fraction in a ratio (of trypsin to the amount of the proteins in the factor fraction) of about 1:7. After incubation at 25°C for 60 min, the digestion was stopped by the addition of an excessive amount of trypsin inhibitor.*

[c]*The factor fraction was heated at 65°C for 5 min and centrifuged to precipitate insoluble materials.*
(Redrawn from Ishiguro et al. (15) by permission of J. Cell Biol.)

AMP-dependent reactivation emerged at the position of the peak of protein kinase activity.

When the active fraction of the DEAE-cellulose chromatography was applied to a hydroxyapatite column, the activity for reactivating Triton models was completely separated from that of protein kinase (Fig. 7). The fractions containing reactivation activity were combined and condensed by ammonium sulfate precipitation and applied to gel filtration columns.

Figure 8 shows the pattern of the gel filtration. The activity to reactivate Triton models in the presence of purified protein kinase emerged in a peak (Fig. 8).

Here this component which is necessary for sperm model reactivation along with cyclic AMP-dependent protein kinase is simply called the factor. Apparent molecular weight of the factor was 58,000 as estimated by gel filtration.

When the factor fraction was treated with trypsin or was heated at 65°C for 5 min, followed by addition of purified protein kinase, the activity of the factor to reactivate Triton models was completely destroyed (TABLE I). Therefore it can be concluded that this factor is a protein.

There are two possibilities with respect to the function

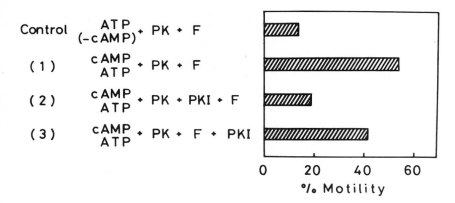

Fig. 9.   Effect of preincubation of the factor with
protein kinase and protein kinase inhibitor on the activity
of stimulating the motility of Triton models. After mixing
of ATP, cyclic AMP and protein kinase (hydroxyapatite fraction), the factor (hydroxyapatite fraction) and protein kinase
inhibitor were added at intervals of 30 sec in the order indicated on the left side of the figure. Percent motility was
assayed after the final addition of Triton model suspension.
ATP and cyclic AMP concentrations in the preincubations were
1 mM and 10 μM, respectively. Abbreviations used in this
figure are: PK, protein kinase (2.6 units); F, the factor
(1.2 μg of protein); PKI, protein kinase inhibitor (30.7
units). (Redrawn from Ishiguro et al. (15) by permission of
J. Cell Biol.)

of this factor in the reactivation process. One is that this
factor is phosphorylated by the protein kinase and the phosphorylated factor is active in the reactivation of the sperm
models.   Another possibility is that the substrates of the
protein kinase exist in the Triton model itself, and that the
phosphorylation of some components in the Triton model along
with the addition of the factor to the phosphorylated model
is necessary for the reactivation.

     To examine these possibilities, the specific protein
kinase inhibitor mentioned above was used.· Protein kinase,
the factor and the kinase inhibitor were mixed and incubated
for 30 sec in different successions, and the mixture of these
materials was added to the Triton models followed by measurement of % motility of the sperm models (Fig. 9).

     The first row in Fig. 9 shows the % motility in the pres-
ence of the mixture of ATP, protein kinase and the factor
without cyclic AMP. The second row shows the value obtained
in the presence of cyclic AMP. Cyclic AMP-dependent reacti-

vation of the models was quite clear.

When cyclic AMP, ATP, protein kinase and protein kinase inhibitor were preincubated, followed by addition of the factor, little reactivation of Triton models occurred. In this case, the activity of protein kinase should have been inhibited before mixing with the factor and then with Triton models.

But when the factor was preincubated with cyclic AMP, ATP, and protein kinase, followed by addition of protein kinase inhibitor as shown in the last row, fairly high reactivation was observed. In this case, phosphorylation of the factor should have occurred before the addition of the inhibitor, and the kinase activity should have been blocked by the kinase inhibitor just before the addition of the mixture to the Triton models.

These data strongly suggest that the phosphorylation of the factor is necessary for the reactivation. It is also suggested that the activity of the protein kinase is not necessary for the reactivation process after the phosphorylation of the factor occurs. In other words, phosphorylation, if any, of some components in the Triton model is not necessary for the reactivation process.

Figure 10 shows that actual phosphorylation of components in the factor fraction occurred being catalyzed by the protein kinase. Purified protein kinase and partially purified factor, high speed liquid chromatography fraction, were incubated with $[\gamma\text{-}^{32}P]$ATP in the presence of cyclic AMP, followed by analysis by SDS-PAGE and autoradiography. Figure 10a is the pattern of the Coomassie Brilliant Blue staining of the

42k 31k 29k

Fig. 10. Analysis of phosphorylated peptides in the factor fraction. The factor fraction (high speed liquid chromatography fraction, 5 μg of protein) and 3 units of protein kinase (high speed liquid chromatography fraction) were incubated in the presence of 0.1 mM $[\gamma\text{-}^{32}P]$ATP and 10 μM cyclic AMP at 25°C for 30 min. Proteins were analyzed by SDS-PAGE (a gradient gel containing 7.5-15 % polyacrylamide) followed by autoradiography. a, protein stained with Coomasie Brilliant Blue. b, autoradiogram.

proteins. An autoradiogram of the same gel is presented in b.
One of the main components and two minor components were phosphorylated. The molecular weights of the phosphorylated components in the factor fraction were estimated to be 42,000,
31,000 and 29,000. It can be considered that one of them or
some of them are the factor or factors which are responsible
for the motility reactivation of the models. Further purification of the factor is necessary to determine this exactly.

VI. DISCUSSION

It has been reported that cyclic AMP activates the motility of mammalian sperm models (5,17). Gibbons and Gibbons
reported that cyclic AMP had no effects on the motility of
sea urchin sperm Triton models (2). We have also noticed
that cyclic AMP, protein kinase and the factor sometimes did
not affect sperm models prepared from very fresh spermatozoa.
However, we suggest that the aging of dry sperm makes their
Triton models susceptible to cyclic AMP-dependent reactivation
for motility. This might be due to dephosphorylation of the
protein factor to some extent during the aging, which may
release more proteins during preparation of Triton models and
leads to immotility of the models.
In the case of mammalian sperm, it was reported that aging
of the sperm made their Triton models more susceptible to
cyclic AMP-dependent motility activation (17). The basic
story on the roles of cyclic AMP in the flagellar motility
of sea urchin and mammalian spermatozoa may be the same.
Further characterization of the factor and detailed
studies on the function of the phosphorylated factor in the
Triton model motility should be performed in order to establish the control mechanism of sperm motility.

REFERENCES

1. Summers, K. E., and Gibbons, I. R., *Proc. Natl. Acad.
    Sci. U.S.A. 68,* 3092-3096 (1971).
2. Gibbons, B. H., and Gibbons, I. R., *J. Cell Biol. 54,*
    75-97 (1972).
3. Naitoh, Y., and Kaneko, H., *Science 176,* 523-524 (1972).
4. Morton, B., Harrigan-Lum, J., Albagli, L., and Jooss, T.,
    Biochem. Biophys. Res. Commun. 56, 372-379 (1974).
5. Mohri, H., and Yanagimachi, R., *Exp. Cell Res. 127,* 191-
    196 (1980).
6. Hoskins, D. D., Casillas, E. R., and Stephens, D. T.,

Biochem. Biophys. Res. Commun. 48, 1331-1338 (1972).

7. Garbers, D. L., First, N. L., and Lardy, H. A., *J. Biol. Chem. 248,* 875-879 (1973).

8. Lee, M. Y. W., and Iverson, R. M., *Biochim. Biophys. Acta 429,* 123-136 (1976).

9. Murofushi, H., *Ph. D. Thesis, The University of Tokyo* (1975).

10. Gray, J. P., Drummond, G. I., Luk, D. W. T., Hardman, J. G., and Sutherland, E. W., *Arch. Biochem. Biophys. 172,* 20-30 (1976).

11. Tang, F. Y., and Hoskins, D. D., *Biochem. Biophys. Res. Commun. 62,* 328-335 (1975).

12. Brandt, H., and Hoskins, D. D., *J. Biol. Chem. 255,* 982-987 (1980).

13. Murofushi, H., *Biochim. Biophys. Acta 327,* 354-364 (1973).

14. Murofushi, H., *Biochim. Biophys. Acta 370,* 130-139 (1974).

15. Ishiguro, K., Murofushi, H., and Sakai, H., *J. Cell Biol.* in press.

16. Walsh, D. A., Ashby, C. D., Gonzalez, C., Calkins, D., Fischer, E. H., and Krebs, E. G., *J. Biol. Chem. 246,* 1977-1985 (1971).

17. Lindemann, C. B., *Cell 13,* 9-18 (1978).

CHAPTER 17

THE DYNAMICS OF MICROTUBULE SLIDING
IN FLAGELLA[1]

Keiichi Takahashi
Shinji Kamimura

Zoological Institute
University of Tokyo
Tokyo, Japan

I. INTRODUCTION

It is now well established that the bending movement of
eukaryotic cilia and flagella is caused by active sliding between the doublet microtubules (1-3). The force for the sliding
is thought to be generated by the dynein arms of the A-tubule
interacting with the B-tubule of the adjoining doublet. Although the nature of the force-generating reaction is unknown,
recent studies have indicated that the dynein arms act in a
manner similar to the myosin cross-bridges in the skeletal muscle. Thus, models consistent with both morphology and enzymology of dynein have been proposed for the cross-bridge cycle in
which each arm makes intermittent attachment to the B-tubule of
the adjoining doublet while repeating conformational changes
so as to exert a shearing force between the two cross-bridged
doublets (4-6).
To understand the mechanochemistry of dynein-tubulin interaction and to correlate it with the motile behavior of the
whole flagellum, it is important to know certain mechanical
parameters of the force-generating mechanism. For over fifty
years, the dynamics of muscular contraction — such as the isotonic and isometric transients, length-tension and force velocity relations — have been extensively studied and provided a

[1]*Supported by Grants-in-Aid for Scientific Research (Nos.
511213 and 534032) from the Ministry of Education, Science and
Culture, Japan*

Biological Functions of Microtubules
and Related Structures
177

wealth of information which cannot be ignored in any discussion
of the mechanism of contraction. In contrast, the study of the
dynamics of microtubule sliding in cilia and flagella has been
hampered by the small size and the two- or three-dimensional
nature of their movement which make both direct determination
of mechanical properties and interpretation of obtained results
difficult. Nevertheless, attempts have been made to estimate
the force of microtubule sliding from the bending moment generated by the whole organelle (7, 8). Also, by using trypsin-
treated axonemal fragments (2), the velocity of microtubule
sliding has been determined as a function of ATP concentration
(9, 10).

Recently, it has been shown that the force of microtubule
sliding can be measured by attaching fine glass needles to a
trypsin-treated flagellar axoneme (11). With this new technique,
which may be applied to other areas of cell motility as well,
it is possible to determine not only the force generated by a
unit length of the doublet microtubule, but also the relationship between the force and the velocity of microtubule sliding
in a more direct way than has previously been possible.

II. THE FORCE OF MICROTUBULE SLIDING

The experimental procedure used to measure the force of
microtubule sliding was as follows (11):

Spermatozoa of the sea urchin (*Hemicentrotus pulcherrimus*,
Anthocidaris crassispina, or *Pseudocentrotus depressus*) were
demembranated with the extracting solution containing 0.15 M
KCl, 4.0 mM $MgSO_4$, 0.5 mM EDTA, 2.5 mM $CaCl_2$, 0.04 % Triton X-
100, 1.0 mM dithiothreitol, 10 mM Tris (pH 8.0) at room temperature (20°C) and stored for up to 1 h in the reactivating solution without ATP containing 0.15 M KCl, 2.0 mM $MgSO_4$, 2.0 mM
EGTA, 1.0 mM dithiothreitol, 10 mM Tris (pH 8.0) at 0°C. Under
a dark-field microscope, a single sperm was mounted between
two glass microneedles whose tips had been coated with poly-L-
lysine (Fig.1). One of the needles (the flexible needle) was
more flexible than the other (the holder needle) and was used
to measure the sliding force exerted by the doublets. The
needles had been calibrated for their compliances by a series
of cross-calibrations with a needle of a known compliance (12).
The needles were separately mounted on micromanipulators. In
each experiment, the sperm was first perfused with the reactivating solution containing ATP and then with the solution containing both ATP and elastase (10 µg/ml). When this was done,
the sperm was first reactivated to beat, but as digestion with
elastase proceeded, the beating ceased and, after a while the

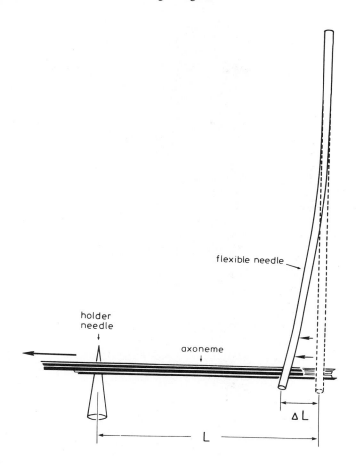

Fig. 1. The principle of the experiment. A demembranated flagellum was attached at two sites to a pair of polylysine-coated microneedles, a stiff holder needle and a flexible needle. The displacement ΔL of the flexible needle, caused by microtubule sliding within the flagellar axoneme was measured. L is the initial distance between the two needles. From Kamimura and Takahashi (11).

axoneme underwent a sliding disintegration (13). It has been suggested that elastase preferentially disrupts structures, possibly the nexin links, which normally bind the doublets together to limit the amount of interdoublet sliding. Trypsin has long been known to have a similar effect (2), but was not used in our experiments because it decreased the adhesiveness of polylysine-coated needles to a considerable extent.

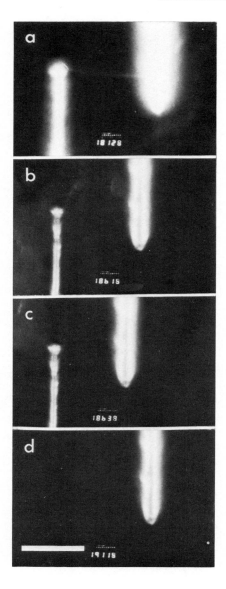

Fig. 2. Movement of the flexible needle. a, *Before the movement. Photographed with increased illumination to show the axoneme.* b,c, *The flexible needle (right) is pulled toward the holder needle.* d, *The holder needle was withdrawn, breaking the axoneme to show the zero-force position of the flexible needle. The clock at the bottom of each frame gives time in 0.1 s. Scale bar in* d, *20 μm. From Kamimura and Takahashi (11).*

    With the onset of sliding disintegration, the flexible
needle was often pulled toward the holder needle (Fig. 1, 2).
The movement was not always smooth, but the needle often stopped or moved back and forth on its way. The movement of the
needle was recorded with a 16 mm cine camera and analyzed frame
by frame on a film motion analyzer.
    To calculate the force generated by a unit length of the
microtubule, we simply assumed that the force was generated
along the length of one doublet between the two needles. Thus,
the force was estimated by the equation

$$f = k \, \Delta L / (L - \Delta L)$$

where k is the elastic coefficient of the flexible needle, L
the initial distance between the two needles and ΔL the displacement of the flexible needle toward the holder needle (Fig.
1). The maximum force was determined from the maximum displacement shown by the needle when it stopped in a fully pulled
position. This has been done in four different concentrations
of ATP and are summerized in Table I which gives the sliding
forces as those generated by a unit length of doublet. Here,
although the average seems to be larger in higher concentrations of ATP, it is possible that the largest values obtained
in each ATP concentration is more reliable because our original
assumption that the force is generated along the entire length
of a doublet between the needles would hold only under most
favorable conditions. It is likely that in many cases we over-
estimated the length of the active region and thus underesti-
mated the force per unit length of doublet. For example, microtubules were often observed to loop out from the axoneme during
the sliding process. If this occurs the true length of the
force-generating region would be less than $L - \Delta L$. It is interesting in this respect that in ATP concentrations between 20
and 500 μM the largest values were about the same (80 - 90 pN).

TABLE I.   Sliding Force in Various Concentrations of ATP

ATP Concentration (μM)	Force ± s.d. (pN/μm)	n	Range (pN/μm)
4	20 ± 7	21	8.5 – 41
20	28 ± 18	22	5.6 – 88
200	35 ± 18	28	11 – 89
500	57 ± 17	5	38 – 83

(From Kamimura and Takahashi, ref. 11.)

Since the dynein arms occur at 24-nm intervals (14) and there are two rows of arms on each doublet, the number of arms per µm of doublet is about 83. From this the maximum time-averaged force generated by a single dynein arm is estimated to be about 1 pN.

III. THE FORCE-VELOCITY RELATION

In the above experiment, the load on the sliding micro-tubules slowly increased with the bending displacement of the flexible needle. This is similar to the condition known to muscle physiologists as auxotonic, and by analyzing the movement of the needle the relation between the load and the veloc-ity of microtubule sliding can be determined, assuming that the relation itself does not change as a result of sliding. Fig. 3a

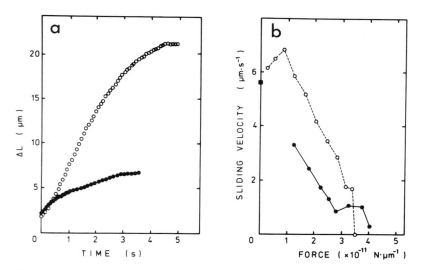

Fig. 3. a, *Displacements (ΔL) of the flexible needle in two experiments are plotted against time. Filled circles, re-sult obtained with a needle which had an elastic coefficient, k, of 180 pN/µm. The initial length of the axoneme between the needles (L) was 38 µm. Open circles, result with a more flexi-ble needle (k = 27 pN/µm, L = 36 µm). ATP concentration, 200 µM. b, Relationship between the sliding velocity and the load given to a unit length of the microtubule. Symbols are as in a. The velocity of unloaded microtubule sliding in trypsin-treated axonemes at 200 µM ATP is shown by a filled square.*

shows two examples of needle displacement plotted against time. From this the sliding velocity was calculated by applying the least-squares method to every five successive points, and the result was replotted against force per unit length of microtubule in Fig. 3b. Although crude, this is the first force-velocity curve ever obtained for the tubulin-dynein system and shows that the sliding velocity decreases with the increasing load. The filled square on the ordinate axis, to which the curves seem to extrapolate, shows the velocity of unloaded sliding we have determined in the same ATP concentration (10).

IV. ISOMETRIC RECORDING

We have so far dealt with experiments in which the isometric force was developed only after the doublets had slid over a considerable distance. This distance, which typically was between 5 and 20 μm, is two orders of magnitude larger than the maximum distance of interdoublet sliding that would take place in the normally beating flagellum. After such an extensive sliding, it is not certain whether all or most of the dynein arms are capable of normal interaction with the B-tubule. Moreover, as mentioned above, microtubules were often seen to loop out from the axoneme during the measurement, causing ambiguity about the length of the force-generating region. These difficulties would be avoided at least partly by reducing as much as possible the extent of sliding, that is, by approaching the true isometric condition. This is much more difficult than the auxotonic method and the results we have obtained are to be regarded as preliminary.

We have developed two types of experimental setup. In the first device, the flexible needle is mounted on an electromechanical transducer consisting of a piezoelectric ceramic. The position of the needle in the microscopic field is detected by a pair of phototransistors and a feedback control is used to compensate for the needle's movement so as to keep its position constant. Thus, if the axoneme pulls the needle, it causes an imbalance of the light impinging on the transistors. The signal is amplified and fed to the piezoelectric ceramic in such a way that the flexible needle mounted on it is moved back to its original position. The voltage applied in this way to the ceramic measures the force exerted on the flexible needle by the axoneme. The performance of this system is limited by the nonlinearity of the sensors' response and also by the mechanical hysteresis of the ceramic which is common to the piezoelectric materials.

Fig. 4 shows a typical record obtained with this method.
As ATP-elastase solution was perfused, tension was developed
between the two needles. The movement of the needle was very
small as the lower trace indicates. The noisy appearance of
the tension record is largely an artifact caused by the insta-
bility of the feedback system, although some part of it may
indicate inherent fluctuation of the sliding force. The plateau of isometric force was terminated by withdrawing the
holder needle so that the tension returned to the zero level.

In the second device, a relatively stiff needle is used as
the flexible needle. As the compliance of the needle is much
smaller than the previously used ones, the sliding force of
the axoneme causes only a very small displacement of its tip.
This displacement is detected by a very sensitive system in
which the light from the dark-field image of the needle is
split by a wedge mirror and sensed by two photodiodes. The
output from this system is amplified and recorded as the isometric force. An example of the record is shown in Fig. 5.
It shows repeated force transients prior to the plateau. This
type of response has been observed in many isometric records.
Again the plateau was terminated by a withdrawal of the holder
needle.

As might have been expected, the force measured with these
two types of isometric devices was somewhat larger than that
obtained with the auxotonic setup.

V. DISCUSSION

It is interesting to compare the present results with previous estimates of the force generated by a dynein arm. By
using the data of Yoneda (12) who measured the force exerted
by a giant compound cilium of *Mytilus* by arresting its movement with a calibrated microneedle, Hiramoto estimated the
sliding force generated by a single dynein arm at about 2 pN
(7). This might have been an underestimation since it was assumed that all the dynein arms in the cilia were contributing
force at the same time. This seems unlikely today since it has
been found that the force generation by the arms is unidirectional and polarized with respect to the base-to-tip axis of
the cilium (15). It is therefore more likely that the maximum
number of cross-bridges that are active at any moment is half
the total number of arms. Moreover, the axial spacing of the
arms along the A-tubule was assumed to be 17 nm, as compared
with the currently used value of about 24 nm. Indeed, using
the same experimental data, Brokaw (8) estimated the corresponding force at 8.8 pN. This value is not unrealistically

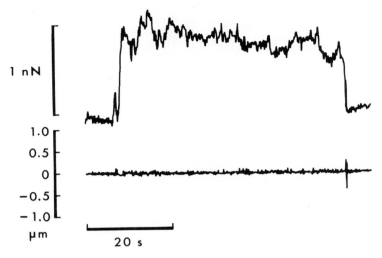

Fig. 4.   *Isometric force recorded with a servo-controlled device. Demembranated sperm of* Anthocidaris crassispina. *The distance between the needles was 25 μm. 100 μM ATP. Upper trace, force record. Lower trace, displacement of the flexible needle. Temperature, 27°C.*

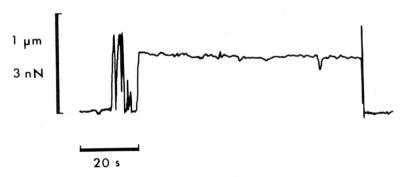

Fig. 5.   *Isometric force recorded with a relatively stiff needle. Note the train of force transients prior to the stable force generation. The spike at the end of the plateau was caused by a deliberate movement of the holder needle which broke the axoneme. Demembranated sperm of* Pseudocentrotus depressus. *The distance between the needles was 25 μm. 20 μM ATP. Temperature, 25°C.*

large compared with our present estimate of about 1 pN per
dynein arm, if we take into account the fact that the latter
value was obtained *in vitro* with demembranated, elastase-
digested axonemes and after an extensive sliding had occurred.
Indeed, our preliminary measurements with isometric recording
devices indicate that the force may be somewhat larger than
1 pN per arm.

By analogy with the force-velocity relation in the striated muscle, we may expect the force to be somewhat smaller in the
beating cilia. Thus, Brokaw (8) estimated the time-average
force in the active region of a swimming sperm of *Lytechinus
pictus* at 0.35 pN. This again seems to be compatible with the
force-velocity relation obtained in the present study although
a strict comparison cannot be made between the two experimental
systems.

It is interesting that the isometric force generated by a
dynein arm is of the same order of magnitude as the values
estimated for the myosin cross-bridge of the striated muscle.
For example, the force per myosin cross-bridge in frog muscle
during a maximal tetanus has been estimated at 2 to 5 pN (15,
16). This might be a mere coincidence and our data are still
insufficient for quantitative analyses. It is hoped that by
further refinement of the technique, it will become possible
to make a more precise study of the dynamics of microtubule
sliding which will give us more insight into the mechanism of
motility in the tubulin-dynein system.

REFERENCES

1.   Satir, P., *J. Cell Biol*. 39, 77 (1968)
2.   Summers, K. E., and Gibbons, I. R., *Proc. Natl. Acad.
     Sci. U.S.A*. 68, 3092 (1971).
3.   Shingyoji, C., Murakami, A., and Takahashi, K., *Nature*
     265, 269 (1977).
4.   Sale, W. S., and Gibbons, I. R., *J. Cell Biol*. 82, 291
     (1979).
5.   Satir, P., *in* "The Spermatozoon" (D. W. Fawcett and J. M.
     Bedford, eds.), p. 81 Urban and Schwarzenberg, Baltimore-
     Munich, (1979).
6.   Satir, P., Wais-Steider, J., Lebduska, S., Nasr, A., and
     Avolio, J., *Cell Motility* 1, 303 (1981).
7.   Hiramoto, Y., *in* "Cilia and Flagella" (M. A. Sleigh, ed.),
     p. 177, Academic, London, (1974).
8.   Brokaw, C. J. *in* "Molecules and Cell Movement" (S. Inoué
     and R. E. Stephens, eds.), p. 165. Raven, New York, (1975).
9.   Yano, Y., and Miki-Noumura, T., *J. Cell Sci*. 44, 169 (1982).

10. Takahashi, K., Shingyoji, C., and Kamimura, S., *in* "Pro-
     karyotic and Eukaryotic Flagella" p. 159. Cambridge
     University Press, Cambridge, (1982).
11. Kamimura, S., and Takahashi, K., *Nature* 293, 566 (1981).
12. Yoneda, M., *J. Exp. Biol.* 37, 461 (1960).
13. Brokaw, C. J., *Science* 207, 1365 (1980).
14. Linck, R. W., *in* "The Spermatozoon" (D. W. Fawcett and
     J. M. Bedford, eds.), p. 99. Urban and Schwarzenberg,
     Baltimore-Munich, (1979).
15. Sale, W. S., and Satir, P., *Proc. Natl. Acad. Sci. U.S.A.*
     74, 2045 (1977).
16. Oplatka, A., *J. Theor. Biol.* 34, 379 (1972).
17. Hill, T. L., *Prog. Biophys. Molec. Biol.* 28, 267 (1974).

CHAPTER 18

ROTATION OF THE CENTRAL-PAIR MICROTUBULES
IN *CHLAMYDOMONAS* FLAGELLA

Ritsu Kamiya

Institute of Molecular Biology
Nagoya University
Nagoya

Reiko Nagai

Department of Biology
Osaka University
Toyonaka, Osaka

Shogo Nakamura

Department of Biology
Toyama University
Toyama

I. INTRODUCTION

It is now established that the bending motion of cilia and flagella is based on the active sliding interactions between outer-doublet microtubules (1,2), but the role of the central-pair microtubules is still unclear. Omoto and Kung demonstrated by electron microscopy that the central pair of *Paramecium* cilia rotates 360° in accordance with the ciliary beat cycle (3). Furthermore, Omoto and Witman have shown that *Micromonas pusilla,* a marine alga, has a flagellum with unusually extended central tubules, and that the cell swims by rotating the helical central pair, like a bacterium swimming by rotating helical flagellar fibers (4). While several other studies also have indicated the central-pair rotation in different kinds of cilia and flagella (5,6), Tamm and Tamm

Biological Functions of Microtubules
and Related Structures

have reported that the orientation of the central pair never
changes in two-dimensionally beating metazoan cilia (7). Thus
the generality and importance of the central-pair rotation
remain to be investigated. Here we report direct observation
of rotating central pairs in *Chlamydomonas* flagella that beat
two-dimensionally either in a cilia-type pattern or in a
flagella-type pattern. Together with works by others (8-10),
our present study leads to the conclusion that the central
pair and its rotation are dispensable for the generation of
bending motion in *Chlamydomonas* flagella, but may be important
for fine control of the wave-form and/or the propagation of
bending. Parts of this study have been reported (11).

II. MATERIALS AND METHODS

A wild-type strain C-239 (BIU-90, mt⁻) and a backward-
swimming mutant RL-10 (12) of *Chlamydomonas reinhardii* were
grown by the method of Sager and Granick (13) under constant
illumination. Cells were washed twice by centrifugation in 10
mM HEPES (N-2-hydroxyethyl-piperazine-N'-2-ethane sulfonic
acid) of pH 7.3. Demembranation and reactivation of axonemes
were carried out at room temperature by mixing about 0.02 ml
of cell suspension with 1 ml of demembranating/reactivating
solution containing 30 mM HEPES, 1 mM DTT, 5 mM $MgSO_4$, 25 mM
KCl, 2% polyethylene glycol, 0.04% Nonidet P40, 1 mM ATP and
a calcium-buffer system consisting of variable amounts of
$CaCl_2$ and EDTA (or EGTA). The pH value was adjusted to 7.3.
These solvent compositions were after ref. 14.
An Olympus microscope with a dark-field condenser (N.A.
1.2 - 1.33), an Apo 40X objective (N.A. 1.0) and a light
source consisting of a 100 W high-pressure mercury arc lamp
was used. The techniques of dark-field microscopy have been
described (15). With this system, we can observe single microtubules from bovine brain (16). The images were recorded with
a TV camera equipped with an SIT (silicon intensified target)
tube (CTC-9000, Ikegami Tsushinki Co., Tokyo) and a SONY J9
video recorder. Video images were filmed with a Bolex 16M
cine camera from an 11 inch monitor.
For ultra-thin section studies, cells were instantaneously
fixed by the method of Tamm with some modifications (3).
Cells were fixed with 2% OsO_4 and 2% glutaraldehyde in 50 mM
sodium cacodylate buffer of pH 7.3 for 15 min and post-fixed
with 1% OsO_4 for 1 h. After being dehydrated, the samples were
embedded in Spurr's resin (17), thin-sectioned, and stained
with uranylacetate and lead citrate. A JEM 100C microscope was
used.

III. RESULTS

A. *Extrusion of Helical Central Pairs*

When wild-type cells were suspended in the demembranating/ reactivating solution of low Ca^{2+} concentrations (less than 10^{-6} M), their flagella displayed ciliary type beating for about 30 min. Under these conditions, most of the flagella remained attached to the cell body. When 10^{-4} M Ca^{2+} was present, the flagella became detached from the cell body and beat with a symmetrical, flagella-type pattern for about 10 min. The behavior of reactivated axonemes at different concentrations of Ca^{2+} was as reported by previous authors (14,18). After periods of such normal beating, the flagella tended to stop or disintegrate. For many axonemes undergoing disintegration, the first event in the disintegration process was extrusion of a helical fiber from the tip of the axoneme (Fig. 1). This fiber, an approximate left-handed helix with a pitch of 3 - 5 µm and a diameter of 1.5 - 2.5 µm, could be easily distinguished from an outer-doublet microtubule or a bundle of outer doublets which had much smaller curvatures. Nearly one pitch of helix was contained in the whole length of this fiber. Nakamura and Kamiya (8) have concluded that this helix is the central pair, because in electron micrographs, curved filaments projecting from the axoneme were always pairs of microtubules with the characteristic striations of the central pair (19,20)(Fig. 2). In agreement with this conclusion, we have found that such a helical fiber was absent from frayed axonemes of central-pairless mutants, *pf-15* and *pf-19* (21,22). The extrusion of the central pair occurred more frequently in

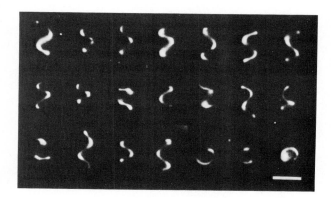

FIGURE 1.   *Helical central pairs released in the medium. Dark-field micrographs. Bar, 5 µm.*

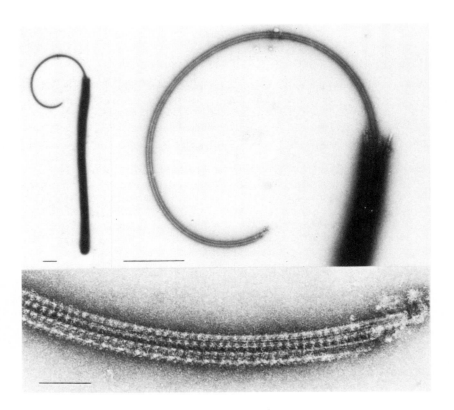

FIGURE 2. *Electron micrographs of an axoneme with a projecting central pair. Negatively stained with 1% uranylacetate. Bars are 0.5 μm in the upper two photographs and 0.1 μm at the bottom.*

the presence of 10^{-4}M Ca^{2+} than in 10^{-8}M Ca^{2+}, and was rarely observed when isolated axonemes were used in place of the whole cells. Hence some factors from the cell body, such as a Ca^{2+}-dependent protease, might be involved in this reaction.

B. *Rotation of the Central Pair*

The central pair seemed to rotate around the axis of the axoneme when it was partially extruded (Fig. 3). The rotation was as rapid as 30 Hz and smooth and continuous just after the central pair projected out. As the extrusion proceeded, however, it gradually slowed down, tended to rest for short periods, and stopped completely in 3 to 120 sec. When the

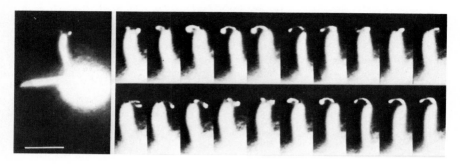

FIGURE 3.   Rotation of a partially extruded central pair
under conditions of low $(10^{-8}$ M) Ca^{2+} concentration. A series
of TV recordings taken at intervals of 1/30 sec for 2/3 sec
(from upper left to lower right). The left photo shows an
entire view of the cell carrying this axoneme. Bar, 10 μm.

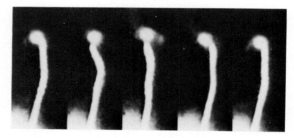

FIGURE 4.   Rotation of a central pair with a bound polystyren latex bead. Taken with an interval of 1/30 sec from
left to right. Note that the positions of the bead on the
second and the fourth frames are opposite with respect to the
axonemal axis.

rotation became slow, there seemed to be one to three angular
positions at which the central pair tended to rest. We have
found no constant relationship between the degree of the projection and the total number of revolution of the central
pair, although the extrusion was always accompanied by rotational movements. The direction of rotation was counter-clock-
wise as viewed from the distal end. This direction was the
same as in *Paramecium* cilia and *Micromonas* flagella (3,4).
When polylysin-coated polystyren beads (23) were added to the
sample, they preferentially became attached to the central
pair. On some occasions, beads attaching near the axonemal tip
alternated their lateral position with respect to the axonemal
axis as the central-pair rotated (Fig. 4). This observation
evidenced real rotation of the central pair as opposed to

propagation of helical waves without true rotation of the
fiber. The central-pair rotation was observed under conditions
of both low (10^{-8} M) and high (10^{-4} M) Ca^{2+} concentrations,
although fewer axonemes displayed the central-pair extrusion
in the former conditions than in the latter. Under conditions
of high Ca^{2+} concentrations, the rotation of the central pair
was observed in detached axonemes sticking to the glass surface by the proximal end (Fig. 5). We have so far not observed
any Ca^{2+}-dependent change in either the direction of rotation
or the helical shape of the central pair. Since the central
pair behaved similarly whether the axoneme was attached to the
cell body or to the glass surface, the possibility that the
whole axoneme rotates, rather than the central pair alone
does, seems highly unlikely.

C. Bending of Axoneme without Central-Pair Rotation

The extrusion and rotation of the central pair was always
accompanied by a marked decrease in the beating amplitude of
the axoneme. In contrast, some axonemes exhibited a slow but
large bending motion while the partially extruded central pair
was not rotating (Fig. 6). Bending was observed even in the

FIGURE 5.   Rotation of the central pair on a detached
axoneme in a high (10^{-4} M) concentration of Ca^{2+}. Interval,
1/30 sec, from left to right.

FIGURE 6.   Bending motion of an axoneme without rotation
of the central pair. Interval, 1/15 sec, from left to right.
Note that the proximal portion is bending, despite the lack
of the central pair. From ref. 11.

proximal portion of the axoneme, from which the central pair
had slipped off. In all of these cases, propagation of the
bend was not observed. Actually, small groups of outer-doublet
microtubules in frayed axonemes were able to exhibit bending
movements (8). Our observations so far have strongly suggested
that groups of two outer doublets could generate a bending
motion.

D. Central-Pair Rotation in Vivo

It was not certain from the above observations whether the
central pair was rotating before projecting out from an axoneme. Ringo reported that, in electron micrographs showing
basal portions of the two flagella of a cell in cross-section,
the central pair was always aligned roughly parallel to the
line connecting the two flagella, i.e. to the plane of flagellar bending (24). Thus his observation apparently did not support the view of in vivo rotation of the central pair. However,
certain degrees of deviation in the orientation of the central
pair can be seen in his published micrographs. Thus we thought
that further studies were needed to prove or disprove the in
vivo rotation, and reexamined the central-pair orientation in

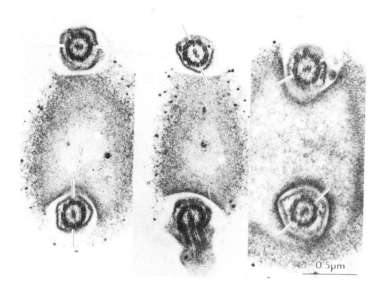

FIGURE 7.   Three sections which show proximal portions of
the two flagella of single cells. The orientation of the
central pairs are indicated with lines.

cross-section, using an 'instantaneous' fixation procedure (3, 25). When five volumes of fixative (2% OsO_4 + 2% glutaralde-hyde) was added to a suspension of vigorously swimming *Chlamydomonas,* all the organisms stopped swimming instanta-neously, and flagella were 'frozen' in various wave-forms corresponding to natural phases of the beat cycle, as in the case of *Paramecium* cilia (3,25).

We have found much greater variation in the central-pair orientation in cross-section than has been reported (24). Some central pairs were aligned perpendicular to the flagellar bending plane, while some others showed a nearly parallel orientation as reported by Ringo (24)(Fig. 7). Since the plane of cross-section was not strictly normal to the apparent two-fold axis of the cell, the observed variation in central-pair orientation must have suffered some artificial modification, and thus requires caution to interpret. However, the occur-rence of both parallel and perpendicular orientations is interpretable only when we assume a rotational movement of the central pair. We have not yet been certain whether these cross-sections are as viewed from base to tip or from tip to base, nor can we distinguish the two central microtubules. Thus the angular variation in the central-pair orientation in these photographs might well be of 90°, 180° or 360°. There-fore, further studies are necessary to prove *in vivo* rotation of the central pair, but, together with the *in vitro* evidence, these electron microscopic observations should strongly sup-port the idea that the central pairs are rotating in intact flagella of *Chlamydomonas.*

We also examined the central-pair orientation in a backward-swimming mutant RL-10 (12). The flagella of this mutant always beat with a symmetrical, flagella-type pattern, while the wild-type flagella beat mostly with an asymmetrical, cilia-type pattern. We have observed a similar degree of variation in the central-pair orientation in both kinds of flagella. Hence it is likely that the central pair rotates irrespective of the bending pattern of the flagella, as indi-cated by the *in vitro* observations conducted at different concentrations of Ca^{2+}.

IV. DISCUSSION

Several models could be thought of for the mechanisms of the central-pair rotation. For example, we could imagine that some specific interactions between the central pair and the radial spokes cause the rotation. However, it is unlikely that the central pair is rotated only at the base of an axoneme, since the central pair can rotate after a certain length of it has been extruded from the axoneme. The observation that the extrusion and rotation of the central pair was accompanied by a marked decrease in the bending amplitude of the axoneme suggests some close coupling between the central-pair rotation and the bending of the axoneme. A possible mechanism is that the rotation is caused by mechanical interactions between axonemal bending and the central pair with a tendency to assume helical forms. If the central pair is free to rotate in an axoneme, it will orient itself (to minimize elastic energy) to match its inherent curvature with any bend produced in the axoneme, and will be rotated as the bend is propagated distally. This simple 'passive rotation' model also explains why the central pair is extruded, as well as rotated. If this is actually the mechanism for rotation, the tendency of the central tubules to assume helical forms is crucial to the rotation. It is interesting to note that in all cases where the central-pair rotation was demonstrated, helical forms or structural twists were observed in the central pair (3-5).

In electron micrographs of negatively stained specimens (Fig. 2), the pair of central tubules appeared to lie side by side, forming concentric arcs. This implies that the two microtubules are slightly different in intrinsic length. If the center-to-center distance between the two tubules is 35 nm and the radius of curvature 1 μm, this difference should be about 3.5% on the average. If the central pair tends to have a curvature in the plane comprising the two tubules, and if the above model for rotation is correct, then the plane of the central-pair orientation will coincide with the plane of the flagellar bending. Parallel orientation of the central pair to the plane of flagellar bending has been observed in *Paramecium* cilia (3), whereas central pairs are aligned perpendicular to the bending plane in many kinds of cilia and flagella (for example, see ref. 7). This discrepancy might be due to the difference in morphological or mechanical properties of the central pair. According to the above model, the plane of flagellar bending should be determined independently of the central pair. The recent work by Goldstein has supported the idea that the central pair is not involved in determining the beating plane of *Chlamydomonas* flagella: he showed that the flagella of a central-pairless mutant (*pf-18*) became transiently motile under some conditions, and that those flagella beat in the same plane as the bending plane of the wild-type flagella.

We have shown that the *Chlamydomonas* axoneme can bend without rotating the central pair. In fact, the bending motion was observed in axonemal portions which apparently lacked the central pair, and even in small groups of outer doublets in frayed axonemes. These observations are consistent with recent work by Luck and Goldstein, both of which showed that the

central pair is dispensable for the generation of bending in
Chlamydomonas flagella (9,10). However, most of these flagellar movements exhibited in the absense of the central pair had
some type of abnormality, such as low frequencies, abnormal
wave-forms, or irregular bend-propagation. Thus the central
pair and its rotation might be important in controlling
subtle coordination within the flagellar machinaries.

REFERENCES

1.   Satir, P., *J. Cell Biol. 39,* 77 (1968).
2.   Summers, K. E., and Gibbons, I. R., *Proc. Nat. Acad. Sci.
     U.S.A. 68,* 3092 (1971).
3.   Omoto, C. K., and Kung, C., *J. Cell Biol. 87,* 33 (1980).
4.   Omoto, C. K., and Witman, G. B., *Nature 290,* 708 (1981).
5.   Jarosch, R., and Fuchs, B., *Protoplasma 85,* 285 (1975).
6.   Tamm, S. L., and Horridge, G. A., *Proc. Roy. Soc. Lond.
     Ser. B. 175,* 219 (1970).
7.   Tamm, S. L., and Tamm, S., *J. Cell Biol. 89,* 495 (1981).
8.   Nakamura, S., and Kamiya, R., *Cell Struct. Funct. 3,* 141
     (1978)
9.   Luck, D. J. L., *Cell Motil.* in press.
10. Goldstein, S. F., *Cell Motil.* in press.
11. Kamiya, R., *Cell Motil.* in press.
12. Nakamura, S., *Exptl. Cell Res. 123,* 441 (1979).
13. Sager, R., and Granick, S., *Ann. N.Y. Acad. Sci. 56,* 831
     (1953).
14. Bessen, M., Fay, R. B., and Witman, G. B., *J. Cell Biol.
     86,* 446 (1980).
15. Kamiya, R., and Asakura, S., *J. Mol. Biol. 106,* 167 (1976).
16. Miki-Noumura, T., and Kamiya, R., *Exptl. Cell Res. 97,*
     (1976).
17. Spurr, A. R., *J. Ultrastruct. Res. 26,* 31 (1969).
18. Hyams, J. S., and Borisy, G. G., *J. Cell Sci. 33,* 235
     (1978).
19. Hopkins, J. M., *J. Cell Sci. 7,* 823 (1970).
20. Allen, C., and Borisy, G. G., *J. Mol. Biol. 90,* 381
     (1974).
21. Warr, J. R., McVittie, A., Randall, J., and Hopkins, J.
     M., *Genet. Res. 7,* 335 (1966).
22. Witman, G. B., Plummer, J., and Sander, G., *J. Cell Biol.
     76,* 729 (1978).
23. Brown, S. S., and Spudich, J. A., *J. Cell Biol. 80,* 499
     (1979).
24. Ringo, D. L., *J. Cell Biol. 33,* 543 (1967).
25. Tamm, S. L., *J. Cell Biol., 55,* 250 (1972).

CHAPTER 19

LOCALIZATION OF FLUORESCENTLY LABELED CALMODULIN IN LIVING SAND DOLLAR EGGS DURING EARLY DEVELOPMENT[1]

Yukihisa Hamaguchi

Biological Laboratory
Tokyo Institute of Technology
Tokyo

Fuyuki Iwasa

Department of Biology
College of General Education
University of Tokyo
Tokyo

I. INTRODUCTION

It is well known that calmodulin plays a central role in regulating many cellular activities in various eukaryotic cells as an intracellular intermediary for calcium ions (1, 2). One of these cellular activities may be mitosis because specific localization of calmodulin in the mitotic apparatus has been found by microinjection of a fluorescent conjugate of cal-modulin into sand dollar eggs (3), and by immunofluorescence in mammalian cells (4-6) during mitosis. Calmodulin has also been found in association with microtubules by immunoelectron micro-scopy in mammalian cells during mitosis (7). In this study, we have investigated the effects of calmodulin-specific inhibi-tors, a microtubule-specific inhibitor, low temperature and calcium on the localization of calmodulin during mitosis.

[1]This work was supported by Grants-in-aid for Scientific Research from the Ministry of Education, Science and Culture Japan.

II. MATERIALS AND METHODS

Gametes of the sand dollar, *Clypeaster japonicus,* were the experimental material.

Calmodulin was purified from porcine brain and labeled with the fluorescent dye, N-(7-dimethylamino-4-methylcoumarinyl)-maleimide (DACM) in molar ratio of 1 : 1 as described elsewhere (3). A solution containing 2.5 mg/ml DACM-calmodulin was used for microinjection. As controls, 2.3 mg/ml DACM-α-lactalbumin (DACM-LA), 6.0 mg/ml DACM-bovine serum albumin (DACM-BSA) and 4.9 mg/ml DACM-rabbit muscle troponin C (DACM-TnC) were used. 100 mM trifluoperazine (TFP), 20 mM N^2-dansyl-L-arginine-4-*t*-butyl-piperazine amide (TI 233), 50 mM N-(6-aminohexyl)-5-chloro-naphthalene-sulfonamide (W-7), and 3 or 30 mM colchicine were used for microinjection. The CaHEDTA buffer consisted of 1 M N-hydroxyethylethylenediaminetriacetic acid (HEDTA)/ 0.5 M $CaCO_3$/ 10 mM piperazine-N,N'-*bis*(2-ethanesulfonic acid) (PIPES)-KOH buffer (pH 7.0). EGTA solution consisted of 1 M ethyleneglycol-*bis*(β-aminoethylether)-N,N'-tetraacetic acid (EGTA)/ 10 mM PIPES-KOH (pH 7.0).

Ca^{2+}-dependent cyclic nucleotide phosphodiesterase was prepared and the activity of the enzyme was measured by the method of Teo *et al.* (8) as described elsewhere (3). Polyacrylamide gel electrophoresis in the presence of sodium dodecyl sulfate (SDS) was performed according to Laemmli (9) except that the gels contained 7% (v/v) glycerol.

Microinjection was carried out by the method of Hiramoto (10) using a braking micropipette. The volume of microinjected protein solutions per egg was usually 20-40 pl, corresponding to 3-5% of the egg volume. Therefore, the final concentration of DACM-calmodulin in the egg cytoplasm was 0.07-0.13 mg/ml (4.2-7.8 µM).

Low temperature treatment of the cells was performed by perfusing cold sea water cooled on ice through the injection chamber.

FIGURE 1.   15% SDS-polyacrylamide gel electrophoresis of DACM-calmodulin.   Ten micrograms of DACM-calmodulin was electrophoresed in the presence of $CaCl_2$ (the upper lane) and in the presence of EGTA (the lower lane) in the sample buffer. This fluorescence pattern was obtained by ultraviolet trans-illumination.

Mitotic apparatuses were isolated from eggs preloaded with DACM-calmodulin by squirting isolation medium containing 0.1% Triton X-100 (cf. 11) onto the eggs through a glass capillary.

Eggs, blastomeres and mitotic apparatuses were observed with an epifluorescence microscope (Nikon XF-EF) equipped with differential interference optics and a Brace-Köhler compensator. This microscope allowed the use of three types of optics—fluorescence, differential interference, and polarization. These three kinds of micrographs could be taken within 15-20 sec.

III. RESULTS AND DISCUSSION

A. *Specific Localization of Calmodulin in the Mitotic Apparatus*

DACM-labeled calmodulin retained the property of a calcium-dependent change of mobility on an SDS-polyacrylamide gel (Fig. 1). No difference was observed in phosphodiesterase stimulating activity between DACM-labeled and unlabeled calmodulin (Fig. 2). Thus, it is concluded that these properties of calmodulin do not change through the labeling procedures. Calmodulin in sea urchin eggs closely resembles that in mammalian brain as shown by amino acid analysis (12) and a Ca^{2+}-dependent

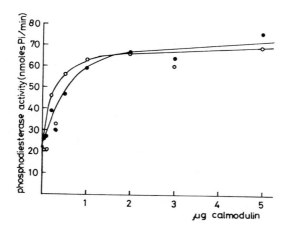

FIGURE 2. *Activation of brain Ca^{2+}-dependent phosphodiesterase by DACM-calmodulin. 0.36 μg of phosphodiesterase was mixed with 0.02 units of 5'-nucleotidase and 1 mM cyclic AMP. The reaction mixture was incubated at 30°C for 30 min. ○, DACM-calmodulin; ●, Unlabeled calmodulin. (From Hamaguchi and Iwasa (3) by the permission of Biomed. Res.)*

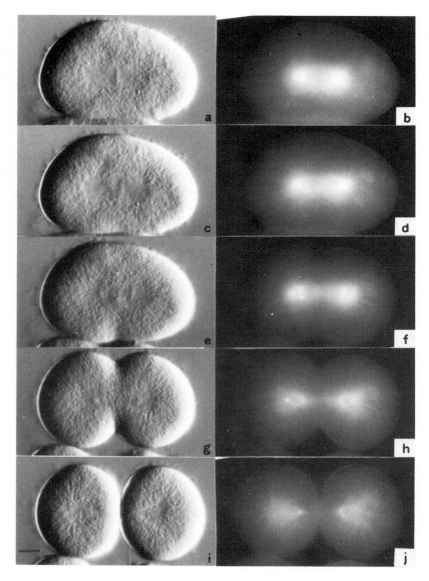

FIGURE 3. *Calmodulin localization in a blastomere during*
the 8 cell stage. a, c, e, g, i, Differential interference
micrographs; b, d, f, h, j, Fluorescence micrographs. Differ-
ential interference and fluorescence micrographs were taken
mutually at one- to two-minute intervals in alphabetical order.
The bar is 10 μm. (From Hamaguchi and Iwasa (3) by the
permission of Biomed. Res.)

inhibitory effect on brain microtubule assembly (13). Therefore, injected DACM-calmodulin is expected to function similarly to intrinsic calmodulin in the eggs.

DACM-calmodulin uniformly diffused in the cytoplasm of
unfertilized and fertilized eggs within a few min after microinjection. Although the total content of calmodulin in the
injected eggs increased by a factor of 1.4-1.8 after injection
in the present study (see in detail, ref. 3), this excess
calmodulin did not perturb the physiological activities of the
eggs, such as fertilization, mitosis or cleavage.

After fertilization, intense fluorescence of DACM-
calmodulin appeared at the center of the sperm aster. When
the aster became larger, fluorescence was distinctly observed
at the astral rays. In cases of DACM-LA and DACM-BSA, intense
fluorescence was not observed in the aster (see in detail,
ref. 3).

After the nuclear membrane disappeared, intense fluorescence was observed in the region of the mitotic apparatus in
the eggs or blastomeres injected with DACM-calmodulin (Fig. 3).
At metaphase (Fig. 3b), the fluorescence of the spindle poles
and the area around the poles became the most intense. The
fluorescence around the poles became diffuse except in the
intensely fluorescent central parts, i.e. centrosomes, and
intense fluorescence was observed along elongating astral rays
during anaphase (Fig. 3d). During telophase (Fig. 3f and h),
the fluorescence was the most intense in the centrosomes, and
both the astral rays and the interzone distinctly fluoresced.
Such a change in fluorescence localization was repeated during
subsequent mitoses until 4 hr or more after injection, when
the embryos developed to blastulae.

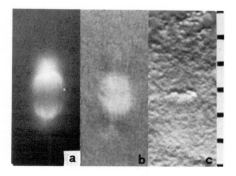

FIGURE 4. Calmodulin localization in an isolated mitotic
apparatus at metaphase. a, Fluorescence micrograph; b, Polarization micrograph; c, Differential interference micrograph.
One division is 10 μm. (From Hamaguchi and Iwasa (3) by the
permission of Biomed. Res.)

Uniform distribution of fluorescence was observed in the region free from yolk granules of the egg injected with DACM-BSA, DACM-LA and DACM-TnC. This phenomenon may be ascribed to the fact that the light path length for all the proteins dissolved in the egg cytoplasm is substantially long in this region because large membrane-bound organelles are excluded from the mitotic apparatus.

Fig. 4 shows a mitotic apparatus isolated at metaphase from an egg which had been loaded with DACM-calmodulin. Calmodulin remained associated with the isolated mitotic apparatus through the isolation procedure with the medium practically free of calcium ions. It is noted that fluorescent fibrous structures projecting toward, but not reaching the chromosomes were observed originating near the spindle poles (Fig. 4a). This fluorescence distribution in the spindle was obviously different from the distribution of birefringence (cf. Fig. 4b); the birefringence was found near the equator in the spindle, but not near the spindle poles. The birefringence is mainly due to microtubules oriented in parallel in the mitotic apparatus (13, 14). It has been reported that the localization of calmodulin is different from that of tubulin as determined by immunofluorescence (4, 5). Calmodulin was reported not to show a Ca^{2+}-dependent inhibitory effect on the assembly of microtubules from sea urchin eggs (15). These facts suggest that calmodulin may be associated with microtubules in a Ca^{2+}-independent manner, and that calmodulin does not directly affect assembly and disassembly of microtubules in the mitotic apparatus.

In the present study, it has been found that DACM-calmodulin is a good molecular probe for studying the intracellular localization of calmodulin. Calmodulin localization during mitosis in the sand dollar eggs is similar to the localization in mammalian fixed cells by immunocytochemistry (4-7). Specific localization as described above suggests that calmodulin plays a significant role in cell motility during fertilization and mitosis.

B. *Effects of Calmodulin-Specific Inhibitors on Calmodulin Localization in the Mitotic Apparatus*

In order to understand the role of calmodulin in mitosis, we examined the effects of several calmodulin-specific inhibitors, TFP, TI 233 and W-7, on mitosis and on the localization of calmodulin in fertilized sand dollar eggs. These drugs are reported to produce a 50% inhibition of calmodulin-mediated processes at a concentration of about 10 μM *in vitro* (16-18). When the inhibitors were injected into the eggs at metaphase or anaphase at intracellular concentrations of more than 100 μM, chromosome movement was not inhibited. When W-7 was injected into blastomeres preloaded with DACM-calmodulin, at an intra-cellular concentration of more than 100 μM, it did not disturb the localization of calmodulin. The localization of TFP and TI 233 in the egg cytoplasm could be observed by fluorescence microscopy because they are fluorescent. After injection of TI 233 at an intracellular concentration of about 100 μM, it was concentrated in the asters, except in the centrosomes. With time, the TI 233 became concentrated in the granular structures of the egg cytoplasm but not in the mitotic appara-tus. In a fluorescent micrograph (Fig. 5a) taken 4 min after injection of TFP at an intracellular concentration of about 1 mM, an amorphous fluorescent region of TFP appeared, in addition to the fluorescence of DACM-calmodulin. TFP neither diffused out from the injected region, nor disturbed the localization of calmodulin. Anaphase chromosome movement and cleavage were not inhibited. These facts indicate that TFP and TI 233 fail to alter the distribution of calmodulin in the egg cytoplasm because they are trapped or concentrated in organelles other than the mitotic apparatus.

C. Effects of Colchicine and Low Temperature on Calmodulin Localization

In order to investigate the relationship between micro-tubules and calmodulin, colchicine was injected into blasto-meres preloaded with DACM-calmodulin. Soon after the injection

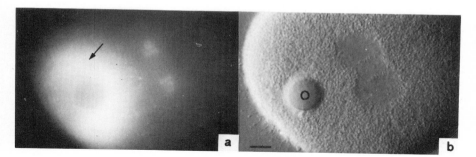

FIGURE 5. Localization of TFP in a blastomere at the two cell stage, preloaded with DACM-calmodulin. a, Fluorescence micrograph; b, Differential interference micrograph. TFP was injected into the blastomere at metaphase at an intracellular concentration of 1.03 mM, and these micrographs were taken 4 min after injection of TFP. Arrow indicates TFP fluorescence. O is an oil drop introduced at the time of injection. The bar is 10 μm.

of colchicine at an intracellular concentration of 0.36 mM, the birefringence of the mitotic appatatus disappeared and the intense fluorescence of calmodulin in the mitotic apparatus became disorganized. The disappearance of birefringence appeared to be faster than the disorganization of calmodulin localization (Fig. 6). At a lower colchicine concentration, i.e. at about 0.03 mM, which is near the threshold for inhibition of cleavage (19), birefringence decreased but did not completely disappear, and the spindle shortened. Calmodulin fluorescence was still observed in the shortened mitotic spindle and the asters.

It is well known that low temperature reduces the birefringence of the mitotic spindle (20), but does not completely

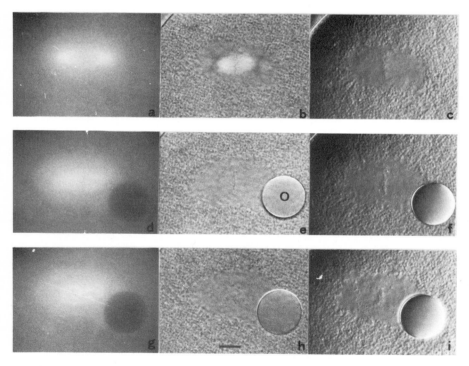

FIGURE 6. *Effect of colchicine at high intracellular concentration on localization of DACM-calmodulin and birefringence. a, d, g, Fluorescence micrographs; b, e, h, Polarization micrographs; c, f, i, Differential interference micrographs. a, b, c, Before colchicine injection; d, e, f, 1 min after injection of colchicine at an intracellular concentration of 0.36 mM; g, h, i, 4 min after injection of colchicine. 0 is an oil drop introduced at the time of injection. The bar is 10 μm.*

disrupt the microtubules (21-24). In the present study, it was observed that spindle birefringence diminished considerably and sometimes disappeared completely within 1 min of the perfusion of cold sea water, and the spindle became shortened. However, the intense fluorescence of calmodulin in the mitotic apparatus was clearly observed in the shortened spindle and the centrosomes (Fig. 7). Our results concerning the effect of low temperature on calmodulin localization are consistent with those obtained by Welsh *et al.* (6) in fixed mammalian cells using immunofluorescence. The localization of calmodulin at low temperature resembled the localization of calmodulin obtained by the injection of colchicine at lower intracellular concentrations. These facts suggests that calmodulin is associated with cold-stable microtubules, but not with cold-labile microtubules, and that cold-stable microtubules are less sensitive to colchicine than cold-labile microtubules. This can explain the difference between the distribution of spindle birefringence and that of calmodulin.

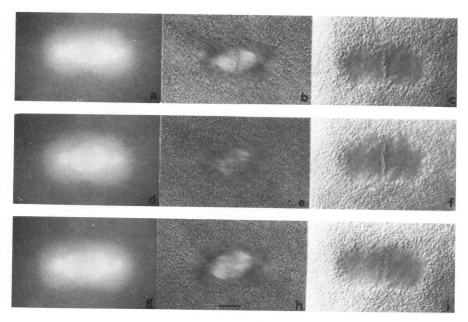

FIGURE 7. Effect of low temperature on localization of DACM-calmodulin and birefringence. a, d, g, Fluorescence micrographs; b, e, h, Polarization micrographs; c, f, i, Differential interference micrographs. a, b, c, Before perfusion of cold sea water; d, e, f, 1 min after perfusion of cold sea water; g, h, i, 1.5 min after stopping the perfusion. The bar is 10 μm.

D. *Effects of Changes in Calcium Ion Concentration*
   in the Egg Cytoplasm on Calmodulin Localization

Calcium has been suggested to be an intracellular regulator
of microtubule assembly and disassembly because it inhibits
microtubule assembly and disrupts microtubules both *in vivo*
(25) and *in vitro* (13, 26). We have investigated the effects

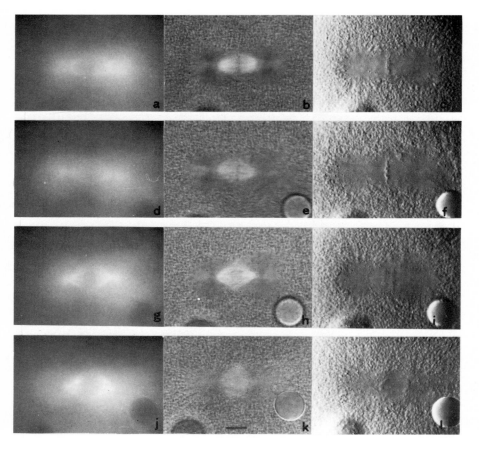

FIGURE 8. *Effect of injection of CaHEDTA buffer on local-*
ization of DACM-calmodulin and birefringence. a, d, g, j,
Fluorescence micrographs; b, e, h, k, Polarization micrographs;
c, f, i, l, Differential interference micrographs. a, b, c,
Before injection; d, e, f, 1 min after injection of CaHEDTA
buffer (CaHEDTA/HEDTA = 1) at an intracellular concentration
of 11 mM. g, h, i, 3 min after injection (anaphase); j, k, l,
9.5 min after injection (telophase). The bar is 10 μm.

of the change in intracellular calcium concentration on the
calmodulin localization and microtubule assembly and disassembly by injecting calcium buffers into blastomeres
preloaded with DACM-calmodulin. CaHEDTA solution may increase
the intracellular calcium ion concentration up to 4 µM when
injected, as calculated in a previous paper (27). Shortly
after injection of CaHEDTA solution, birefringence of the
mitotic apparatus decreased and the spindle became shortened,
but the fluorescence of DACM-calmodulin remained localized in
the shortened mitotic apparatus (Fig. 8d and e). Within a few
min after injection, the birefringence of the mitotic
apparatus began to increase and simultaneously, the fluorescence of DACM-calmodulin in spindle became more intense than
that before injection (Fig. 8g and h). Subsequently, anaphase
chromosome movement was observed though spindle elongation
never occurred (Fig. 8g-1).

EGTA solution depletes calcium ions from the egg cytoplasm,
which may cause a decrease in the intracellular calcium ion
concentration. After injection of EGTA solution, birefringence
of the mitotic spindle appeared to increase, but the intensity
and distribution of DACM-calmodulin fluorescence in the mitotic
apparatus scarcely changed. EGTA injection at an intracellular
concentration of 10 mM did not inhibit chromosome movement.
Thus, the localization of calmodulin may be regulated by the
level of free calcium in a manner somewhat different from the
calcium regulation of microtubule assembly and disassembly.

ACKNOWLEDGMENTS

We express our gratitude to Professors Y. Hiramoto and
H. Mohri for their invaluable advice and for reading the manuscript. We wish to thank Dr. M. Maruyama of Mitsubishi-kasei
Institute of Life Sciences and Yoshitomi Pharmaceutical Indus-
tries Ltd. for their generous gifts of TI 233 and TFP, respectively, and we also wish to thank Misaki Marine Biological
Station for supplying the sand dollars used as materials.

REFERENCES

1.   Kakiuchi, S., *Taisha 17,* 233 (1980).
2.   Means, A. R. and Dedman, J. R., *Nature 285,* 73 (1980).
3.   Hamaguchi, Y. and Iwasa, F., *Biomed. Res. 1,* 502 (1980).
4.   Andersen, B., Osborn, M. and Weber, K., *Cytobiologie 17,*
     354 (1978).

5. Welsh, M. J., Dedman, J. R., Brinkley, B. R. and Means,
    A. R., *Proc. Natl. Acad. Sci. USA 75,* 1867 (1978).
6. Welsh, M. J., Dedman, J. R., Brinkley, B. R. and Means,
    A. R., *J. Cell Biol. 81,* 624 (1979).
7. De Mey, J., Moeremans, M., Geuens, G., Nuydens, R., Van
    Belle, H. and De Brabander, M., *in* "Microtubules and
    Microtubule Inhibitors" (M. De Brabander and J. De Mey,
    eds.), p. 227. Elsevier, Amsterdam, (1980).
8. Teo, T. S., Wang, T. H. and Wang, J. H., *J. Biol. Chem.*
    248, 588 (1973).
9. Laemmli, U. K., *Nature 227,* 680 (1970).
10. Hiramoto, Y., *Exp. Cell Res. 87,* 403 (1974).
11. Sakai, H., Shimoda, S. and Hiramoto, Y., *Exp. Cell Res.*
    104, 457 (1977).
12. Head, J. F., Mader, S. and Kaminer B., *J. Cell Biol. 80,*
    211 (1979).
13. Sato, H., Ellis, G. W. and Inoué, S., *J. Cell Biol. 67,*
    501 (1975).
14. Hiramoto, Y., Hamaguchi, Y., Shôji, Y., Schroeder, T. E.,
    Shimoda, S. and Nakamura, S., *J. Cell Biol. 89,* 121
    (1981).
15. Nishida, E. and Kumagai, H., *J. Biochem. 87,* 143 (1980).
16. Levin, R. M. and Weiss, B., *Biochim. Biophys. Acta 540,*
    197 (1978).
17. Hidaka, H., Yamaki, T., Naka, M., Tanaka, T., Hayashi, H.
    and Kobayashi, R., *Mol. Pharmacol. 17,* 66 (1980).
18. Maruyama, M. and Takagi, K., *Biomed. Res.* (in press).
19. Hamaguchi, Y., *Develop. Growth Differ. 17,* 111 (1975).
20. Inoué, S., Fuseler, J., Salmon, E. D. and Ellis, G. W.,
    Biophys. J. 15, 725 (1975).
21. Brinkley, B. R. and Cartwright, J., *Ann. NY Acad. Sci.*
    253, 428 (1975).
22. Bershadsky, A. D., Gelfand, V. I., Svitkina, T. M. and
    Tint, I. S., *Cell Biol. Int. Rep. 3,* 45 (1979).
23. Salmon, E. D. and Begg, D. A., *J. Cell Biol. 85,* 853
    (1980).
24. Rieder, C. L., *Chromosoma 84,* 145 (1981).
25. Kiehart, D. P., *J. Cell Biol. 88,* 604 (1981).
26. Salmon, E. D. and Segall, R. R., *J. Cell Biol. 86,* 355
    (1980).
27. Hamaguchi, Y. and Hiramoto, Y., *Exp. Cell Res. 134,* 171
    (1981).

CHAPTER 20

ANALYSIS OF D_2O EFFECT ON *IN VIVO* and
IN VITRO TUBULIN POLYMERIZATION AND
DEPOLYMERIZATION

Hidemi Sato
Toyoki Kato
T. Choku Takahashi
Tomohiko Ito

Sugashima Marine Biological Laboratory
Nagoya University
Sugashima, Toba, Mie

I. INTRODUCTION

After the successful isolation of deuterium by Urey *et al.*
in 1932 (1), biological application of heavy water (D_2O) immediately became a fascinating research topic for many biolo-
gists. Water is the essential solvent for the organic compounds in living cells. Thus the partial substitution of
heavy water for water should introduce some isotope effects on
both *in vivo* and *in vitro* cell physiology. Many papers have
been published, and the basic facts about the biological
action of deuterium are known although not completely understood (2-4).

As pointed out by Thomson (4) and Kritchevsky (3), high
cost and the impurity of heavy water considerably retarded
further investigations until a technical improvement which was
achieved in the early 1950s made available a quantity supply
of heavy water at a markedly reduced price. This evoked
research interest in heavy water again in the bio-medical
sciences. For instance, the toxic effect of heavy water was
re-examined (3), induction of male sterility in mice was found
(5), isotope effect on the molecular structure was surveyed (6),
and differential effects of heavy water on mitosis and development of sea urchin embryos were studied by various authors (7-
13).

In general, heavy water is considered to be a toxic for

211

mitotic events, including cytokinesis in eukaryotes, and is
defined as a mitotic retarder. However, the molecular mechanism of heavy water must be more carefully examined. In this
article, we shall describe briefly some of our published data
as well as new data concerned with the D_2O effect on *in vivo*
and *in vitro* tubulin polymerization and depolymerization in
fertilized sea urchin eggs.

II. EFFECT OF HEAVY WATER ON THE MITOTIC SPINDLES

A. *Isotope Effect of D_2O*: Heavy water (D_2O) enhances spindle
volume and birefringence (BR) (14, 15). Using the metaphase-
arrested meiosis I spindle of the oocyte of *Pectinaria gouldi*,
Inoué and Sato (14) found an 8-fold increase in spindle
volume and practically a doubling in retardation. The
increase of volume and retardation depends on the concentration
of D_2O and the stage of mitosis. The maximum increase is
obtained by applying 45% D_2O during metaphase to onset of
anaphase. This particular concentration of D_2O is common for
metaphase spindles in various organisms. The changes are
rapid, the new state of dynamic equilibrium being reached
within 2 min in *Pectinaria* oocytes, 90 sec in developing
Japanese sea urchin eggs of *Anthocidaris crassispina, Hemicentrotus pulcherrimus, Mespilia globulus* or *Pseudocentrotus
depressus,* and 5 min in *Pisaster ochraceus* (16-19). The
D_2O effect is completely reversible and repeatable on
the same spindle.

$H_2{}^{18}O$ has no effect, whereas HDO and $HD^{18}O$ have half the
effect of D_2O. pD also has no effect. In many respects, the
D_2O effect is quite similar to that of elevating temperature
within an optimum range. However, the spindle will over-sta-
bilize or freeze with the application of a high concentration
of D_2O (9, 20).

B. *Thermodynamic Approach*: To compare the effect of temperature and D_2O, we analyzed the association-dissociation reaction of the spindle with a thermodynamic approach (21-23).
Metaphase-arrested meiosis I spindle of the mature oocyte of
Pisaster ochraceus was used as the material.
Thermodynamic parameters were calculated from retardation
measurements at various temperatures with or without D_2O (20,
23). In sea water we obtained ΔH = 58.9 Kcal/mol, ΔS =
205.9 eu, and ΔF = -1.1 Kcal/mol. In 45% D_2O the values were
ΔH = 29.55 Kcal/mol, ΔS = 106.3 eu, and ΔF = -0.9 Kcal/mol.
These values are similar to those obtained for *Chaetopterus*
spindle (22) and *Pectinaria* spindle (24). Both association
and dissociation processes appear to follow first-order

kinetics. E_{act} for the dissociation reaction is 45 Kcal/mol
and 39 Kcal/mol with D_2O. E_{act} for the dissociation reaction
on removing D_2O is 15 Kcal/mol. These data support the
hypothesis that the spindle reaction is the reversible association of tubulin dimers into the linearly aggregated polymers
in a first-order reaction. However, the difference of both
ΔH and ΔS in control and D_2O spindles and their similar E_{act}
suggests that D_2O in fact elevates the concentration of polymerizable tubulin dimers.

When D_2O sea water is applied to the metaphase spindle in
developing sea urchin zygotes *Mespilia globulus, Hemicentrotus
pulcherrimus,* and *Pseudocentrotus depressus,* spindle size and
birefringence are increased depending on the applied concentration. Maximum isotope effect of D_2O is achieved at the concentration of 45%; then spindle birefringence decreases beyond
this limitation even though the spindle holds its typical
barrel shape. D_2O-dependent association reaction is completed within 90 sec to 2 min and is totally reversible and repeatable. This reaction is endothermal, and D_2O is thought to
increase the amount of polymerizable tubulin dimer. Thermodynamic parameters are calculated in the developing egg of
Mespilia as ΔH = 20.9 Kcal/mol, ΔS = 71.9 eu in sea water and
ΔH = 18.6 Kcal/mole, ΔS = 64.2 eu in 45% D_2O (25, 26).
Indirect tubulin immunofluorescence clearly reveals the
increase in the number of microtubules in the spindle.

C. *Number of Spindle Microtubules*: Spindle birefringence (BR)
clearly reflects the number of orderly aligned microtubules
(27). In the case of *Pisaster* oocytes, we know that the
spindle microtubules increase in number and elongate when
D_2O is applied to the dividing cells (Fig. 1). However, the
coefficient of BR (n_e - n_o) of both control and D_2O spindle
remains constant at 5 X 10^{-4}. This means there is a significant increase in microtubules in D_2O spindle with no distur-
bance of original population density. From electron micrographs, we found the total number of spindle microtubules
increased from 4,200 in the control to 10,000 in D_2O spindle,
holding an average population density of 106. The length of
each spindle microtubule was also increased by D_2O. The
estimated rate of tubulin association was calculated as 1 X
10^2 monomers/sec microtubule in *Pisaster* oocyte (17, 20). The
comparison of D_2O spindle and control spindle is summarized in
Table 1.

D. *Tissue Culture Cells*: D_2O effect on the mitotic spindle
was thought to occur only in specialized cells such as mature
oocytes or unfertilized eggs, which hold a large amount of
polymerizable tubulin. Arguments were based on lack of a
sufficient tubulin source in the somatic cells which have to

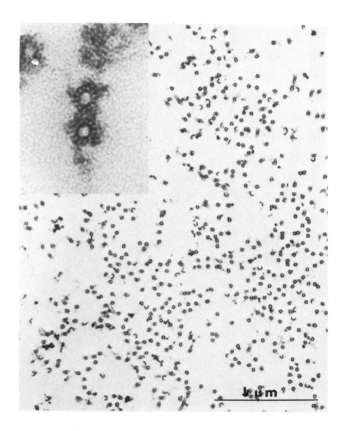

Fig. 1.   *Removal of D_2O or slight alteration of environ-
mental pH immediately induces microtubular dissociation as
seen in this photograph.   28% of total microtubules are splits
to "C" shape. Electron microscopy. Material: meiosis 1
spindle of the mature oocyte of* Pisaster ochraceus. *Left
shoulder photograph shows tubulin subunit.*

synthesize enough polymerizable tubulin during a cell cycle to
support the spindle assembly. However, this assumption was
wrong, and we found that D_2O effect can be clearly demonstrated even in a tissue culture cell such as the dividing rat-
Kangaroo Pt-K2 cell. Knowing that the artificially induced
tubulin paracrystals by the vinblastine can represent almost
all available tubulin in a cell (28), we calculated tubulin
concentration photometrically. We estimate that at most 18%
of tubulin is used for spindle assembly in a dividing Pt-K2
cell (20). The same value also obtained in a dividing tissue-
cultured salamander lung epithelial cell. The order of

TABLE I. *Comparison of D_2O-spindle versus Control Spindle Material:* Pisaster ochraceus

Condition of isolation	13°C,12%HG	45%D_2O,13°C,12%HG
Pol. M.		
Retardation(BR),nm	3.8	5.4
Spindle diameter,µm	8.0	12.4
Coefficient of BR(n_e-n_o)	5 X 10^{-4}	5 X 10^{-4}
E. M.		
Dimension of microtubule		
OD and ID	24nm,15nm	24nm, 15nm
Density of microtubules/µm^2	106	106
Average number of microtubules		
in spindle core/µm^2	130	130
Total number of microtubules		
per spindle	4,200	10,000
Coefficient of BR(n_e-n_o)	5 X 10^{-4}	5 X 10^{-4}

magnitude of tubulin consumption for spindle assembly in these cells is quite comparable to the case of developing zygotes such as fertilized sea urchin eggs, in which about 5-8% of cytoplasmic tubulin reserve is used for each mitosis.

III. CAN D_2O AFFECT ANAPHASE CHROMOSOME MOVEMENT?

A. In Vivo Observations: Mitotic spindle is a motile organella organized *de novo* in eukaryotes in metaphase to aid mutual segregation of replicated chromosomes. Its major structural components are the oriented spindle microtubules, and we postulated the existence of a dynamic equilibrium between the microtubules and tubulin dimers (9, 13) and, further, that the microtubules were responsible for the anaphase chromosome movement (9, 17). Now a question arises: Can D_2O affect anaphase chromosome movement? D_2O increases the number of spindle microtubules as well as their length. If D_2O is acting in favor of the tubulin association, then it may well be unfavorable for tubulin dissociation. Thus, D_2O may retard the anaphase chromosome movement in dividing cells. To examine this possibility, we used developing sea urchin eggs of *Mespilia globulus* and *Hemicentrotus pulcherrimus* because of their synchronous mitotic events and rather exact time schedule. As shown in Figs. 2 and 3, we found that the

mitotic process of developing sea urchin eggs was greatly
extended by immersing them in D_2O sea water. However, the
retardation of mitotic process mainly occurred in prophase to
metaphase, and no anaphase chromosome movement was disturbed
by D_2O. In fact, neither promotion nor retardation of the
speed of anaphase chromosome movement was detected. These
observations imply that D_2O may give the isotope effect for
elevating tubulin concentration but has no significant effect
on tubulin dissociation *in vivo* (8).

B. *Measurements in Spindle Models*: Various attempts have been
made to stabilize viable sea urchin spindles of different
species retaining the calcium-labile nature and the cold-
temperature sensitivity of the living spindle. Glycerol seems
to be an important prerequisite for successful stabilization
(29, 30). Methods developed in our Sugashima MBL to preserve

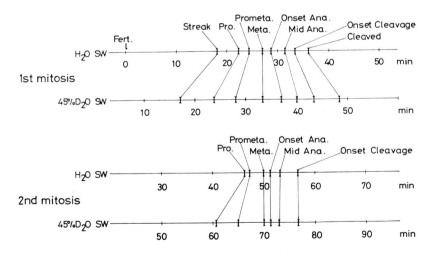

*Fig. 2. Time course of mitosis of Japanese summer sea
urchin,* Mespilia globulus, *immersed in H_2O or D_2O.*

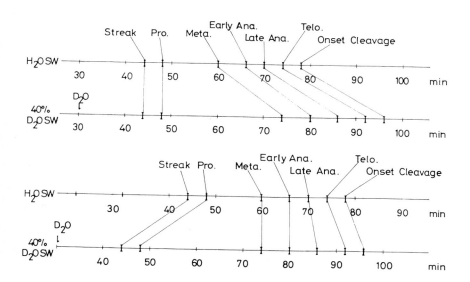

Fig. 3. *Time course of mitosis of Japanese winter sea urchin,* Hemicentrotus pulcherrimus, *immersed in H_2O or D_2O.*

Japanese sea urchin spindles are shown below.

A) *Stabilization medium for Japanese autumn and winter sea urchins,*

Pseudocentrotus depressus   Hemicentrotus pulcherrimus.

Glycerol	1 M		0.8 to 1 M
MES buffer	10 mM	PIPES buffer	10 mM
DTT	3 mM		2.5 mM
EGTA	2 mM		2 mM
MgSO₄	0.6 mM		1 mM

pH 6.2 to 6.3   pH 6.8

Addition of 0.6 to 0.7% of Nonidet P-40 is quite helpful to disperse yolks and other granules.

B) Stabilization medium for Japanese summer sea urchin,
   Anthocidaris crassispina.

Glycerol	*20 to 25%*
PIPES buffer (pH 6.9)	*10 mM*
DTT	*2.5 mM*
EGTA	*2 mM*
MgSO$_4$	*1 mM*
Nonidet P-40	*1 %*

Salinity, molarity of glycerol, pH and temperature must be carefully adjusted and controlled for the stabilization of each specimen, and no standardized preservation medium is available. Preserved spindles of *Hemicentrotus pulcherrimus* and *Anthocidaris crassispina* respond very well to μM Ca^{2+} perfusion: spindle birefringence disappears within 3 to 5 min at room temperature. On the other hand, spindles isolated from *P. depressus* shows rather poor response to μM Ca^{2+} perfusion or temperature shift. Fig. 4 shows some examples of stabilized spindles of *H. pulcherrimus* in metaphase (A, B1, control, C1, C2, with 45% D_2O) and in anaphase (B2), observed with a polarizing microscope.

Anaphase chromosomal movements are carefully followed with these spindle models using ATP, μM Ca^{2+} and lowered temperature as the possible triggers. It is confirmed that: a) μM to mM ATP perfusion cannot initiate any chromosome movement in spindle models even with minute alteration of temperature; b) Ca^{2+} can splits chromosomes in early anaphase and increases interzonal distance, but the moving speed of the chromosomes is 10 times slower than that of controls; and c) lowering temperature significantly reduces overall spindle birefringence but never ignites chromosomal separation or migration.

To determine the accurate speed of anaphase chromosome movement, mass spindle stabilization has been performed following the time course of mitosis, and interkinetochore distances in specific time were plotted using polarizing and differential interference optics. We found the average speed of anaphase chromosomal movement of *H. pulcherrimus* was 2.6 μm/min at 19°C. This value was surprisingly consistent in both control and D_2O spindles, and we concluded that D_2O has neither acceleration nor retardation effect on anaphase chromosome movement.

C. Can D_2O Effect Microtubule Depolymerization? There is a possibility that D_2O may strengthen the hydrophobic bond of

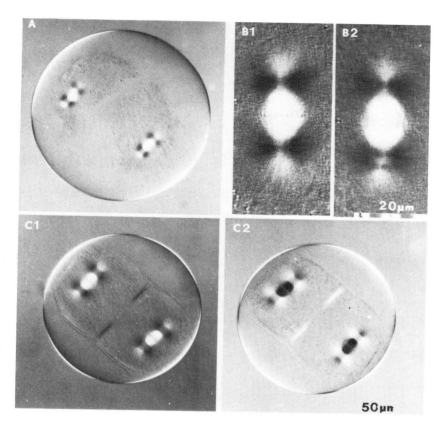

Fig. 4.   Stabilized spindles of H. pulcherrimus.
*Polarizing microscopy.   A: control.   Cl and C2:   45% D₂O spin-
dle in white and black contrast.   Bl and B2: metaphase and
anaphase.*

biological polymers (6, 9, 12, 13, 21).   If D_2O intensifies
the binding property of tubulin molecules, it will retard the
disassembly of spindle microtubules *in vitro*.   From this point
of view, the stability of isolated mitotic apparatuses of sea
urchin eggs was examined under conditions with or without D_2O.
Spindles of *M. globulus* were isolated by immersing the developing eggs in the isolation medium containing 1 M glycerol, 1 mM
EGTA, 10 mM DTT, 1 mg/l soybean trypsin inhibitor, 10 mM MES,
pH 6.2.   Then isolates were transferred to glycerol buffer (1
M glycerol, 1 mM EGTA , 10 mM MES, pH 6.2) with or without D_2O,
and stored at 4°C.   The breakdown of spindle microtubules was
followed measuring the decay of spindle birefringence with

rectified polarizing optics (27). As shown in Fig. 5, no
differences in the fashion of decrease in spindle birefringence
were found in spite of the presence of D_2O. Thus D_2O has
little effect on the improvement of binding strength among
tubulin molecules in a polymerized form.

IV. D_2O EFFECT ON *IN VITRO* TUBULIN ASSEMBLY

    In contrast to the above observations, *in vitro* polymeri-
zation and depolymerization of tubulin occur quite differently
than in *in vivo* or model systems.
    Microtubule protein (MTP) was prepared from fresh bovine
brain by three cycles of the assembly-disassembly method of
Shelanski *et al.* MAP-free tubulin dimer (PCT) was isolated
from MTP by phosphocellulose ion-exchange chromatography by
the method of Weingarten *et al.* (31).
    Turbidimetric assays of microtubule assembly were perform-
ed at 350 nm using a Hitachi 200-20 spectrophotometer.
Assembly was initiated by the addition of GTP to a concentration of 1 mM followed by the rapid raising of temperature from

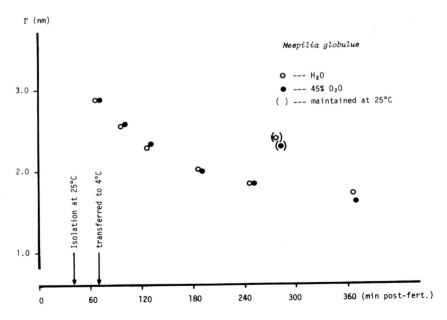

Fig. 5. *Stability of isolated mitotic apparatus under
conditions with or without D_2O at 4°C.*

4°C to 25 or 37°C within buffer made in H_2O and D_2O in varying proportions. As in D_2O, the pK_a values of week acids are increased by about 0.55 pH unit. To maintain pD values in the reaction mixture as constant, we calculated the pD values from the relation

    pD = pH reading + ΔpH,

where $\Delta pH = 0.3314n + 0.0766n^2$ and n is the mole fraction of D_2O.

    As shown in Figs. 6 and 7, D_2O enhances the initial rate of tubulin polymerization and the final extent of the polymers. Following the time course of turbidity change, we found that the total number and the length of microtubules increased depending on the given concentration of D_2O. The maximum

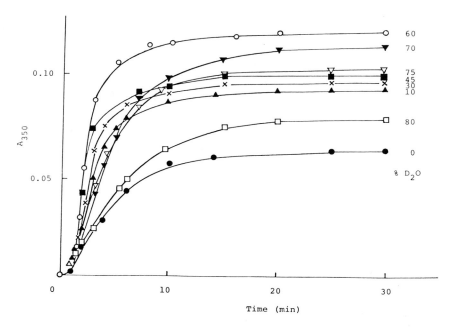

Fig. 6   The effect of D_2O on the polymerization of MT-protein. 0.5 ml of the reaction mixture containing 1.0 mg/ml MTP, 0.1 M Pipes, pD 6.9, 0.5 mM $MgSO_4$, 1 mM EGTA, was pre-warmed at 37°C for 5 min and polymerization was started by adding 10 μl of GTP. The final concentration of GTP was 1 mM. The turbidity was followed by the absorbance at 350 nm.

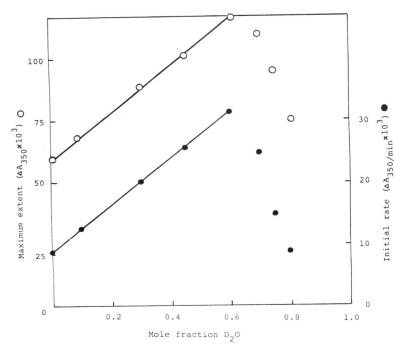

*Fig. 7. Maximum extent and initial velocity of polymer-
ization as a function of the mole fraction of D_2O.
Microtubules were formed as in Fig. 1 for each D_2O concentration.*

amount of polymers obtained at 60% D_2O and yield declined at
higher concentrations. Plotting of initial rates of polymerization within one minute reveals the existance of linearity of
the tubulin polymerization up to the 60% D_2O (Fig. 8). Higher
concentration of D_2O shows stronger inhibitory effect on the
microtubule dissociation. It reveals the definite existence
of D_2O dependency of tubulin polymerization and depolymerization.

Finally, we examined D_2O dependency of *in vitro* tubulin
polymerization using the notion of molecular configuration.
Circular dichroism (CD) spectra were recorded on a Jasco J-40
spectrophotopolarimeter, using the data processor D-DPY. A
1-mm cell was used with thermocontrolled jacket. The protein
concentration used for CD was 0.1 mg/ml in buffer containing
0.1 m KCl, 0.5 mM $MgSO_4$, 1 mM EDTA, 50 mM phosphate, pD 6.4.
Spectra were routinely recorded from 265 to 200 nm at

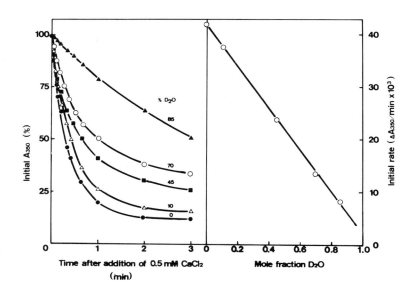

Fig. 8. *Depolymerization time course after addition of*
0.5 mM CaCl₂ to per formed microtubules (left). The 100%
point is the A₃₅₀ value at zero time corrected for the
absorbance of the reaction mixture before polymerization.
Right shows rate of depolymerization as a function of the
mole fraction of D₂O.

sensitivity of 2×10^{-3} deg/cm. The mean residue weight was
taken as 115. Protein concentration was determined by the
method of Lowry, using bovine serum albumin as the standard.

Fig. 9 shows the CD spectra of PC-tubulin in the presence
and absence of D_2O. These data indicate the definite occurrence of conformation change of tubulin molecules from α
helix to the state of random coil in the presence of 20% or
higher concentrations of D_2O. Critical concentration of
tubulin polymerization is also decreased with the presence of
D_2O, suggesting an apparent increase of polymerization nuclei
within the solution.

This discrepancy existing in between *in vivo* and *in vitro*
systems must be examined more thoroughly. Further investigations are in progress in our laboratory.

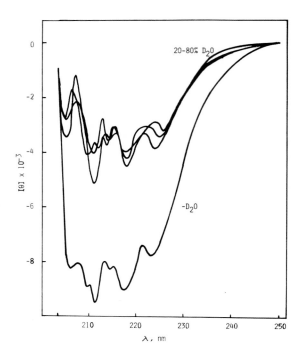

Fig. 9. CD spectra of PC-tubulin in the presence and absence of D_2O. 0.1 mg/ml of PC-tubulin in the buffer containing 0.1 M KCl, 0.5 mM $MgSO_4$, 1 mM EGTA, 1 mM GTP, 50 mM phosphate, pD 6.4 was subjected to CD study at 25°C.

ACKNOWLEDGEMENT

This paper was supported by the Grant-in-aid for Scientific Research Nos. 344074, 438037, 434038 and Special Project Research No. 511214 from the Ministry of Education, Science and Culture of Japan.

REFERENCES

1. Urey, H. C., *Science, 78,* 566 (1933).
2. Lucké, B., and Harvey, E. N., *J. Cell. and Comp. Physiol., 36,* 473 (1935).
3. Kritchevsky, D.,*Ann. N. Y. Acad. Sci., 84,* Art. 16, 573 (1960).

4. Thomson, J. F., "Biological Effect of Deuterium", Pergamon Press Inc., New York, N. Y. (1963).

5. Bennet, E. L., Holms-Hansen, O., Hughes, A. M., Longberg-Holm, K., Moses, V., and Tolbert, B. M., *Science, 128,* 1142 (1958).

6. Berns, D. S., *in* "Biological Macromolecules Series, Vol. 5, Part A. Subunits in Biological Systems" (S. N. Timasheff and G. D. Fasman, eds.), p. 105. Marcel Dekker, New York, N. Y. (1971).

7. Gross, P. R., and Spindel, W., *Ann. N. Y. Acad. Sci., 84,* 745 (1960).

8. Gross, P. R., and Spindel, W., *Ann. N. Y. Acad. Sci., 90,* Art. 2, 500 (1960).

9. Inoué, S., and Sato, H., *J. Gen. Physiol., 50,* 259 (1967).

10. Marsland, D., and Hiramoto, Y., *J. Cell Physiol., 67,* 13 (1966).

11. Marsland, D., and Zimmerman, A. M., *Exptl. Cell Res., 30,* 23 (1963).

12. Marsland, D., and Zimmerman, A. M., *Exptl. Cell Res., 38,* 306 (1965).

13. Stephens, R. E., *Biol. Bull., 142,* 145 (1972).

14. Inoué, S., Sato, H., and Ascher, M., *Biol. Bull., 129,* 409 (1965).

15. Inoué, S., Sato, H., and Tucker, R. W., *Biol. Bull., 125,* 380 (1963).

16. Bryan, J., and Sato, H., *Exptl. Cell Res., 59,* 371 (1970).

17. Sato, H., *in* "Aging Gamates" (R. Blandau, ed.), p. 19, Karger, A. G., Basel, Switzerland, (1975).

18. Sato, H., Takahashi, T. C., and Sato, Y., *European J. Cell Biol., 22,* 310 (1980).

19. Sato, H., Kato, T. and Takahashi, T. C., *Tracer, 6,* 2 (1981).

20. Sato, H., Ohnuki, Y., and Sato, Y., *in* "Cell Motililty: Molecules and Organization" (S. Hatano, H. Ishikawa and H. Sato , eds.), p. 551. Univ. Tokyo Press, Tokyo, (1979).

21. Inoué, S., *in* "Biology and the Physical Sciences." (S. Devons, ed.), p. 139. Columbia Univ. Press, New York, N.Y. (1969).

22. Inoué, S., and Morales, M. F., *Rev. Mod. Phys. 31,* 402 (1959).

23. Sato, H., and Bryan, J., *J. Cell Biol., 39,* 118a (1968).

24. Carolan, R. M., Sato, H., and Inoué, S., *Biol. Bull., 129,* 402 (1965).

25. Takahashi, T. C., and Sato, H., *Cell Str. and Funct., 3,* 391 (1978).

26. Takahashi, T. C., and Sato, H., *Cell Str. and Funct., 4,* 448 (1980).

27. Sato, H., Ellis, G. W., and Inoué, S., *J. Cell Biol., 67,* 501 (1976).

28. Strahs, K. R. and Sato, H., *Exptl. Cell Res., 80,* 10 (1973).
29. Sakai, H., and Kuriyama, R., *Dev., Growth & Diff., 16,* 123 (1974).
30. Salmon, E. D., and Segal, R. R., *J. Cell Biol., 86,* 355 (1980).
31. Weingarten, M. D., Lockwood, A. H., Hwo, S. Y., and Kirschner, M. W., *Proc. Nat. Acad. Sci.* USA, *72,* 1858 (1975).

CHAPTER 21

SPINDLE STRUCTURE AFTER CHROMOSOME MICROMANIPULATION[1]

R. Bruce Nicklas, Donna F. Kubai, and Thomas S. Hays

Department of Zoology
Duke University
Durham, North Carolina

I.   INTRODUCTION

We have combined micromanipulation of living cells with
electron microscopy to reveal some otherwise obscure features
of spindle organization. The major goal was to make visible
the structures responsible for the mechanical attachment of
chromosomes to the spindle. Chromosome attachment to the
spindle is the basis of directed chromosome movement in
mitosis, which results in equal numbers and kinds of
chromosomes in the daughter cells. _Melanoplus differentialis_
(a grasshopper) spermatocytes were manipulated as shown in
Fig. 1: a chromosome was displaced laterally, stretched, and

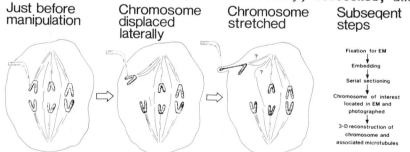

Fig. 1.   The experiment in outline.

then the cell was fixed for electron microscopy followed by
the three-dimensional reconstruction of spindle microtubules.
The rationale is two-fold: 1. The KMts (kinetochore
microtubules) of the displaced chromosome should be shifted

[1]Supported in part by NIH Grants GM 13745 and GM 27569 from
the Institute of General Medical Sciences, and by NSF
Grant PCM 79-11481 from the Division of Physiology,
Cell and Molecular Biology.

outside the confusing thicket of spindle microtubules and
hence should be identifiable. 2. The tension before fixation
on the displaced chromosome will be felt by all the
microtubules involved in the chromosome's attachment to the
spindle. This should produce shifts in position and/or bends
in these microtubules and thus permit their identification.

II. RESULTS

The results will be described in full in a publication
now in preparation. Here, a composite diagram based on
observations from seven cells will be used to illustrate the
chief findings. The diagram (Fig. 2) shows

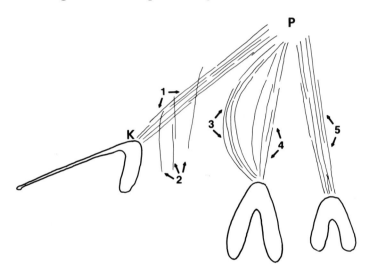

Fig. 2. A sketch of the chromosomes and microtubule
configurations observed in cells in which one chromosome was
displaced to the left and stretched before fixation. The
numbers are explained in the text. K: the kinetochore of the
manipulated chromosome; P: the pole, recognized by the
presence of pericentriolar material.

a manipulated chromosome on the left and two nearby
chromosomes to the right. It is important to emphasize that
the micromanipulation needle touched only the one chromosome:
 the microtubule configurations observed result from tension
on that chromosome, not from stirring the spindle itself with
the needle. With one exception, all the chromosomes continued
normal anaphase movement during the experiment, but at a
reduced rate. The chief findings are as follows: 1. KMts of
the manipulated chromosome ("1", Fig. 2) are displaced as

expected and run straight toward the pole. They encounter
large numbers of other spindle microtubules only near the
pole. 2. A few non-KMts (non-kinetochore microtubules) enter
the KMt bundle closer to the kinetochore ("2", Fig. 2).
However, the mechanical association of these non-KMts with
microtubules of the KMt bundle is weak, at best. Thus, they
usually pass straight through the bundle, showing no sign of
attachment to KMts. The significance of these microtubules
for chromosome movement is dubious but not ruled out. 3. KMts
of the chromosome nearest the manipulated chromosome are bent
("3", Fig. 2) in the direction of the manipulated chromosome's
KMts. The effect diminishes with increasing distance. If
some KMts of the nearest chromosome are unbent, they are those
furthest from the manipulated chromosome ("4", Fig. 2). The
KMts of chromosomes further away are invariably unaffected by
the tension and run more or less straight toward the pole
("5", Fig. 2), as in control cells.
4. Comparably bent non-KMts have never been observed. 5.
Sometimes a few KMts of the manipulated chromosome can be
traced throughout their length (not illustrated in Fig. 2).
In this material (after micromanipulation!), fewer than half
of the KMts extend all the way from the kinetochore to the
pole. Some of those that do not are associated with a second
microtubule -- the two overlap in the middle and together are
long enough to span the whole kinetochore to pole distance.
This result contrasts with recent, convincing reports in other
materials that most KMts extend without interruption from
kinetochore to pole (1, 2).

III.   CONCLUSIONS ON SPINDLE STRUCTURE

We conclude first that microtubules and whatever links
them together almost certainly are responsible for chromosome
attachment to the spindle and for the mechanical coherence of
the spindle as a whole. The evidence for this is indirect,
but (a) no other spindle structures so far observed could fill
the bill, and (b) the microtubule configurations seen after
manipulation invariably are consistent with that expected if
microtubules and linkages among them bear the load imposed by
tension on one chromosome before fixation. The earlier
studies of Begg and Ellis (3, 4) on micromanipulated cells
observed by polarized light microscopy provide important
additional evidence implicating microtubules in determining
the mechanical properties of the spindle.

Second, chromosome attachment to the spindle is mediated
by KMts firmly linked to those of the rest of the spindle only
in the immediate vicinity of the pole. This is exactly what
was expected from earlier, light microscopic studies of
micromanipulated cells (3, 4, 5).

Third, while the thicket of microtubules near the pole precludes a characterization of microtubule-to-microtubule linkages in that region, the bulk of the spindle is open to such characterization after micromanipulation. In the analyzable region, we have no evidence whatever for tight, microtubule-by-microtubule linkages of the sort mediated by dynein, for instance. Especially intriguing are the bent KMts of chromosomes adjacent to the manipulated one. As shown in Fig. 2, these bent microtubules are not close to other microtubules in the vicinity of the bend: there is no sign of the microtubule/microtubule conjunction seen in cilia and flagella (reviewed in ref. 6) or in some parts of the spindle (reviewed in ref. 7). Our tentative conclusion is that the spindle microtubules we can analyse are loosely, but significantly, associated by embedment in a gel of some sort (cf. ref. 8). Mechanical association via a gel would transmit the stress on the micromanipulated chromosome and its KMts to other, nearby microtubules. The nearby microtubules would be shifted in position as groups, rather than as individual microtubules tightly coupled to a particular KMt of the manipulated chromosome. We also have evidence for the same sort of loose mechanical association among the KMts in one KMt bundle and between the KMts and the non-KMts which enter the bundle.

McIntosh (8) has suggested that a gel or "net" investing spindle microtubules merits serious consideration as a source of chromosome movement. The storage of energy in a stretched gel would permit chromosome movement without concurrent ATP hydrolysis (8, 9). Such a mechanism is now a serious contender to explain chromosome to pole movement in anaphase because of Cande's (9) evidence, not yet confirmed, that chromosome to pole movement, unlike spindle elongation, does not require ATP. Our results, of course, provide no evidence whatever on the molecular source of chromosome movement, but they do suggest the presence of a gel which provides significant mechanical linkage among spindle microtubules. This feature of spindle organization is at least consistent with McIntosh's proposal.

IV.   INTERDEPENDENT CHROMOSOME MOVEMENT

The mechanical association observed between the KMts of adjacent chromosomes may provide a structural basis for one example of interdependent chromosome movement. Chromosomes ordinarily move as individuals, with a variable degree of co-ordination among different chromosomes depending on the stage of mitosis. The structural basis for this measure of independence is provided by the separate attachment of each chromosome via its KMts to the spindle as a whole. Exceptions

to this rule provide important tests of the adequacy of models
for mitosis. A good example is the so-called "hinge anaphase"
described by Bĕlař (10) and Ris (11). When a chromatin bridge
prevents the separation of one pair of chromosomes, the
separation of adjacent chromosomes is also reduced, even
though they are not themselves stuck together. The effect is
graded across the spindle -- the separation achieved by pairs
of chromosomes increases with increasing distance from the
pair stuck together. Now compare this graded effect on
velocity with the graded effect on structure observed when one
chromosome is stretched with a microneedle before fixation.
The bending of KMts observed decreases with increasing
distance from the stretched chromosome (Fig. 2). In brief, we
suggest that when a chromosome is stretched by a manipulation
needle or by the spindle (due to a bridge which prevents free
separation), the resulting disruption of spindle organization
is not localized. Instead, adjacent regions of the spindle
are affected because they are mechanically associated and
hence share the stress. A likely consequence of the resulting
disruption is a loss in the capacity of the spindle to bear
the load imposed by chromosome movement -- the Newtonian
counter-force toward the equator associated with chromosome to
pole movement. Thus, a gradient of structural disruption
extends from stress on one chromosome, leading to a gradient
in the rate or extent of chromosome movement.

REFERENCES

1.   Rieder, C.L., Chromosoma 84, 145 (1981).
2.   Witt, P.L., H. Ris, and G.G. Borisy, Chromosoma
        83, 523 (1981).
3.   Begg, D.A., and G.W. Ellis, J. Cell Biol. 82, 528 (1979).
4.   Begg, D.A., and G.W. Ellis, J. Cell Biol. 82, 542 (1979).
5.   Nicklas, R.B., and C.A. Staehly, Chromosoma 21, 1 (1967).
6.   Warner, F.D., in "Microtubules" (K. Roberts and
        J.S. Hyams, eds.),p.359. Academic Press, London (1979).
7.   McIntosh, J.R., in "Microtubules" (K. Roberts and
        J.S. Hyams, eds.), p. 381. Academic Press,
        London (1979).
8.   McIntosh, J.R., in "International Cell Biology
        1980-1981" (H.G. Schweiger, ed.), p. 359. Springer-
        Verlag, Berlin (1981).
9.   Cande, W.Z., Cell 28, 15 (1982).
10.   Bĕlař, K., Wilhelm Roux'Arch. Entwicklungsmech.
        118, 359 (1929).
11.   Ris, H., Biol. Bull. (Woods Hole) 96, 90 (1949).

CHAPTER 22

MECHANICS OF ANAPHASE B MOVEMENT

Paul J. Kronebusch[1]
Gary G. Borisy

Laboratory of Molecular Biology
University of Wisconsin
Madison, Wisconsin

I.   INTRODUCTION

Anaphase is the process where sister chromosomes separate from each other and move towards the poles of the cell. The poles themselves also move and therefore the overall process has been characterized as consisting of two components: the movement of chromosomes relative to a pole (anaphase A) and the movement of poles relative to each other, carrying the chromosomes with them (anaphase B) (1-3).

Most of the experimental literature on anaphase has focused on anaphase A. It is clear from the micromanipulation experiments that chromosomes are pulled to the pole and that the force pulling them is transmitted along a fiber mechanically connected to the pole at one end and to a kinetochore at the other. A chromosome in anaphase behaves as if it were connected to the pole by a fine piano wire; that is, it may be displaced readily towards the pole or in an arc centered on the pole but displacement radially away from the pole meets with great resistance (4,5). The force transmitting fiber is known to consist primarily of a bundle of microtubules (MTs), but it is not known whether the MTs generate the force or merely transmit a force generated by a separate traction element (6,7).

[1]Supported by NIH grant 25062 to GGB. PJK was an NIH Predoctoral Trainee.

Biological Functions of Microtubules
and Related Structures

233

In contrast, experimental investigation of anaphase B has been neglected. Older studies on the relative sensitivity of anaphase A and B to drug inhibition (2, 8) have suggested that the two processes proceed by different mechanisms and recent experiments with in vitro models have supported this conclusion (9, 10). Based on morphological analyses of MT distributions between separating chromosomes, models have been proposed suggesting that the force driving separation of the poles is generated in the equatorial zone where MTs emanating from opposite poles overlap each other (11-14).

Although the models differ in detail concerning MT polarity and the nature of microtubule interaction, they all have in common the feature that the force is generated in the overlap zone. This predicts that the poles in anaphase B are essentially being pushed apart.

Alternatives to pushing might be considered. The poles might be pulled apart and the sliding of MTs in the overlap region would then be a consequence rather than a cause of the pole motion. The poles might move autonomously; that is, they might generate a traction force relative to a supporting structure in the local cytoplasm and effectively crawl through the cytoplasm.

We have sought to test the prediction that the poles are pushed apart. Basically our plan has been to place an obstacle in the equatorial zone of a PtK1 cell in anaphase and to observe the effects on subsequent motion of the poles. The obstacle we have devised is a hole which completely prevents mechanical interaction via overlapping MTs between the two halves of the spindle. We have found that such obstacles do not prevent separation of poles and we therefore conclude that, in PtK1 cells, contrary to the commonly accepted view, poles are not pushed apart in anaphase B. These findings have prompted us to consider new models for the mechanism of anaphase B movement.

II. TESTS OF ANAPHASE B MODELS

Figure 1 shows, in schematic fashion, the three general possibilities for the movement of poles in anaphase B. The first two mechanisms may be considered passive; that is, the poles are either pushed or pulled by forces acting external to them. In the third mechanism the poles are considered to be active; that is, the forces acting on them are generated in their immediate locality.

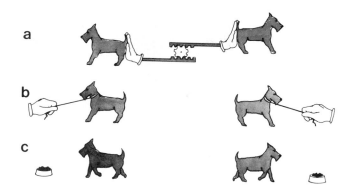

Figure 1. Diagrammatic representation of three mechanisms for separating the spindle poles. The dogs represent the poles of the spindle. (a) The poles are pushed apart by forces generated within the spindle. (b) The poles are pulled apart by forces generated outside the spindle. (c) The poles are able to propel themselves and are aligned in order to insure separation.

As indicated before, most current models have consistently neglected the second and third mechanisms shown in Figure 1 and have focused on the pushing mechanisms. In order to put these pushing hypotheses to a critical test we decided to break the MT connections between the separating spindle poles to see whether anaphase B movement stopped or continued (15).

The spindle in the cell line, PtK$_1$, undergoes a two-fold elongation during anaphase. This elongation may be quantitated readily by phase microscopy. To sever the central spindle of MTs we have used a fine glass needle attached to a micromanipulator. By pinching the cell across the equator with the needle during anaphase we could fuse the dorsal and ventral cell membranes together, severing the MT connections in the process. The result of such an operation was an anaphase cell with a membrane-bound hole through the interzone region. An example of such a cell is shown in Figure 2. The spindle poles and the attached chromosomes continue to move away from each other after the central spindle has been cut. The arrows in Figure 2b point to cytoplasmic markers which the chromosomes move past, indicating that this is indeed a mitotic movement and not some induced bulk flow of cytoplasm.

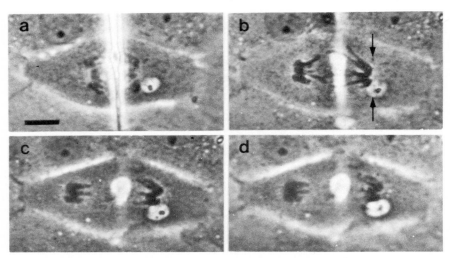

Figure 2.   Series of photographs demonstrating continued
spindle elongation in a living PtK$_1$ cell after a hole has
been introduced through the interzone.   (a) Glass needle
pinches down on anaphase interzone.   (b) 1.0 minutes after
needle has been removed.   The central spindle and several
chromosome arms attached to the right pole have been cut as
the dorsal and ventral cell membranes have been fused.
Arrows point to a large vacuole and a dark granule in the
cytoplasm which chromosomes move past.   (c) 4.7 minutes.
The boundary of the hole is more obvious.   (d) 5.2 minutes.
Bar = 10 μm.

    We have found that the pole movement continues at a rate
that is actually faster than in cells with intact central
spindles.   Figure 3 compares the movements of the poles in
two representative cells.   For each experimental cell we
calculated the rate at which the poles moved away from each
other during the first three minutes after the hole had been
introduced.   We also measured the rate at which each pole
moved away from the boundary of the hole and summed the
values for the two poles to calculate the pole-hole rate for
each cell.   In some cells the hole expanded late in anaphase
and this measurement subtracts any movement that may be due
solely to the expansion of the hole.   The average rates for
a number of experiments are presented in Table I.   In many
experiments cells were treated with cytochalasin B.
Initially this was done because we thought pole movement
might be linked to cleavage movements.   We found that cytochalasin B did not prevent pole movement, but the treatment

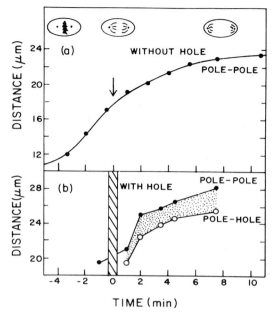

Figure 3.   Effect of introducing a hole through the
spindle on pole separation. (a) An unmanipulated cell.
Arrow indicates the stage of anaphase at which needle would
have been inserted. (b) A cell with a hole introduced at
time 0. Filled circles represent total distance between
poles. Open circles represent distance between poles minus
the diameter of the hole. Both cells treated with 20 μm/ml
cytochalasin B at least 20 minutes prior to metaphase.
From reference 15.

did result in a greater proportion of cells surviving the
operation and it also inhibited hole expansion.
        In order to determine whether the introduction of the
hole had actually severed all the MT connections between the
two half spindles, we stained cells with an antibody against
tubulin and observed the immunofluorescence distribution of
fibers. Figure 4 shows that the bundles of MTs usually seen
at the equator of normal cells are absent from the two
bridges of cytoplasm still connecting the two half spindles.
The MT distribution in the rest of the cell does not seem to
have been altered by the operation, however.
        In summary, we have severed the MT connection between
the two poles and have observed continued movement at a rate
that exceeds normal. This observation deals a serious blow
to the pushing models for anaphase B, since it appears that

TABLE I.   Anaphase B Movement

		Rate of Separation	
Treatment	Number of Experiments	Pole-Pole μm/min	Pole-Hole μm/min
No drug addition			
without hole	7	0.89 ± 0.23[a]	
with hole	9	1.80 ± 0.82	1.20 ± 0.52
Cytochalasin B (20 μg/ml)			
without hole	11	0.41 ± 0.19	
with hole	8	2.53 ± 0.58	1.83 ± 0.71

[a]Standard deviation.

the central spindle connecting the two poles may actually slow their movement rather than generate any propelling force.

Similar results have been obtained by Aist and Berns for the intranuclear division of the fungus Fusarium solani (16). After breaking the central spindle with a laser microbeam, the daughter nuclei separated at nearly three times their normal speed. Thus, in both animal and fungal cells, there is now reason to consider alternatives to pushing models for the pole motion in anaphase B.

III.   ANAPHASE B MOVEMENT AND ASTER MIGRATION

In fungal cells the nucleus-associated body (NAB) and its cytoplasmic MTs appear to play a causal role in most nuclear movements. The migrating interphase nucleus is led and apparently pulled by the NAB which displays an array of cytoplasmic MTs during migration (17-21). Similarly, it lies at the spindle pole during mitosis and anaphase nuclear movements. In light of the results of Aist and Berns it is no longer reasonable to propose that the NAB pulls on the nuclear envelope during interphase, but pushes on it during mitosis. One movement merges into the next in the cell and the evidence now indicates that a similar pulling force is operating during both.

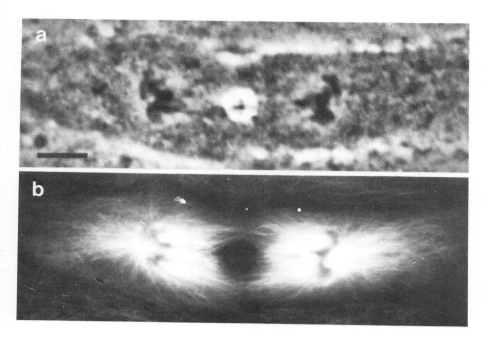

Figure 4.   PtK$_1$ cell fixed in glutaraldehyde 4 minutes
after the hole was introduced through the interzone.
(a) Phase contrast. (b) Anti-tubulin immunofluorescence
pattern.   Bar = 10 μm.   From reference 15.

We propose that in animal cells the centrosome and aster
MTs play a role similar to that of the NAB and cytoplasmic
MTs in fungi.   A causal relationship between the centrosome
aster movement and interphase nuclear movement is not as
well documented in animal cells as it is in the fungi.
However, it has been observed that when directed cell migration is induced the centrosome moves to the leading edge of
the nucleus (22, 23).

Many other observations have described the motile properties of asters in animal cells and all of these motile
functions are inhibited by MT poisons (24–31).   Asters can
remain stationary and transport other particles toward them
as in the saltatory movement of vacuoles and organelles
(25, 26) and the rapid aggregation of pigment granules in
specialized cells (27, 28).   Asters themselves are capable
of migration through the cytoplasm as in prophase centrosome separation (32, 33) and during egg fertilization (29).
Spindles, which consist of two connected asters, are
capable of rotation and migration through the cytoplasm.

This is especially evident during the movement toward an
eccentric position in the cell prior to unequal division
(34, 35). In certain polarized cells the spindle will
reorient to its original position after being rotated by
micromanipulation (35).

In the early divisions of an insect egg Wolf has
observed a case in which there is aster mediated migration
of both spindles and daughter nuclei in addition to polarized transport of yolk particles (31). Nuclear divisions
are followed by nuclear migration without ensuing cleavage.
During migration each nucleus is preceded by an enlarged
polar aster. While the nucleus is migrating the neighboring
cell membrane shows indentations repeatedly, while underlying
yolk particles move in a saltatory manner toward the aster
center. During mitosis there is a sequential activation of
asters along the length of the elongated insect egg.
Anaphase movements begin at the same time as particle transport toward the aster. When one center is activated before
the other it pulls the entire mitotic apparatus until the
other center is activated. Then, net nuclear movement
ceases until mitosis is completed and each daughter nucleus
follows its aster. Particle transport continues throughout
mitosis and during the subsequent nuclear migration. From
the observation of these aster-nuclei migrations Wolf concluded that astral rays adhere to peripheral egg structures
and exert traction forces toward the aster's center. These
combined forces pull the aster and accompanying nucleus
through the cytoplasm.

Considering these observations, we propose that anaphase
B movement is a special case of aster migration involving
similar microtubule dependent mechanisms as other aster
migrations in cells.

IV. MECHANISMS FOR ANAPHASE B MOVEMENT

If we assign the motile functions to the asters we are
obliged to find an alternative role for the central spindle
during anaphase. What is the purpose of the central spindle
if it is not responsible for pushing the poles apart? Pole
movement was accelerated in both fungal and animal cells
when the central spindle was cut during anaphase B. Aist
and Berns (16) suggested that the central spindle served as
a mechanical governor that limited the rate of spindle
elongation. However, it is not immediately obvious what
cellular catastrophe would result from having an ungoverned
rate of spindle elongation. Perhaps there is a more crucial function that it performs.

An important role that the central spindle might play is restricting the directions in which the spindle poles may move. The overlapping MTs of the central spindle may form a framework which only allows the poles to move in directions 180° away from each other. It might function like a telescoping sleeve upon which the asters pulled. The retardation of pole movement may be due to the resistance of interacting MTs of the spindle sliding past each other. Bajer observed that prophase asters migrating away from each other without any spindle MTs between them were capable of rotating and approaching each other again (36). The role of the spindle may be to prevent such accidents, since such freedom of movement for anaphase asters would be disastrous for the cell. In the fungal system, when the central spindle was broken in the middle, the two remaining bundles each immediately rotated so that they were no longer parallel to the original spindle axis (16). This reaction provides evidence that while intact, the central spindle was capable of resisting this bending force.

Thus, by requiring that the poles move in opposite directions the central spindle would play a crucial role in mitosis. It would insure maximal separation of poles and sister chromatids even if it did not provide the motive force itself.

The proposed involvement of astral MTs rather than interzonal MTs in the spindle elongation requires new models to explain the movement. If spindle poles do not push off each other, on what are the pulling forces anchored? One might propose that the MTs are either anchored at the cell membrane or attached to some other supporting matrix in the cytoplasm. These two possibilities are diagramed in Figure 5.

In the studies we have discussed in which pulling by astral MTs has been implicated, MTs or astral rays have been observed to extend all the way to the cell membrane (15, 16, 31). In other systems it has been shown that the position of asters within the cell somehow affects the arrangement of the cell cortex to accurately position the contractile ring of the cleavage furrow (35, 37). This ability implies some sort of connection or interaction between the aster and the surface. Such a membrane attachment model is attractive for explaining situations in which divisions in a polarized cell always occur in line with one cell axis (35). It could allow the cell to program the direction of aster movements by localizing the MT attachment points to certain regions of the cell surface.

A weakness of this model as a general mechanism is that it does not provide a way for asters to migrate away from the cell surface, such as in the movement and fusion of

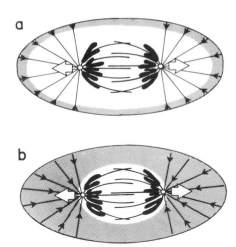

Figure 5. Two models for anchoring the aster in the cell so that a pull may be exerted on the spindle pole. (a) Astral fibers anchored at the cell surface. (b) Astral fibers anchored in a cytoplasmic matrix. Small black arrows represent attachment points which resist the pull of astral rays. The gray area represents the domain in which astral attachment points are anchored. The large white arrow represents the resultant force on the spindle pole as a consequence of the forces exerted along astral rays.

pronuclei occurring after fertilization (29, 38). Similarly, in large syncytia with many synchronously dividing nuclei, not all the poles seem to be moving toward a cell surface (39,40).

An alternative model for anchoring the pulling forces on the spindle pole is one in which astral MTs are attached to anchored structures distributed throughout the cytoplasm. The observed transport of cytoplasmic particles along the length of astral rays provides evidence that such attachments are possible (25, 31). Even though the mechanism for these movements is not known, their existence allows us to propose a model. Suppose these particles or attachment sites were anchored in position so that they could not move. If the same force generating systems were activated, the aster itself might now be pulled through the cytoplasm.

This model would allow an aster to crawl about the cytoplasm in an autonomous fashion. The direction of its movement would be determined by the sum of the vector forces acting upon it. It could accommodate the examples that were problems for the previous model. In addition, it

could accommodate the observations of Bajer in which
monoasters with associated chromosomes tend to move in the
direction of the most developed array of MTs and that on
occasion would migrate to the margin of the cell, rotate,
and then migrate back towards the center of the cell (36).
     We have seen how a mechanism which exerts traction
forces along astral MTs accommodates many of the movements
observed within the cell. It allows anaphase B movement to
be explained by the same mechanism in such diverse systems
as the intranuclear spindle of the fungi and the open
astral spindle of the animal cell. By modulating the
proposed astral pulling forces one can also explain such
puzzling phenomena as the alignment of the spindle in a
predetermined orientation, the movement of the spindle to an
eccentric position prior to unequal division, and the
migrations and oscillations of spindles that are sometimes
observed in cells. In addition, it allows a common mechanism to be responsible for aster movements throughout all
stages of the cell cycle. Interphase migrations, prophase
pole separation and anaphase spindle elongation can all be
accommodated by such a model.
     Experiments have yet to be performed which specifically
interrupt astral MTs or prevent their interaction with
possible anchorage sites. In order to fully embrace this
attractive mechanism, future experiments must show that
disruption of astral MTs stops anaphase B movement in cases
where disruption of interzone MTs does not.

REFERENCES

1. Ris, H., Biol. Bull. 85:164-178 (1943).
2. Ris, H., Biol. Bull. 96:90-160 (1949).
3. Carlson, J.G., Chromosoma 64:191-206 (1977).
4. Begg, D.A., and Ellis, G.W., J. Cell Biol. 82:528-541
    (1979).
5. Nicklas, R.B., and Staehly, C.A., Chromosoma 21:1-16
    (1967).
6. Begg, D.A., and Ellis, G.W., J. Cell Biol. 82:542-554
    (1979).
7. Forer, A., in "Cell Cycle Controls" (G.M. Padilla,
    I.L. Cameron and A. Zimmerman, eds.) pp. 319-336.
    Academic Press, Inc., New York, 1974.
8. Oppenheim, D.S., Haushka, B.T., and McIntosh, J.R.,
    Exp. Cell Res. 79:95-105 (1973).
9. Cande, W.Z., McDonald, K., and Meeusen, L., J. Cell Biol.
    88:618-629 (1981).

10. Cande, W.Z., J. Cell Biol., in press (1982).
11. McIntosh, J.R., Hepler, P.K., and Van Wie, D.G., Nature 224:659-663 (1969).
12. McIntosh, J.R. and Landis, S.C., J. Cell Biol. 49:468-497 (1971).
13. Brinkley, B.R., and Cartwright, J., J. Cell Biol. 50: 416-431 (1971).
14. Margolis, R.L., Wilson, L., and Kiefer, B.I., Nature 272:450-452 (1978).
15. Kronebusch, P.J., and Borisy, G.G., J. Cell Biol., in press (1982).
16. Aist, J.R., and Berns, M.W., J. Cell Biol. 91:446-458 (1981).
17. Wilson, C.L., and Aist, J.R., Phythopathology 57:769-771 (1967).
18. Gibardt, M., in "Aspects of Cell Motility". Twenty-second Symposium of the Society for Experimental Biology, Oxford. Cambridge University Press, pp. 249-259 (1968).
19. Poon, N.H., and Day, A.W., Can. J. Microbiol. 22:495-506 (1976).
20. Nakai, Y., and Ushiyama, R., Can. J. Bot. 56:1206-1211 (1978).
21. Heath, I.B., Int. Rev. Cytol. 69:191-221 (1981).
22. Malech, H.L., Root, R.K., and Gallin, J.I., J. Cell Biol. 75:666-693 (1977).
23. Gotlieb, A.I., May, L.M., Subrahmanyan, L., and Kalnin, V.I., J. Cell Biol. 91:589-594 (1981).
24. Brinkley, B.R., and Stubblefield, E., J. Ultrastruct. Res. 19:1-18 (1967).
25. Rebhun, L.I., Int. Rev. Cytol. 32:93-137 (1972).
26. Bajer, A.S., and Mole-Bajer, J., in "Spindle Dynamics and Chromosome Movement". Int. Rev. Cytol. (Suppl. 3), pp. 1-271. Academic Press, Inc., New York (1972).
27. Murphy, D.B., and Tilney, L.G., J. Cell Biol. 61:757-779 (1974).
28. Porter, K.R., in "Locomotion of Tissue Cells". Ciba Foundation Symposium, pp. 149-169 (1973).
29. Bestor, T.H., and Shatten, G., Develop. Biol. 88:80-91 (1981).
30. Gaulden, M.E., and Carlson, J.G., Exp. Cell Res. 2:416-433 (1951).
31. Wolf, R., Develop. Biol. 62:464-472 (1978).
32. Molé-Bajer, J., Cytobios. 13:117-140 (1975).
33. Rattner, J.B., and Berns, M.W., Cytobios. 15:37-43 (1976).
34. Longo, F.J., and Anderson, E., J. Ultrastr. Res. 33: 495-527 (1970).

35. Kawamura, K., Exp. Cell Res. 21:1-18 (1960).
36. Bajer, A.S., Debrabander, M., Molé-Bajer, J., DeMay, J.,
     Paulaitis, S., and Geuens, G., in "Microtubules and
     Microtubule Inhibitors" (M. Debrabander and J. DeMay,
     eds.), pp. 399-425 Elsevier/North-Holland Biomedical
     Press, Amsterdam (1980).
37. Rappaport, R., Int. Rev. Cytol. 31:169-213 (1971).
38. Hamaguchi, M.S., and Hiramoto, Y., Develop., Growth &
     Differ. 22:517-530 (1980).
39. Fullilove, S.L., and Jacobson, A.G., Develop. Biol. 26:
     560-577 (1971).
40. Koevenig, J.L., and Jackson, R.C., Mycologia 58:662-667
     (1966).

CHAPTER 23

LOCATION OF THE MOTIVE FORCE
FOR CHROMOSOME MOVEMENT
IN SAND-DOLLAR EGGS[1]

Y. Hiramoto
Yôko Shôji

Biological Laboratory
Tokyo Institute of Technology
Tokyo, and
National Institute for Basic Biology
Okazaki

I. INTRODUCTION

Although echinoderm eggs have been used by many investigators in studies of cell division, especially cytokinesis and
physico-chemical properties of the mitotic apparatus,
chromosome movement has scarcely been investigated in these
materials because chromosomes were hardly observable in the
living state by conventional light microscopy. Recently,
Hiramoto et al. (1) found that chromsomes can be observed in
living eggs of the sand-dollar, *Clypeaster japonicus* with a
differential interference microscope. This finding has enabled us to use this material for experimental studies of the
mechanism of chromosome movement in the living state. In the
present study, we have analyzed the location where the motive
force for anaphase chromosome movement is generated by means
of micro-operations on the mitotic apparatus and by destroying
a pre-selected region of the mitotic apparatus by local
application of Demecolcine.

[1]This work was supported by Grants-in-Aid for Scientific
Research nos. 348016, 548015 and 511214 from The Ministry of
Education, Science and Culture of Japan awarded to Y.H.

Biological Functions of Microtubules
and Related Structures

II. MICROMANIPULATION OF THE MITOTIC APPARATUS.

Throughout the present study, fertilized eggs at the first division or blastomeres at the second division of the sand-dollar, *Clypeaster japonicus* were used as materials. The cells were so transparent that chromosomes could be observed by differential interference microscopy as described else-where (1). Eggs deprived of both fertilization membrane and hyaline layer were put into a manipulation chamber described by Hiramoto (2). At metaphase or anaphase of mitosis, a part of the mitotic apparatus was removed by sucking it out with a braking micropipette (cf. ref. 2) inserted into the cell, while observing with a Nikon differential interference micro-scope. The movement of chromosomes in the operated cells and their further development were observed with the same micro-scope.

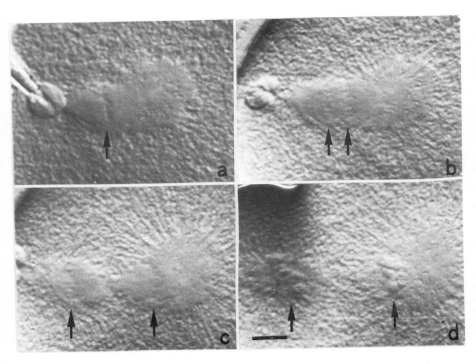

Fig. 1. Chromosome movement in a mitotic apparatus from which an aster including the centrosome has been removed by sucking through a micropipette during metaphase. In this and subsequent figures, arrows indicate positions of chromatids and bars indicate 10μm scales.

In the first experiment, one of the asters of the mitotic apparatus was removed by sucking it out with a micropipette during metaphase (Fig. 1). As a result of this operation, the remainder of the mitotic apparatus in the cell often moved toward the tip of the micropipette (Fig. 1a). When anaphase started, chromatids migrated poleward at almost normal speeds (Fig. 1b and c) and then a daughter nucleus was formed near each end of spindle. A similar result was obtained when a similar operation was made during anaphase. In these operations, it seems that the centriole at the center of the removed aster and a small portion of the spindle attached to the aster were removed together with the remaining region of the aster during the above operations. Therefore, it may be concluded that neither the centrosome at the center of the aster nor the polar part of the spindle participates in anaphase chromosome movement.

When a considerable amount of protoplasm was removed from the region between the equatorial plane and a pole of the spindle during metaphase by sucking with a micropipette, the arrangement of chromosomes at the equatorial

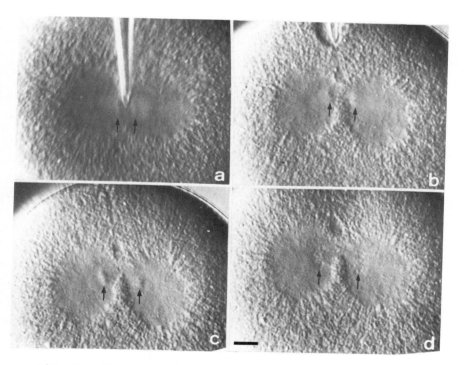

Fig. 2.   Chromosome movement in spindle in which the interzonal region has been removed during early anaphase.

plane was disturbed more or less passively by the distortion
of other regions of the spindle. When anaphase started, the
movement of some chromatids, especially those in the un-
operated half spindle, was normal while movements of other
chromatids were inhibited to varying degrees; in consequence,
chromatids changed into chromosomal vesicles at various
positions in the spindle during telophase.

When the protoplasm at the interzonal region of the
spindle was sucked into a micropipette placed at one side of
the spindle, granular cytoplasm at the other side was drawn
into the region between chromatid groups moving toward the
pole (Fig. 2a and b). Chromatids continued to move poleward
after the above operation and subsequently daughter nuclei
were formed near the poles of the spindle (Fig. 2). The separation of the poles during anaphase was significantly smaller
than the separation of poles in normal cells. The above results suggest that the interzonal region of the spindle does
not participate in the generation of the motive force responsible for anaphase A (cf. ref. 3), and that it does participate in anaphase B.

In the next experiment, a very thin glass plate about
10 μm wide and 1 μm thick at the tip was inserted into the
interzonal region of an anaphase spindle in order to cut it
into two parts (cf. Fig. 3a) In this case, poleward movement
of chromatids continued, while the separation of poles during
anaphase (anaphase B) was more or less inhibited. This result
resembles the result of the previous experiment in which the
interzonal region was removed; together they suggest the motive force responsible for poleward movement of chromatids
(anaphase A) is not located at the interzonal region of the
spindle.

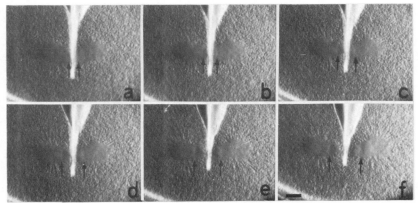

*Fig. 3. Chromosome movement in a spindle which has been
cut into two parts by inserting a glass plate into the interzone during early anaphase.*

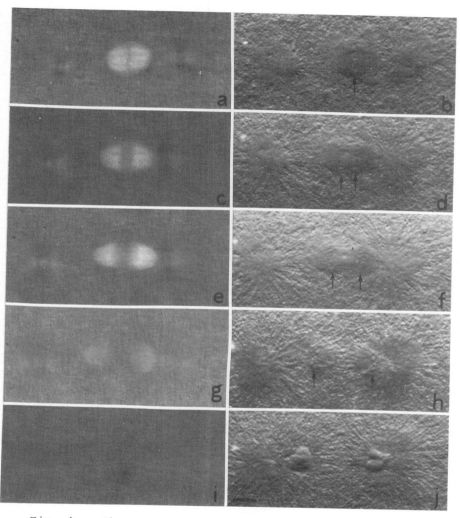

*Fig. 4.   Chromosome movement in a mitotic apparatus from
which an aster has been separated by rapidly compressing the
cell during metaphase. Pictures in the right column indicate
differential interference contrast micrographs showing positions of chromatids (arrows), and those in the left column
polarization micrographs showing birefringent structures due
to the oriented microtubules.*

III. CUTTING THE SPINDLE INTO TWO PARTS
     BY RAPIDLY COMPRESSING THE CELL

In *Clypeaster* eggs, it was possible to cut the mitotic
spindle into two parts at various sites along the spindle by
compressing the egg rapidly between a glass slide and a coverslip during metaphase. Fig. 4 shows an example from experiments in which one of the asters, including its centrosome,
was separated from the main part of the mitotic apparatus.
Pictures in the right column are differential interference
contrast micrographs showing chromosomes, and those in the
left column are polarization micrographs showing birefringent
structures of the spindle and asters. Neither the velocity
nor the extent of chromosome movement was different from those
in a normal spindle. This fact implies that the centrosome at
the pole of the spindle is unnecessary for the anaphase chromosome movement.

When the spindle was cut at a site between a pole and
center of the spindle, chromatids moved during anaphase from
the equatorial plane toward the cut edge of the spindle and
stopped a few micrometers in front of the cut edge.

These results with compressed eggs indicate that neither
the centrosome attached to the pole of the spindle nor the
polar region of the spindle is necessary for anaphase chromosome movement.

IV. DESTROYING A SELECTED PART OF THE SPINDLE
    BY COMBINED DEMECOLCINE TREATMENT
    AND REGIONAL IRRADIATION WITH ULTRAVIOLET RAYS

N-methyl-N-desacetylcolchicine (Demecolcine, Sigma Chemical Company, St. Louis, Mo.) is a compound which destroys
microtubule structures of the mitotic apparatus. Aronson and
Inoué (4) reported that the spindles in sea urchin eggs which
had been destroyed by treatment with N-methyl-N-desacetylcolchicine (Colcemid, Ciba Pharmaceutical Company, Fairlawn,
N. J.) were re-formed when the eggs were irradiated with
ultraviolet (UV) rays of 366 nm wavelength. It is considered
that Colcemid (Demecolcine) is changed into a compound which
does not destroy spindle structure.

We have attempted to destroy spindle structure in a limited region by combined Demecolcine treatment and regional irradiation with ultraviolet rays as follows. Eggs stuck to a
slide with polylysine were put on the stage of a microscope,
which could be converted between a differential interference
contrast and a polarization optics (cf. Fig. 5). Changing the

optical systems from one to the other was achieved by inserting or removing Nomarski prisms (N_1 and N_2). A mica plate (M)
could be inserted into the light path to add an appropriate
retardation to the background when the microscope was used as
a polarizing microscope. The cell could be irradiated with UV
rays of 366 nm wavelength arising from a high-pressure mercury
arc (L_1), passing through a window (W) and filters (F_1, F_2, F_3
and F_4), and reflecting by a dichroic mirror (DM). The area
in the cell to be irradiated was controlled by adjusting the
position, shape and size of the window (W) so that its image
was formed at the plane of the cell on the stage by the

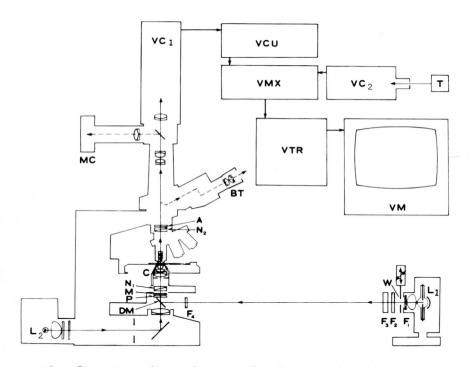

*Fig. 5.   Experimental setup for locally irradiating the
cell with UV rays while the cell was observed with a differential interference microscope as well as a polarization microscope. A, analyzer; BT, binocular tube of the microscope; C,
condenser lens; DM, dichroic mirror; F_1, F_2, F_3 and F_4, filters; L_1, light source for UV rays; L_2, light source for visible rays; M, mica plate; MC, microscope camera; N_1 and N_2,
Nomarski prisms; P, polarizer; T, timer; VC_1 and VC_2, video-
cameras; VCU, video-camera contral unit; VM, video-monitor;
VMX, video-mixer; VTR, video-tape recorder.*

condenser lens (C) of the microscope. The whole microscopic
field was illuminated with visible rays arising from the
built-in light source (L_2) and passing through the dichroic
mirror (DM). The eggs could be observed through the binocular
tube of the microscope (BT) or a video-monitor (VM) because
the video-camera(VC_1) was sensitive to the UV rays. Positions
of chromatids in the spindle during experiments were checked
by differential interference microscopy, and the birefringence
of the spindle, which may be closely related to the number of
microtubules, (cf. ref. 1) was checked by polarization microscopy from time to time.

When the eggs on the glass slide developed to early metaphase of mitosis, they were continuously treated with sea
water containing 10^{-5} M Demecolcine by perfusion, and at the
same time the eggs were irradiated with UV rays of 366 nm
wavelength over the entire area of the mitotic apparatus in

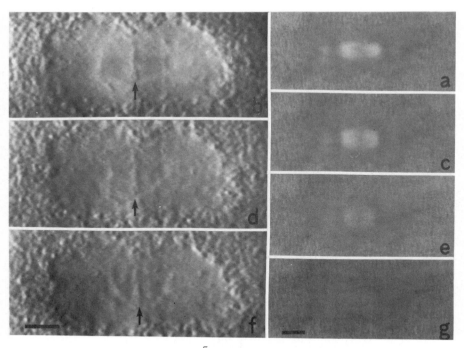

*Fig. 6.   Effects of 10^{-5} M Demecolcine on the structure of
the mitotic apparatus.   a and b, before application of Demecolcine; c, d, e, f and g, after application.   b, d and
f are differential interference contrast micrographs and a, c,
e and g are polarization micrographs.*

a cell. At late metaphase or anaphase, irradiation was con-
fined to various regions of the mitotic apparatus during continued perfusion with Demecolcine sea water in order to destroy the spindle structure at the shaded area (i.e. unirradiated area).

In preliminary experiments, it was found that mitosis and
cleavage proceeded normally in a cell when the irradiation
over the entire mitotic apparatus was unintercepted, while
both mitosis and cleavage were blocked in all other cells on
the same glass slide when UV irradiation was intercepted.
When the UV rays were intercepted over the entire mitotic
apparatus during and after mid-metaphase, birefringence in the
asters and spindle decreased, the spindle shortened, and
finally both structures disappeared (cf. Fig. 6). Chromosome
movement was never observed. When the UV rays were intercepted during and after early anaphase, chromatids stopped
moving shortly afterwards. The birefringence of the spindle
and asters disappeared and the elongation of the spindle
during anaphase was not observed.

When the UV rays were intercepted during and after mid-
metaphase (cf. Fig. 7) over one "side" of the mitotic apparatus (with reference to the spindle axis), the spindle structures in the shaded area were destroyed and movement of
chromatids during anaphase was inhibited, while the movement
was almost normal in the irradiated "side" of the spindle
(cf. Fig. 7). This fact may signify that the chromatids move
almost independently of one another. It also demonstrates

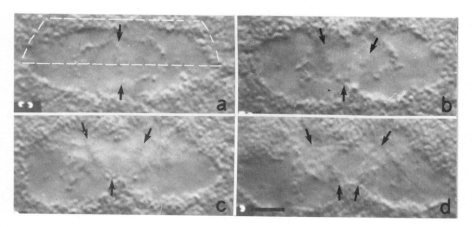

Fig. 7. Chromosome movement in a spindle in which micro-
tubules are preserved from destruction by inactivating
Demecolcine in the medium with UV irradiation of one "side"
of the mitotic apparatus (with reference to the spindle axis).
The UV-irradiated area is indicated by a dashed line.

that the boundary between the irradiated area and the shaded
one was fairly sharp in the present experimental system.

When the UV rays were intercepted over an aster and a
small portion of the spindle attached to the aster, fibrous
structures in these regions were destroyed, but poleward
movement of chromatids during anaphase was scarcely affected.
A similar result was obtained when the interception of UV rays
was started after the onset of anaphase. These results seem
to exclude the participation of the centrosome and, possibly
polar part of the spindle, in anaphase chromosome movement;
they thus support the conclusion obtained by the micromanipulation and compression experiments mentioned above.

When the UV rays were intercepted over one half of the

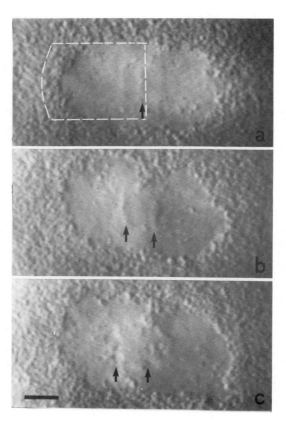

*Fig. 8. Chromosome movement in a spindle in which micro-
tubules are preserved from destruction with UV irradiation of
one half of the mitotic apparatus (with reference to the
equatorial plane). The UV-irradiated area is indicated by a
dashed line.*

spindle (with reference to the equatorial plane) during and
after mid-metaphase (cf. Fig. 8), chromatids moved poleward
in the irradiated half-spindle in almost normal fashion during anaphase, while they moved with lower speeds and then
stopped midway in the shaded half-spindle.   When the interception of UV rays was started during early anaphase, normal
poleward movement of chromatids continued in the irradiated
half-spindle, while chromatids stopped midway in the shaded
half-spindle.   Birefringence decreased markedly in the shaded
half of the mitotic apparatus.
     When the UV rays were intercepted over the whole mitotic
apparatus except for one aster during and after mid-metaphase,
the spindle was shortened but chromatids moved poleward in the
shortened spindle during anaphase.   Because weak birefringence
was detected in the spindle at regions near the chromatids in
the present case, it seems likely that motile machinary for

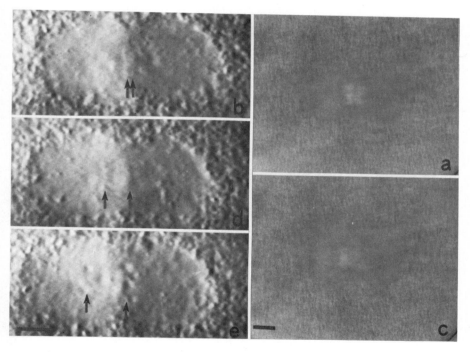

*Fig. 9.   Chromosome movement in a half-spindle which is
formed by UV irradiation of one half of the spindle treated
with Demecolcine.   b, d and e are differential interference
contrast micrographs showing positions of chromatids, and a
and c are polarization micrographs showing birefringent
structures in the spindle.*

chromatids still operated within the shortened half-spindle area.

As originally reported by Aronson and Inoué (4), we found that birefringence of the spindle which had disappeared by Demecolcine treatment increased by irradiation with UV rays over the whole mitotic apparatus. Chromosome movement was observed in such irradiated spindle. When the Demecolcine-treated spindle was irradiated over an area covering half of the spindle (with reference to the equatorial plane), birefringence of the spindle and asters incresed in the irradiated area and chromatids moved toward the pole in this half-spindle; in the unirradiated half, recovery of birefringence was not detected and chromatids remained near the equator (cf. Fig. 9). This observation indicates that chromatids can move in one half of the spindle independently of movement of sister chromatids in the other half of the spindle.

V. CONCLUSIONS

From results of the above experiments, we interpret the location of motive force for anaphase chromosome movement in sand-dollar eggs to be as follows:

1. The motive force for chromosome movement is located at and/or near the chromatids in the half-spindle. The centrosome and the polar region of the spindle do not participate in the generation of the motive force for chromosome movement.

2. The interzonal region of the spindle probably participates in the spindle elongation during anaphase but it does not participate in chromosome movement toward the pole.

3. Each chromatid can move independently of its sister chromatids.

4. Chromosome movement is closely correlated with the presence of fibrous (birefringent) structures-----probably microtubules ----- in the spindle which are reversibly destroyed by Demecolcine.

ACKNOWLEDGMENT

We thank Dr. Thomas E. Schroeder for critical reading of the manuscript.

REFERENCES

1. Hiramoto, Y., Hamaguchi, Y., Shôji, Y., Schroeder, T.E.,
    Shimoda, S., and Nakamura, S., *J. Cell Biol. 89*, 121
    (1981).
2. Hiramoto, Y., *Exp. Cell Res. 87*, 403 (1974).
3. Inoué, S., and Ritter, H., *in* "Molecules and Cell Move-
    ment" (S. Inoué and R.E. Stephens, eds.), p.3. Raven
    Press, New York, (1975).
4. Aronson, J., and Inoué, S., *J. Cell Biol. 45*, 470 (1970).

CHAPTER 24

ASSEMBLY AND DISASSEMBLY OF
ECHINODERM EGG ACTIN[1]

Issei Mabuchi

Department of Biology
College of General Education
University of Tokyo
Komaba, Meguro-ku, Tokyo

I. INTRODUCTION

In many nonmuscle cells, contractile apparatus or cytoskeletons are formed transiently when these structures are
necessary for the cell. A good example of the transient
contractile apparatus is the contractile ring in dividing
cells which forms in a very short time at the onset of cytokinesis (1). This apparatus is composed of numerous actin
filaments oriented in a parallel manner at the equatorial cell
cortex. Myosin (2, 3), tropomyosin (3, 4), and α-actinin (5)
have been observed in the same region using immunofluorescence
techniques, and the necessity of myosin in its formation or
function has been demonstrated physiologically (6). The contractile apparatus is reduced in volume during the progression
of the cleavage furrow, while the population density of the
actin filaments does not seem to change, and it finally disappears after the completion of the cleavage (7). Since actin
filaments are the main components of these structures, it is
speculated that association and dissociation of the pre-
existent actin filaments or polymerization and depolymerization of actin molecules take place in these cells. The former

[1]*Supported by grants-in-aid for Scientific Research and
for Special Project Research from the Ministry of Education,
Science and Culture in Japan, and by a research grant from the
Yamada Science Foundation.*

process has been demonstrated *in vitro* with crude extract or reconstituted systems using purified proteins in gelation or actin bundle formation. Recent findings that many nonmuscle cells contain large pools of monomeric actin (8-13) seem to suggest that polymerization also takes place in the cell. However, only a very few examples have been demonstrated for the existence of polymerization or for the polymerization-depolymerization cycle of actin in the cell. It is still not known whether contractile ring actin filaments are formed from monomeric actin or from filamentous actin.

II. ACTIN IN THE CORTICAL LAYER OF ECHINODERM EGGS

In echinoderm eggs, cell division is the most remarkable motile activity. The contractile ring may be the only contractile apparatus which has been demonstrated to be composed mainly of actin filaments and is believed to be actually contractile. However, little is known about its formation, as described above. It has been speculated that the precursor of the contractile ring actin filament is located in the cortical layer but not in the cytoplasm before its formation since no increase in the amount of actin in the cortex was observed during mitosis of the sea urchin egg (14).

In addition, apparent actin polymerization takes place when cell surface projections are produced. It has recently been found that spike-like surface projections are formed transiently on starfish oocytes treated with 1-methyladenine (15). These projections contain bundles of actin filaments as cores. In sea urchin eggs, microvilli, which also contain bundle of actin filaments as cores, elongate upon fertilization (14, 16-18). A meshwork of microfilaments (presumably actin) which seemed to interconnect rootlets of the microvillar actin core was also observed in the cortical layer (ref. 14, Fig. 1). It has been speculated that these actin filaments are formed through polymerization since no filamentous structures are detectable in the cortical layer of the unfertilized eggs (14, 16-18). The content of actin in the isolated cortices actually increased several-fold after fertilization (14, 17-19), which supported this idea. The trigger for this polymerization has not yet been identified. Begg and Rebhun (17) have done some very interesting work on this problem. First, they isolated cortices from unfertilized sea urchin (*Strongylocentrotus purpuratus*) eggs at pH 7.3-7.5 (which is considered to be the pH of fertilized eggs), and they found that the appearance of the cortical layer was very similar to that of the fertilized eggs -- namely, actin

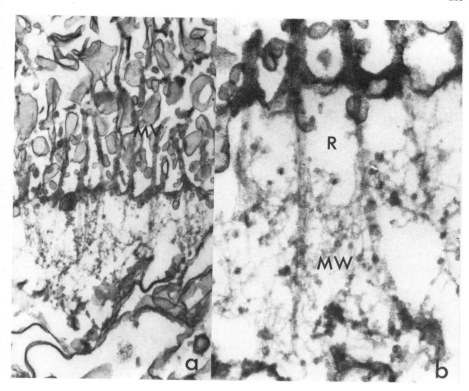

Fig. 1.   Cross-sectional view of the cortex isolated from a fertilized sea urchin (Hemicentrotus pulcherrimus) *egg. MV, microvilli. R. rootlets of the microvillar actin core. MW, microfilament meshwork. Magnification: a, X 40,000; b, X 100,000. From ref. 14.*

filaments were observed in microvilli-like surface projections and in the cortical layer. Secondly, they isolated unfertilized egg cortices at pH 6.5 (which is considered to be close to the pH of unfertilized eggs) and then transferred them into a pH 7.5 medium. They found polymerized actin in these cortical layers. On the basis of these findings they speculated that the trigger may be an increase in the intracellular pH after fertilization.

On the other hand, biochemical studies have shown the presence of monomeric actin in sea urchin eggs; it might be the precursor of organized actin in the cortical layer of fertilized eggs. A previous study with unfertilized sea urchin (*Strongylocentrotus purpuratus*) eggs showed that 1) about 50%

of the total actin is monomeric, 2) this monomeric actin has a high critical concentration for polymerization, and 3) the poor polymerizability of the monomeric actin is due to the presence of other proteins (13, 20).

III. AN ACTIN DEPOLYMERIZING PROTEIN

One of these proteins was purified from starfish (*Asterias amurensis*) oocytes (21; Mabuchi, in preparation) using ammonium sulfate fractionation of 65-90% saturation followed by DEAE-cellulose, hydroxylapatite, and Sephadex G-75 column chromatography techniques. Its molecular weight was determined to be 17,000 by SDS[2]-gel electrophoresis (Fig. 2) for denatured protein, and to be 20,000 by gel filtration chromatography for native protein. The isoelectric point was determined to be about 6 (Fig. 2). This protein can either inhibit the extent of actin polymerization or depolymerize F-actin quickly. After the latter property, this protein will tentatively called *depactin*. Figure 3 showed the time course of actin polymerization induced by addition of salts, monitored by viscometry. In the presence of depactin, the viscosity of actin increased to a certain level and then decreased slightly to attain a plateau. This "overshooting" was reproducibly observed at an appropriate depactin concentration. A half-time $(T_{1/2})$ of the polymerization to the initial viscosity peak was

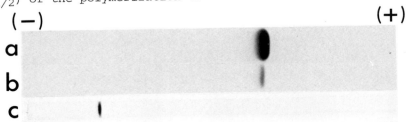

Fig. 2. *SDS-gel electrophoresis and isoelectric focusing of depactin. a and b: SDS-gels of depactin, 6 μg and 1 μg, respectively. c: an isoelectric focusing gel of depactin (1 μg). pH 4-6.5 Pharmalyte was used.*

[2]*Abbreviations used are: SDS, sodium dodecyl sulfate; EGTA, ethyleneglycol-bis(β-aminoethyl ether)-N, N, N', N'-tetraacetic acid; TAME, p-tosyl-L-arginine methylester·HCl; PIPES, piperazine-N, N'-bis(2-ethane sulfonic acid); MOPS, morpholinopropane sulfonic acid; HMM, heavy meromyosin.*

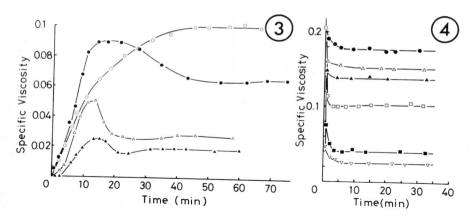

Fig. 3.   Time course of actin polymerization. 0.1 M KCl and 1 mM MgCl$_2$ were added at pH 6.85 to mixtures of actin (5.5 μM) at 25°C and various amounts of depactin (o, 0; •, 1.2; △, 2.9; ▲, 5.6 μM) and the viscosity was measured using an Ostwald-type viscometer.

Fig. 4.   Time course of actin depolymerization. To the mixture of 5.6 μM actin in 0.1 M KCl and 1 mM MgCl$_2$ (pH 6.85) was added various amounts of depactin (•, 0.6; △, 1.2; ▲, 1.7; □, 2.9; ■, 5.6; ▽, 10.9 μM). Other conditions are the same as described in the legend of Fig. 3. From ref. 21.

estimated. Compared to control polymerization in the absence of depactin ($T_{1/2}$ = 9.5 min), depactin shortened the half-time a little ($T_{1/2}$ = 7 to 7.5 min). This may indicate that depactin accelerates nucleation of actin to some extent. When depactin was added to F-actin solution, depolymerization of F-actin took place and was complete in a few minutes (Fig. 4). The extent of depolymerization was dependent upon the amount of added depactin. Electron microscopic examination showed that straight and long actin filaments became flexible and short upon addition of an insufficient amount of depactin. No filament was observed with an equimolar amount of depactin.

IV. EFFECT OF pH ON ACTIN-DEPACTIN INTERACTION

It is important to investigate the conditions under which actin is released from inhibition by depactin in order to establish the role of depactin in actin organization in the cell. The first possible factor I checked was pH of the

solution. As mentioned in section II, Begg and Rebhun (17)
have postulated that change in the intracellular pH after
fertilization would trigger the polymerization of actin in the
cell. They demonstrated by electron microscopy the appearance
of polymerized actin in isolated cortices of unfertilized eggs
after raising the pH. Therefore, the pH of the crude extract
of unfertilized sea urchin eggs was raised from 6.8 to 8.0,
via 7.5 and changes in the G-actin concentration was measured
(13). The G-actin concentration in the crude extract was 1.2
mg/ml at pH 6.8. It did not decrease as the pH of the extract
increased, but rather increased a little. The same was true
for the purified actin-depactin system: when pH of a solution
of actin and depactin in 0.1 M KCl and 1 mM $MgCl_2$ was changed
from 6 to 8.5, the G-actin concentration did not decrease but
rather increased as the pH was increased. Therefore, it may be
that the increase in intracellular pH after fertilization is
not a direct trigger for polymerization of actin in the
cortical layer after fertilization.

V. POLYMERIZATION OF ACTIN FROM A CRUDE MONOMERIC ACTIN
   FRACTION

   When a crude monomeric actin fraction obtained by gel
filtration chromatography was concentrated to about 1 mg
actin/ml in 0.1 M KCl, 2 mM $MgCl_2$, 1 mM EGTA, 0.2 mM ATP, 0.5
mM dithiothreitol, and 20 mM MOPS buffer (pH 6.85), about 5%
actin precipitated as fine needle-like structures of several
μm. The rest of the actin remaining in the solution was in the
monomeric state. These needles were analyzed by electron
microscopy and optical diffraction and were shown to be actin
paracrystals (22). These paracrystals had periodical transverse bands, the spacing of which was one-third of the

   *Fig. 5. Paracrystals of actin formed in a crude monomeric
actin fraction. A crude extract of unfertilized sea urchin*
(Anthocidaris crassispina) *egg was applied to a Sephadex G-150
column, and eluted monomeric actin fraction was pooled.
Precipitates formed upon concentration of this fraction with
an Amicon diaflo cell (UM-2 membrane) were examined by an
electron microscope with a negative staining technique.
Magnification, X 100,000. From ref. 22.*
   *Fig. 6. An optical diffraction pattern of actin para-
crystal. A region on a paracrystal (indicated by an arrow)
shown in Fig. 5 was analyzed. From ref. 22.*
   Fig. 7. SDS-gel electrophoresis of paracrystals. a: crude

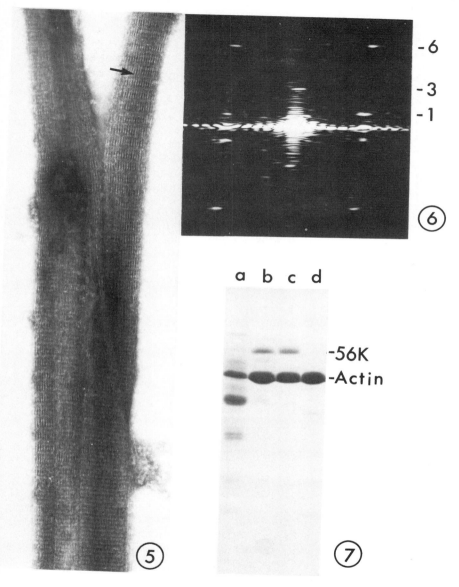

monomeric actin fraction. b: precipitates (paracrystals) formed upon concentration of a. c: precipitates formed upon addition of 0.1 M KCl and 1 mM MgCl₂ after depolymerization of b (purified paracrystals). d: high-speed pellet from a 20,000 x g supernatant (after removal of c) of the salts-supplemented depolymerized b. From ref. 22.

distance between the cross-over points of the two long-pitch
helical strands of the actin filament (Figs. 5 and 6). The
ratio for the so-called sixth and first layer lines was 6.32
(Fig. 6), which is larger than that of the pure actin
filament. These properties of the present paracrystal indicate
that it has a structure very similar to that of a microvillar
actin bundle (23) or an actin bundle in an extract gel of
Hawaiian sea urchin eggs (24). SDS-gel electrophoresis
revealed that the paracrystals are composed only of actin and
a 56,000 molecular weight protein. These paracrystals were
purified by a cycle of depolymerization-polymerization under
conditions applicable to purification of actin. All the 56K
protein present attached to actin to form paracrystals at a
molar ratio of about 5:1 (Fig. 7).

These facts suggest that the 56K protein may be a trigger
for polymerization of monomeric actin. Judging from its
molecular weight, its binding ratio to actin, and the
appearance of the paracrystal, this 56K protein may be the
same as the 58K protein contained in the gel of Hawaiian sea
urchin eggs found by Kane (25) and later called *fascin* (26).

VI. EFFECT OF MYOSIN ON THE ACTIN-DEPACTIN INTERACTION

Within the cell, there may exist a number of proteins
which can interact with actin in various ways. It may be
reasonable to assume that there are some interactions between
such actin-binding proteins and actin-depactin complex, which
may cause actin polymerization by releasing depactin from the
complex in the cell. Along this line, effect of myosin or its
tryptic fragment heavy meromyosin (HMM), alpha-actinin,
tropomyosin, or calmodulin was investigated. Only myosin and
HMM showed a positive effect. When HMM was added to an actin-
depactin mixture in 0.1 M KCl and 1 mM $MgCl_2$ and in the
absence of nucleotides, the G-actin concentration was reduced
with time, indicating that actin polymerization took place.
This effect was dependent on the HMM concentration: nearly
full polymerization was observed with HMM of twice the amount
of depactin (Fig. 8a). In the presence of an appropriate
concentration ATP or ADP, however, HMM did not induce actin
polymerization. On the other hand, 5'-AMP was not effective
(Fig. 8b). This HMM-induced polymerization of actin was
confirmed by electron microscopy and sedimentation studies.
Filaments appeared when HMM was added to the actin-depactin
mixture. All these filaments showed an arrowhead configuration
characteristic of an actin-HMM complex (not shown). In
contrast, filamentous structure was never observed in the
presence of 0.5 mM ADP.

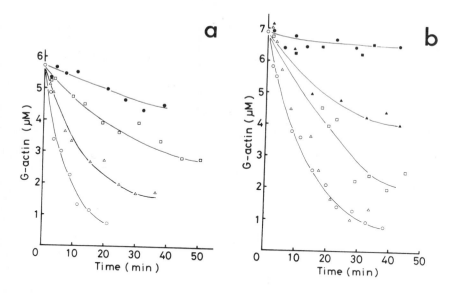

Fig. 8.   HMM-induced polymerization of actin in the pres-
ence of depactin. To a mixture of rabbit skeletal actin and
depactin in 0.1 M KCl, 1 mM MgCl$_2$ and 10 mM MOPS buffer (pH
7.0) at 25°C, HMM was added, and G-actin concentration was
monitored by the DNase I inhibition assay (27) with minor
modifications (13). a: 5.6 μM actin, 10.2 μM depactin and
various amounts of HMM (●, 1.7 μM; □, 3.4 μM; △, 6.6 μM; o,
13 μM). b: 6.8 μM actin, 8.4 μM depactin, 15.0 μM HMM, and
various adenine nucleotides (o, no nucleotide; △, 60 μM 5'-
AMP; □, 10 μM ADP; ▲, 30 μM ADP; ■, 60 μM ADP; ●, 60 μM ATP).

VII. DISCUSSION

Recently, several proteins have been isolated from non-
muscle cells which are capable of depolymerizing actin. Such
proteins are listed in Table I. This table does not include
actin gelling or capping proteins. It shows that depactin
resembles profilin from calf spleen (35) in molecular weight;
however, other properties, including mode of interaction with
actin, are very different. The amino acid composition of these
proteins (not shown) is also different. As far as the effect
of depactin on actin is concerned, this protein is similar to
fragmin from a slime mold *Physarum polycephalum* (32). However,
there is one functional difference, which is the sensitivity
to calcium ions. This point should not be neglected when one

Table I. Proteins That Depolymerize Actin

Proteins	Tissues	M.W.	pI	Rate of actin depolymerization[2]	Inhibition of nucleation	Ca²⁺ requirement	Reversal by
Villin	Intestinal epithelial cells	95K		Fast	No	Yes	
Brain modulator	Porcine brain	94K		Slow	Yes	?	
ADF[1]	Plasma	92K	6 - 6.5	Fast	?	No	
Gc-globulin	Human serum	56K		Medium	?	?	
Fragmin	Physarum	43K x 2	4.7	Fast	No	Yes	
DNase I	Bovine pancreas	31K		Medium	No	No	
Depactin	Starfish oocytes	17K	6	Fast	No	No	Myosin
Profilin	Calf Spleen	15K	9?	Slow	Yes	No	Alpha-actinin

[1]Actin depolymerizing factor. [2]The criteria are as follows. Fast means that the depolymerization of actin caused by the protein will be complete in a few minutes. Medium means in more than ten minutes. Slow means over several hours to a few days. References are as follows: villin, 28; brain modulator, 29; ADF, 30; Gc-globulin, 31; fragmin, 32; DNase I, 33 and 34; depactin, 21 and this report; profilin, 35-37.

considers the function of these proteins in the cell. At
present, we do not find a reasonable relationship among the
proteins listed in this table. Further studies will clarify
whether these proteins are evolutionarily related or not.

The present results suggest that the direct trigger for
the polymerization of the monomeric actin in the cell may be
other proteins. The effect of the pH increase found by Begg
and Rebhun (17) might be indirect, and it is possible that the
pH increase regulates actin-polymerizing activity in these
proteins. Paracrystal formation from the monomeric actin
indicates that a certain concentration of both actin and the
56K protein may be necessary for this process. It may be that
accumulation of the 56K protein near the plasma membrane
after fertilization is a candidate for the trigger for polymerization of the microvillar actin core.

In the case of contractile ring formation, the situation
may be more complex. There are conflicting observations on the
distribution of actin in the cortical layer at cytokinesis.
Sanger (38) reported, using fluorescein-labelled HMM, that
actin is concentrated in the cleavage furrow of chick fibroblasts. Later, Aubin et al. (3) confirmed his observation
using fluorescently labelled antibody against actin or HMM on
PtK2 cells. On the other hand, Herman and Pollard (39)
reported, using similar techniques, that actin is *not* concentrated in the cleavage furrow of PtK2 or HeLa cells. We
also did not find a concentration of actin in the cleavage
furrow of dividing sand dollar eggs when we used microinjected fluorescently labelled phalloidin, although it
accumulated preferentially in the cortical layer (40). Moreover, as already mentioned, the actin content in the isolated
sea urchin egg cortex did not change during mitosis (14). The
former findings (3, 38) indicate that the precursor of the
contractile ring actin filaments migrates to the equatorial
cell cortex from other regions of the cell when the contractile ring is formed. The latter findings (14, 39, 40) indicate
that the precursor is previously present in the cell cortex
and does not change its location. I do not intend to judge
which is more likely. The difference may be attributed to the
species difference. But I should like to point out that, in
the former situation, the contractile ring filaments may be
formed through polymerization. In the latter situation, the
precursor will be a filamentous actin as well as a monomeric
actin bound to some structures. In any case, it is possible
that the contractile ring actin filaments are formed through
polymerization.

On the other hand, myosin (2, 3), tropomyosin (3, 4), and
alpha-actinin (5) are known, by immunofluorescent staining
techniques, to be concentrated in the cleavage furrow region.

Accumulation of a myosin-like component in the cell cortex towards the first cell division of the sea urchin egg has also been noted (14). The present report demonstrated that myosin reversed the inhibition of actin by depactin. In the case of profilin, alpha-actinin has been reported to reverse its inhibition at a substoichiometric concentration (37). These results suggest that those structural proteins which may construct the contractile ring with actin would be candidates for the trigger for the polymerization of the contractile ring actin filaments.

REFERENCES

1. Schroeder, T. E. (1975). *In* "Molecules and Cell Movement" (S. Inoue, and R. E. Stephens, eds.), p. 305. Raven Press, New York.
2. Fujiwara, K., and Pollard, T. D. (1976). *J. Cell Biol.* *71,* 848.
3. Aubin, J. E., Weber, K., and Osborn, M. (1979). *Exptl. Cell Res.* *124,* 93.
4. Ishimoda-Takagi, T. (1979). *Exptl. Cell Res.* *119,* 423.
5. Fujiwara, K., Porter, M. E., and Pollard, T. D. (1978). *J. Cell Biol.* *79,* 268.
6. Mabuchi, I., and Okuno, M. (1977). *J. Cell Biol.* *74,* 251.
7. Schroeder, T. E. (1972). *J. Cell Biol.* *53,* 419.
8. Tilney, L. G., Hatano, S., Ishikawa, H., and Mooseker, M. (1973). *J. Cell Biol.* *59,* 109.
9. Bray, D., and Thomas, C. (1976). *J. Mol. Biol.* *105,* 527.
10. Gordon, D. J., Boyer, J. L., and Korn, E. D. (1977). *J. Biol. Chem.* *252,* 8300.
11. Adelman, M. R. (1977). *Biochemistry 16,* 4862.
12. Merriam, R. W., and Clark, T. G. (1978). *J. Cell Biol.* *77,* 439.
13. Mabuchi, I., and Spudich, J. A. (1980). *J. Biochem.* *87,* 785.
14. Mabuchi, I., Hosoya, H., and Sakai, H. (1980). *Biomed. Res.* *1,* 417.
15. Schroeder, T. E. (1981). *J. Cell Biol.* *90,* 362.
16. Burgess, D. R., and Schroeder, T. E. (1977). *J. Cell Biol.* *74,* 1032.
17. Begg, D. A., and Rebhun, L. I. (1979). *J. Cell Biol.* *83,* 241.
18. Spudich, A., and Spudich, J. A. (1979). *J. Cell Biol.* *82,* 212.
19. Vaquier, V. D., and Moy, G. W. (1980). *Develop. Biol.* *77,* 178.

20.   Mabuchi, I. (1979). *In* "Cell Motility: Molecules and Organization" (S. Hatano, H. Ishikawa, and H. Sato, eds.), p. 147. University of Tokyo Press, Tokyo.
21.   Mabuchi, I. (1981). *J. Biochem. 89,* 1341.
22.   Mabuchi, I., and Nonomura, Y. (1981). *Biomed. Res. 2,* 143.
23.   Spudich, J. A., and Amos, L. A. (1979). *J. Mol. Biol. 129,* 319.
24.   DeRosier, D., Mandelkow, E., Silliman, A., Tilney, L., and Kane, R. (1977). *J. Mol. Biol. 113,* 679.
25.   Kane, R. E. (1975). *J. Cell Biol. 66,* 305.
26.   Otto, J. J., Kane, R. E., and Bryan, J. (1979). *Cell 17,* 285.
27.   Blikstad, I., Markey, F., Carlsson, L., Persson, T., and Lindberg, U. (1978). *Cell 15,* 935.
28.   Bretcher, A., and Weber, K. (1980). *Cell 20,* 839.
29.   Nishida, E. (1981). *J. Biochem. 89,* 1197.
30.   Harris, H. E., and Gooch, J. (1981). *FEBS Lett. 123,* 49.
31.   Van Baelen, H., Bouillon, R., and DeMoor, P. (1980). *J. Biol. Chem. 255,* 2270.
32.   Hasegawa, T., Takahashi, S., Hayashi, H., and Hatano, S. (1980). *Biochemistry, 19,* 2677.
33.   Mannherz, H. G., Leigh, J. B., Leberman, R., and Pfrang, H. (1975). *FEBS Lett. 60,* 34.
34.   Hitchcock, S. E., Carlsson, L., and Lindberg, U. (1976). *Cell 7,* 531.
35.   Carlsson, L., Nyström, L. -E., Sundkvist, I., Markey, F., and Lindberg, U. (1977). *J. Mol. Biol. 115,* 465.
36.   Grumet, M., and Lin, S. (1980). *Biochem. Biophys. Res. Commun. 92,* 1327.
37.   Blikstad, I., Sundkvist, I., and Eriksson, S. (1980). *Eur. J. Biochem. 105,* 425.
38.   Sanger, J. W. (1975). *Proc. Nat. Acad. Sci. U. S. 72,* 1913.
39.   Herman, I. M., and Pollard, T. D. (1978). *Exptl. Cell Res. 114,* 15.
40.   Hamaguchi, Y., and Mabuchi, I. (1982). *Cell Motility 2,* in press.

CHAPTER 25

CAPPING, BUNDLING, CROSSLINKING
THREE PROPERTIES OF ACTIN BINDING PROTEINS

G. Isenberg

Max-Planck-Institute for Psychiatry
8033 Martinsried, Munich

B. M. Jockusch

Department of Developmental Biology
University of Bielefeld
4800 Bielefeld

I. INTRODUCTION

Cytoskeletal proteins represent the structural basis for cellular and subcellular motility, such as locomotion, uni-directional growth, establishment of polarity, membrane mobility and intracellular transport mechanisms. Evidence is accumulating that there are at least three different princi-ples of generating force for such complex phenomena: (i) The structural proteins actin and tubulin themselves may directly create movement by reversible self-assembly or disassembly of microfilaments and microtubules, respectively (1,2), (ii) microfilaments and microtubules may provide structural back-bones to which a motor for mechano-chemical energy trans-duction may be attached (muscle, cilia, this volume), (iii) microfilaments and microtubules are not separate cellular entities but may be interconnected with each other and with cellular organelles (e.g. 3-6). Such interactions must be very important for cytoplasmic architecture, and, as a consequence of that, for cellular shape and intracellular transport.

It is the study of accessory proteins of microfilaments and microtubules which may eventually lead to an understanding of motility processes based on mechanisms (ii) and (iii).

Biological Functions of Microtubules
and Related Structures

So far, it seems that the microfilament system contains many more of such proteins than the microtubules, and much has been learned recently about their influence on actin filament organization from *in vitro* experiments.

Here, we report on three actin binding proteins (capping protein, vinculin, α-actinin) as examples for factors which exert entirely different effects on actin filaments: capping, bundling and crosslinking.

II. RESULTS AND DISCUSSION

A. *Capping protein*

Viscometry provides a convenient method to study the effect of actin binding proteins on actin polymerization and filament interactions *in vitro*. A very sensitive albeit qualitative method for detecting even small changes in the apparent viscosity of actin solutions is the use of a low shear falling ball viscometer (7). In Fig. 1, the apparent viscosity of skeletal muscle F-actin (0.5 mg/ml) is plotted as percent in comparison to a control (100 %) after the addition of increasing amounts of *Acanthamoeba* capping protein (8), chicken gizzard vinculin and skeletal muscle α-actinin (9,10). Capping protein and vinculin, both have an inhibitory effect on the viscosity of actin solutions, whereas α-actinin

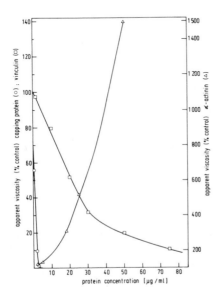

FIGURE 1. *Apparent viscosities as measured under low shear conditions. Actin solutions (0.5 mg/ml) were polymerized in the presence of varying concentrations of capping protein (○), vinculin (□) and α-actinin (Δ) at 25°C. The viscosity is expressed in % relative to control viscosities of actin alone (100 %)*

increases the viscosity several fold, as compared with the control. At the same protein concentration, capping protein reduces the low shear viscosity much more drastically than vinculin. The original data (8) showed that 1-2 µg protein reduces the viscosity of 0.5 mg/ml actin to less than 5 % of the control value. Assuming a native molecular weight of

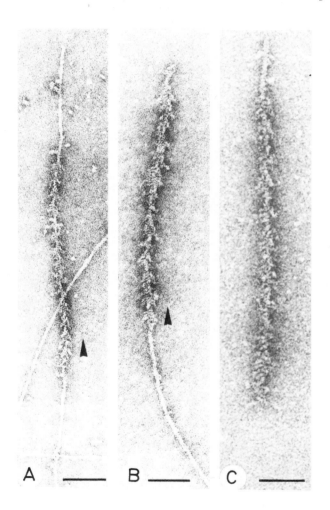

FIGURE 2. *Growth of actin filaments onto S-1 decorated filament fragments. A and B: Bidirectional growth onto S-1 decorated actin filament fragments which have been fixed in the presence of vinculin. Bar = 0.1 µm. C: In the presence of* Acanthamoeba *capping protein which blocks monomer addition at the barbed, fast growing end. Bar = 0.05 µm (8).*

74,000 for capping protein (11), the molar ratio of capping
protein:actin under these conditions was 1:1000. This drastic
effect on viscosity at such a low molar ratio became plausible
when actin:capping protein complexes were analyzed by electron
microscopy: the addition of capping protein inhibits monomer
addition to filament ends by binding to the barbed (fast
growing) end of F-actin [(8) and Fig. 2 C)]. In addition, it
was shown that capping protein promotes filament nucleation,
probably by stabilizing actin dimers during the early phase of
polymerization. These characteristics of capping proteins are
shared by villin (12), gelsolin (13), fragmin (14,15) and
platelet factors (16,17). All these proteins are thought to
"cap" actin filament ends and thus can act as potential length
regulators of actin filaments within the cell (For reviews,
see refs. 18 and 19).

B. *Vinculin*

Vinculin, on the other hand, though having a similar
inhibitory effect on the viscosity during actin polymerization
may act rather differently. The stoichiometry is much higher
than when compared with capping factors: the low shear viscosity is noticeably affected by a molar ratio of 1 vinculin:200
actin molecules, and a 90 % decrease is seen at a ratio of
1:10. Thus, these differences, when compared to the action of
"classical" capping proteins on actin polymerization suggested
a different mode of binding.

There is somewhat more biochemical and structural information on chicken gizzard vinculin than on *Acanthamoeba*
capping protein: vinculin is a globular, monomeric protein of
130,000 (5,20-22). Single vinculin molecules from an electrophoretically homogeneous preparation of chicken gizzard
vinculin can be seen by negative staining and rotary shadowing
in electron micrographs (Fig. 3 A). The protein is roughly
globular with a central depression and shows three or four
lobes. The average diameter in negatively stained preparation
is about 8.5 nm (10,23).

We (9,10) and others (24) have shown by electron microscopy of negatively stained preparations that, in the presence
of vinculin, actin filaments form tightly packed bundles.
These bundles can also be seen in the light microscope (10)
and in angle shadowed preparations (23). Optical diffraction
of negatively stained preparations and computer analysis of
the obtained patterns have shown that these bundles are in
fact paracrystalline structures, with a packing of actin filaments quite similar to Mg-induced paracrystals. However,
unlike other proteins which have been found to induce such
paracrystals [fascin from sea urchin eggs (25,26) and a

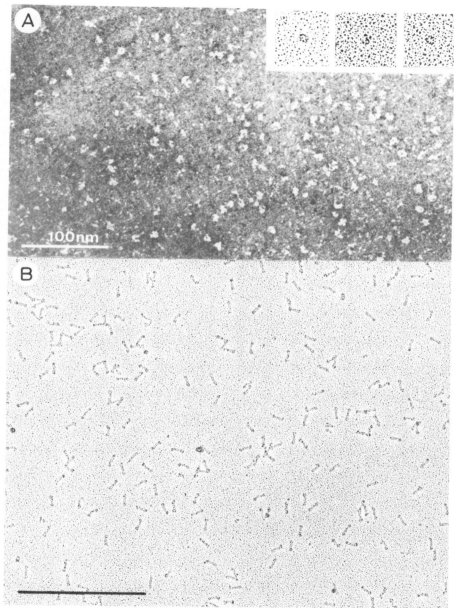

FIGURE 3. A:Single vinculin molecules prepared from chicken gizzard as seen after negative staining and rotary shadowing (insets). The compact particles with a central depression have an average diameter of 8.5nm. Bar=0.1 m. B: A rotary shadowed preparation from chicken gizzard α-actinin. The rod-shaped molecules represent individual α-actinin dimers. Bar=0.25 m.

similar protein from *Limulus* (27)] vinculin does not form
regular cross bridges within the paracrystal.

When actin is polymerized with a high amount of vinculin
present (1:10 up to 1:3 molar ratio), vinculin molecules can
directly be seen bound laterally along actin filaments, over
the entire length of the bundle (Fig. 4 B)(23). Again, and in
agreement with the optical diffraction data, the binding of
vinculin to vinculin-induced bundles lacks any periodicity.
The lateral binding of vinculin to actin bundles was also
confirmed by indirect labelling of vinculin-actin complexes
with ferritin-coupled anti-vinculin [F(ab)' fragments (23)].
In order to test whether vinculin, in addition to binding
laterally, may also associate with actin filament ends, we
incubated S-1 decorated actin filament fragments with vinculin
prior to fixation with 0.1 % glutaraldehyde. Any vinculin
bound at the ends should thus be arrested and inhibit further
growth after the addition of G-actin at one or both ends of
the filament fragments. Quantitative analysis of electorn
micrographs (Fig. 2 A and B), however, showed that growth of
those pretreated filament fragments was identical to that of
controls (23). The same result had been obtained previously
with unfixed filaments in the presence of vinculin (9). Also,
vinculin does not promote nucleation (9,10,24), as do the
typical "capping factors" described above. The parallel
alignment of paracrystalline F-actin can account for the
decreased high and low shear viscosity found for actin polymerizing in the presence of vinculin. *In vivo*, the bundling
of microfilaments by vinculin located closely to the plasma
membrane (20,28) may be a necessary requirement for the interaction of actin-filaments with membrane components.

C. Alpha Actinin

α-Actinin is an example of quite another class of actin-
binding proteins, which were termed "gelation factors". The
observed increase in high (29) and low shear viscosity (9) of
actin polymerizing in the presence of skeletal or smooth
muscle α-actinin can be attributed to an extensive cross-
linking of individual filaments by single α-actinin molecules.
The protein is a dimer of 2 x 100,000, with a particle size of
roughly 4 x 40 nm (30). In rotary shadowed preparations of
chicken gizzard α-actinin, elongated particles are seen which
show a subunit composition of two rod-like structures, and a
knob-like thickening at each end (Fig. 3 B). It is tempting
to speculate that those may represent the two identical polypeptide chains with one binding site each for actin. Thus,
the two identical binding sites are located at different ends
of the rather stiff, rod-like molecule, and, while crosslinking

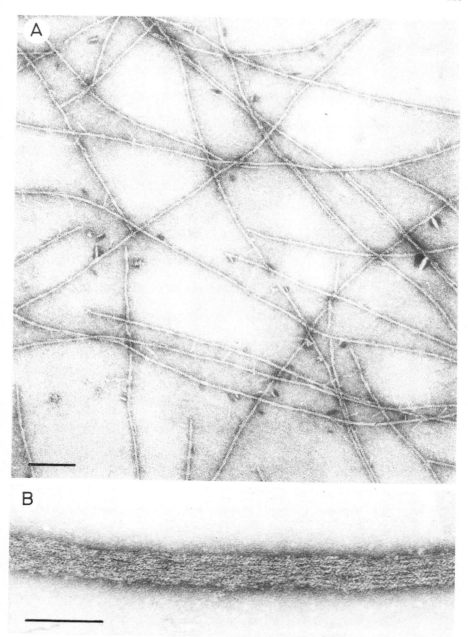

FIGURE 4. A: Micrographs of an actin-α-actinin mixture taken
during the viscosity measurements as shown in Fig. 1. Actin
filaments are crossbridged by α-actinin at various angles and
arranged in a three dimensional network. Bar = 0.1 μm.

B: A paracrystalline actin bundle induced by vinculin (vin-culin/actin molar ratio 1:3) under low salt conditions (20 mM KCl) single vinculin molecules can be seen binding late-rally to actin filaments and decorating the bundle over its entire length in an irregular pattern. Bar = 0.1 μm.

actin filaments, space them out, forming ladder-like structures (Fig. 4 B). *In vivo,* such crosslinking may be an important factor in stabilizing actin filaments against high shear forces in areas of active motility, and also in arranging microfilaments parallel to each other. The location of α-actinin in areas of membrane ruffling in migrating or spreading cells (31) as well as inside microfilament bundles (stress fibers) of stationary cells (32) is consistent with this idea.

From the long and still growing list of actin-capping, -bundling and crosslinking proteins (for a recent review, see ref. 33) we have chosen for this report three typical examples which, from their location inside the cell can be expected to influence the spatial arrangement of actin filaments also *in vivo*. Little is known so far on the fine regulation of the binding of such proteins to actin, of their interactions with each other and of the binding sites for such proteins on the actin molecule. Clearly, we still have to learn a great deal about those processes before we will be able to understand the generation and modulation of force in cellular motility.

ACKNOWLEDGMENTS

The data about *Acanthamoeba* capping protein were obtained in collaboration with Drs. U. Aebi and T. D. Pollard, when G.I. was a visiting scientist at the Dept. of Cell Biology and Anatomy, The Johns Hopkins Medical School, Baltimore, USA. We thank Dr. Brian Burke, EMBL, Heidelberg, for help with the rotary shadowing. The Deutsche Forschungsgemeinschaft supports B.M.J. by grant Jo 55/9 and G.I. by grant Is 25/2-1.

REFERENCES

1.   Tilney, L. G., *J. Cell Biol.* 77:551 (1978).
2.   Oakley, B. R., and Morris, N. R., *Cell* 24:837 (1981).
3.   Griffith, L. M., and Pollard, T. D., *J. Cell Biol. 78:* 958 (1978).

4.   Heggenes, M. H., Simon, M., and Singer, S. J., *Proc. Natl. Acad. Sci. USA. 75*:3863 (1978).
5.   Geiger, B., and Singer, S. J., *Proc. Natl. Acad. Sci. USA. 77*:4769 (1980).
6.   Schliwa, M., and Blerkom, V., *J. Cell Biol. 90*:222 (1981)
7.   McLean-Fletcher, S., and Pollard, T. D., *J. Cell Biol. 85*:414 (1980).
8.   Isenberg, G., Aebi, U., and Pollard, T. D., *Nature 288*: 455 (1980).
9.   Jockusch, B. M., and Isenberg, G., *Proc. Natl. Acad. Sci. USA. 78*:3005 (1981).
10. Jockusch, B. M., and Isenberg, G., *Cold Spring Harbor Symp. Quant. Biol. vol. 46*, in press.
11. Cooper, J. A., Isenberg, G., and Pollard, T. D., *J. Cell Biol. 91*:299a (1981).
12. Glenney, J. R., Kaulfus, Ph., and Weber, K., *Cell 24*: 471 (1981).
13. Yin, H. L., Zaner, K. S., and Stossel, T. P., *J. Biol. Chem. 255*:9494 (1980).
14. Hasegawa, T., Takahashi, S., Hayashi, H., and Hatano, S., *Biochem. 19*:2677 (1980).
15. Hinssen, H., *Europ. J. Cell Biol. 23*:225 (1981).
16. Grumet, M., and Lin, S., *Cell 21*:439 (1980).
17. Wang, L. L., and Bryan, J., *Cell 25*:637 (1981).
18. Pollard, T. D., and Craig, S., *Trends in Biochem.* in press (1982).
19. Craig, S., and Pollard, T. D., *Trends in Biochem.* in press (1982).
20. Geiger, B., *Cell 18*:;93 (1979).
21. Geiger, B., Dutton, A. H., Tokuyasu, K. T., and Singer, S. J., *J. Cell Biol. 91*:614 (1981).
22. Feramisco, J. R., and Burridge, K., *J. Biol. Chem. 255*: 1194 (1980).
23. Isenberg, G., Leonard, K., and Jockusch, B. M., *J. Mol. Biol.* (submitted).
24. Wilkins, J. A., and Lin, S., *Cell* in press (1982).
25. Spudich, J. A., and Amos, L. A., *J. Mol. Biol. 129*:319 (1979).
26. Otto, J. J., Kane, R. E., and Bryan, J. E., *Cell 17*:285 (1979).
27. Tilney, L. G., *J. Cell Biol. 64*:289 (1975).
28. Geiger, B., *Int. Cell Biol. 1980-1981*:761 (1981).
29. Holmes, G. R., Goll, D. E., and Suzuki, A., *Biochim. Biophys. Acta 253*:240 (1971).
30. Suzuki, A., Goll, D. E., Singer, I., Allen, R. E., Robson, R. M., and Stromer, M. H., *J. Biol. Chem. 251*: 6860 (1976).

31. Lazarides, E., *J. Supramolec. Struct.* 5:531 (1976).
32. Lazarides, E., and Burridge, K., *Cell* 6:289 (1975).
33. Schliwa, M., *Cell* 25:587 (1981).

CHAPTER 26

PHOSPHORYLATION
OF MICROTUBULE-ASSOCIATED PROTEINS (MAPs)
CONTROLS BOTH MICROTUBULE ASSEMBLY
AND MAPs-ACTIN INTERACTION[1]

Eisuke Nishida
Susumu Kotani
Tomoyuki Kuwaki
Hikoichi Sakai

Department of Biophysics and Biochemistry
Faculty of Science
University of Tokyo
Tokyo

The microtubules which exist in almost all eukaryotic cells appear to consist of tubulin and microtubule-associated proteins (MAPs). MAPs are demonstrated to be essential for tubulin polymerization *in vitro* and are believed to be distributed over the entire surface of microtubules both *in vivo* and *in vitro* (1,2). There is a possibility that MAPs mediate the interaction of microtubules with other cellular structures. Indeed, Griffith and Pollard demonstrated that the *in vitro* interaction of microtubules with actin filaments requires MAPs (3).

On the other hand, it is well known that MAPs are strongly phosphorylated by cyclic AMP-dependent protein kinase which copurifies with microtubules (4-8). Recently, Caplow and his colleagues (7,9) demonstrated that phosphorylation of MAPs modulates microtubule assembly. We examined the effect of phosphorylation of MAPs on MAPs-actin interaction and found that phosphorylation of MAPs (MAP2 and tau) not only regulates microtubule assembly but also inhibits MAPs-actin association. It is possible that phosphorylation and de-

[1]*This work was supported in part by a Grant-in-Aid for Scientific Research from the Ministry of Education, Science and Culture, Japan (No. 444074).*

phosphorylation of MAPs regulate both microtubule assembly
and MAPs-actin association in cells.

I. MATERIALS AND METHODS

Microtubule protein was isolated from porcine brains by
three cycles of temperature-dependent assembly and disassembly
(10,11). Tubulin was purified from microtubule protein by
phosphocellulose column chromatography (12,13). Tubulin which
was eluted from the column in the unbound fraction was supplemented with $MgCl_2$ (1 mM) and GTP (0.5 mM), and then concentrated by using an Amicon PM30 membrane. MAPs were prepared
by heat treatment of microtubule protein as described (14).
MAP2 and tau proteins were obtained from these MAPs fractions
by gel filtration on Bio-gel A-15m and DEAE-cellulose (DE52)
chromatography as follows. MAPs were, first, gel-filtered
on Bio-gel A-15m equilibrated with 0.1 M MES-0.5 mM $MgCl_2$-1 mM
EGTA (pH 6.8). MAP2 and tau proteins were completely separated from each other at this step. Tau protein fraction was
then adsorbed to DE52 equilibrated with the same buffer.
After extensive washing with the same buffer, purified tau
proteins were obtained by a stepwise elution with the buffer
containing 75 mM NaCl. Both MAP2-rich fraction and purified
tau proteins were concentrated and dialyzed against 0.1 M MES-
0.5 mM $MgCl_2$-1 mM EGTA (pH 6.5). Rabbit skeletal muscle actin
was prepared by the method of Spudich and Watt (15) with a
slight modification (16). Monomeric actin was further purified by gel filtration on Sephadex G-100. Actin concentration
was determined by UV absorption using $A_{290}^{1\%}=6.5$.
Microtubule assembly was initiated by elevating the
temperature of the protein solution from 0 to 35°C and
monitored by measuring turbidity at 350 nm. Interactions of
MAPs with actin filaments were analyzed using the low-shear
falling ball viscometric technique described by MacLean-
Fletcher and Pollard (17). Immediately after MAPs were mixed
with actin, samples were drawn up into capillary tubes and
incubated at 22°C for desired times before measurement of
viscosity. The time required for a stainless ball to fall
between two points was measured and converted into viscosity
in centipoise (cp). Protein concentration was determined by
the method of Lowry *et al*. (18) or that of Bradford (19).
Polyacrylamide gel electrophoresis in the presence of SDS was
carried out by the method of Laemmli (20).

II. INHIBITION OF MICROTUBULE ASSEMBLY BY PHOSPHORYLATION OF MAPs

Jameson *et al.* (7,9) demonstrated that phosphorylation of microtubule protein at low ionic strength inhibits subsequent microtubule assembly which is induced by increasing the ionic strength. In search of other phosphorylation conditions that induce the inhibition of microtubule assembly, we have found that preincubation of microtubule protein with ATP and cAMP in the presence of mM Ca results in marked inhibition of subsequent microtubule assembly which is induced by decreasing Ca concentration on addition of EGTA (data not shown). Since control sample was also supplemented with ATP and cAMP at the onset of polymerization, the inhibition observed can be attributed to a phosphorylation reaction which takes place during preincubation time. Autoradiography experiments showed that the major substrate for phosphorylation is MAP2 even in the presence of Ca and that tau proteins are also phosphorylated although less intense (14). Then, we prepared phosphorylated MAPs and unphosphorylated MAPs by heat treatment of microtubule protein that had been preincubated for 35 min at 30°C either with or without ATP (1 mM) and cAMP (5 µM) in a buffer solution consisting of 17 mM MES (pH 6.8), 5 mM $MgCl_2$, 90 mM KCl, 0.15 mM GTP and 1.1 mM $CaCl_2$. The phosphorylated MAPs were found to have a decreased ability to induce tubulin polymerization as compared with the unphosphorylated MAPs (data not shown). These results indicate clearly that phosphorylation of MAPs by a cyclic AMP-dependent protein kinase which copurifies with microtubules inhibits microtubule assembly. This is consistent with the finding of Jameson *et al.* (7,9).

III. INHIBITION OF MAPs–ACTIN INTERACTION BY PHOSPHORYLATION OF MAPs

We then compared the phosphorylated MAPs with the unphosphorylated ones in their abilities to interact with actin filaments, using a low-shear falling ball viscometric technique in order to determine whether phosphorylation of MAPs affects MAPs–actin interaction. The unphosphorylated MAPs (39 µg/ml) increased markedly the viscosity of actin filaments (0.12 mg/ml) when mixed with monomeric actin before polymerization, as originally observed by Griffith and Pollard (3). When mixed with preformed actin filaments, the unphosphorylated MAPs (78 µg/ml) also induced a great increase in actin viscosity. In contrast, the same concentrations of the phosphorylated MAPs did not increase the viscosities of actin filaments at all in either case. Figure 1 shows the dependence of low-shear viscosity of MAPs–actin mixtures on the

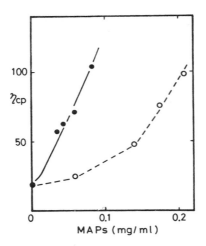

*Fig. 1   Dependence of low-shear viscosity of MAPs-actin
mixtures on the concentrations of phosphorylated or unphosphorylated MAPs.   After mixing various concentrations of MAPs
with monomeric actin in a solution of 25 mM MES-2.25 mM MgCl₂
-50 mM KCl-0.2 mM EGTA-0.08 mM ATP (pH 6.6), samples were
drawn up into capillary tubes and incubated for 30 min at
22°C before viscosities were measured.   ●, actin (0.12 mg/ml)
plus unphosphorylated MAPs;   ○ actin plus phosphorylated MAPs.*

concentrations of MAPs; it indicates that the phosphorylated
MAPs are about 4-fold less potent than the unphosphorylated
MAPs in increasing the actin viscosity.   It is clear from
these experiments that phosphorylation of MAPs inhibits MAPs-
actin interaction.

IV. EFFECT OF pH CHANGE ON MAPs-ACTIN INTERACTION

        Since many studies have suggested that changes in intra-
cellular pH may be an important control mechanism in regulating a variety of cellular functions, we examined the effect
of pH change on the MAPs-actin interaction.   As shown in
Fig. 2, the ability of the unphosphorylated MAPs to increase
the low-shear viscosity of actin filaments is high in the
neutral pH range but falls off sharply at around pH 7.5.   In
contrast, the low-shear viscosity of a mixture of the phosphorylated MAPs and actin filaments is low at neutral and alkaline pH, as described before, but it increases at an acidic
pH (Fig. 2).   Thus, small changes in the physiological range

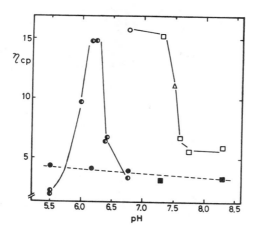

Fig. 2   Effect of pH on the low-shear viscosities of
mixtures of actin filaments with phosphorylated MAPs (◑) or
unphosphorylated MAPs (○□△). After mixing each MAPs (final
78 μg/ml) with preformed actin filaments (0.125 mg/ml) in a
solution of an appropriate buffer 20 mM (◑○● MES, △ HEPES,
■ □ Tris)–1.15 mM MgCl₂–80 mM KCl–0.23 mM EGTA–0.027 mM ATP,
samples were drawn up into capillary tubes and incubated for
10 min.   ● ■ , actin alone;   ◑ , +phosphorylated MAPs;
○△□, +unphosphorylated MAPs.

of pH can regulate MAPs–actin interaction *in vitro*. More
interestingly, the results showed that phosphorylation of
MAPs will modulate pH sensitivity of MAPs–actin association.

V. INTERACTIONS OF ACTIN WITH MAP2 AND TAU PROTEINS

Heat-stable MAPs consist mainly of MAP2 and tau proteins.
We extended our experiments to the interactions of actin with
purified MAP2 or tau proteins. MAP2 and tau proteins were
purified from unphosphorylated MAPs fraction as described in
"MATERIALS AND METHODS". In order to learn whether both MAP2
and tau proteins are capable of interacting with actin, we
mixed MAP2 or tau with actin before polymerization (that is,
G-actin), and after a 30 min-incubation the low-shear viscosity of the MAP2-actin or tau-actin mixture was measured. In
this case, polymerization of actin occurred in a capillary
tube (viscometer) together with MAP2 or tau. The result,
shown in Fig. 3b, demonstrates that both MAP2 and tau increased the actin viscosity in a concentration-dependent but

non-linear fashion, like other actin-crosslinking proteins.
Similar mass concentrations of MAP2 and tau are required to
induce the increase in viscosity. Therefore, MAP2 has a
higher actin-crosslinking activity than tau when expressed on
a molar basis, since the molecular weight of MAP2 is about
5-fold larger than that of tau.

  Next, MAP2 or tau was mixed with preformed actin filaments (F-actin), and after a 10 min-incubation the low-shear
viscosity of each mixture was measured. In this case, the
viscosity of the tau-actin mixture exhibited a biphasic dependence on the tau concentration, as shown in Fig. 3a. In
contrast, MAP2 was found to increase the actin viscosity mo-
notonously, and no sign of decrease in viscosity was observed
within the concentration range examined (Fig. 3a). This
suggests that MAP2 and tau have different effects on the
actin filament interaction.

  The difference between MAP2 and tau in their interaction

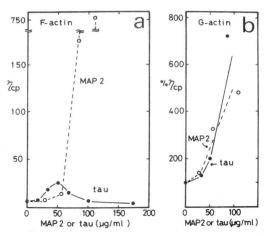

*Fig. 3 Dependence of low-shear viscosities of mixtures
of actin filaments with MAP2 or tau on the concentration of
MAP2 or tau. Unphosphorylated MAP2 and tau were prepared as
described in the text. (a) After mixing MAP2 (○) or tau
(●) with preformed actin filaments (final 0.15 mg/ml) in a
solution of 85 mM MES-0.8 mM MgCl$_2$-0.8 mM EGTA-0.036 mM ATP
(pH 6.5), samples were drawn up into capillary tubes and
incubated for 10 min before viscosities were measured. (b)
After MAP2 (○) or tau (●) was mixed with monomeric actin
(final 0.15 mg/ml) in a solution of 90 mM MES-2.4 mM MgCl$_2$-
0.9 mM EGTA-0.033 mM ATP (pH 6.5), samples were drawn up into
capillary tubes and incubated for 30 min.*

with actin filaments was also evident when turbidity was
measured. When the same mass concentration of MAP2 or tau
(final 110 µg/ml) was mixed with actin filaments (final 150
µg/ml), tau-actin mixture (A_{320}=0.057) showed higher turbidity than MAP2-actin mixture (A_{320}=0.017). It is clear from
these experiments that both MAP2 and tau interact with actin,
but the mode of interaction is somewhat different. Further
experiments are in progress to elucidate the nature of MAP2-
actin and tau-actin interactions.

VI. EFFECT OF PHOSPHORYLATION OF MAP2 AND TAU ON THEIR
ABILITY TO INTERACT WITH ACTIN

In order to examine the effects of phosphorylation of
MAP2 and tau proteins on MAP2-actin and tau-actin interactions, we prepared both phosphorylated and unphosphorylated
MAPs fractions as described before, and then purified from
them phosphorylated MAP2 and tau, and unphosphorylated MAP2
and tau, by the procedure described in "MATERIALS AND METH-
ODS". Figure 4 shows the low-shear viscosity attained by
mixtures of actin filaments with either phosphorylated or un-

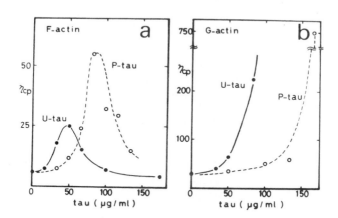

*Fig. 4 Comparison of the low-shear viscosity attained by
mixtures of actin filaments with either phosphorylated (O)
or unphosphorylated (●) tau. Experimental conditions (a,b)
are the same as described in the legend to Fig. 3 (a,b).
Phosphorylated and unphosphorylated tau were prepared as described in the text.*

phosphorylated tau. As illustrated in Fig. 4b, the phosphorylated tau was about 2-fold less potent than the unphosphorylated in increasing the viscosity of actin filaments when
mixed with actin before polymerization. Thus, phosphorylation
of tau weakens the tau-actin association. When the phosphorylated tau was mixed with preformed actin filaments, the
viscosity of the tau-actin mixture showed a biphasic dependence on the tau concentration which is qualitatively the
same as in the case of the unphosphorylated tau (Fig. 4a).
The tau concentration that gave the maximum viscosity was
higher for the phosphorylated tau than for the unphosphorylated, also indicating that phosphorylation of tau tends to
inhibit tau-actin interaction. However, the maximum viscosity attained by the mixtures of actin and phosphorylated tau
was higher than that attained by the unphosphorylated tau-
actin mixtures (Fig. 4a). It is certain from these experiments that phosphorylation of tau modulates tau-actin interaction, although the precise mechanism by which phosphorylation of tau affects tau-actin interaction remains to be elucidated.
     The effect of phosphorylation of MAP2 on MAP2-actin
interaction is shown in Fig. 5, which indicates clearly that

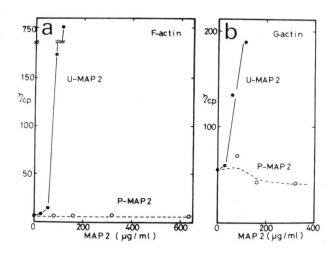

*Fig. 5 Comparison of the low-shear viscosity attained by
mixtures of actin filaments with either phosphorylated (○)
or unphosphorylated (●) MAP2. Experimental conditions (a,b)
are the same as described in the legend to Fig. 3 (a,b).
Phosphorylated and unphosphorylated MAP2 were prepared as
described in the text.*

the phosphorylated MAP2 did not increase the low-shear viscosity of actin filaments at all when mixed with actin either
before or after polymerization within the concentration range
examined.   From the result illustrated in Fig. 5a, the phosphorylated MAP2 was found to be at least 12-fold less potent
than the unphosphorylated MAP2 in increasing actin viscosity.
Therefore, it is not too much to say that phosphorylation of
MAP2 completely inhibits the ability of the protein to cross-
link actin filaments.   Thus, it was found that MAP2-actin
interaction can be controlled by phosphorylation of MAP2.

VII. EFFECT OF PHOSPHORYLATION OF MAP2 AND TAU ON THEIR
      ABILITY TO INDUCE TUBULIN POLYMERIZATION

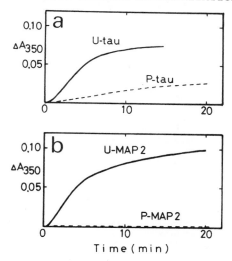

*Fig. 6 Effect of phosphorylation of tau and MAP2 on
their ability to induce tubulin polymerization.
(a) Phosphorylated (----) or unphosphorylated (——) tau was
mixed with phosphocellulose-purified tubulin at 0°C in a
solution of 0.1 M MES-2.5 mM MgCl$_2$-1 mM EGTA-1 mM GTP (pH
6.5), and then microtubule assembly was initiated by elevating the temperature of the mixture from 0 to 35°C.   Final
protein concentrations, tau (phosphorylated or unphosphorylated) 0.12 mg/ml; tubulin 1.4 mg/ml.   (b) Phosphorylated (----)
or unphosphorylated (——) MAP2 was mixed with phosphocellulose-purified tubulin at 0°C, and then microtubule assembly
was initiated as described in (a).   Final protein concentrations: MAP2 (phosphorylated or unphosphorylated) 0.38
mg/ml; tubulin 1.4 mg/ml.   Phosphorylated and unphosphorylated MAP2 and tau were prepared as described in the text.*

   Caplow and his colleagues (7,9) demonstrated that phosphorylation of MAPs inhibits or modulates microtubule assembly. However, since their results were obtained with whole
MAPs, it was not certain whether the activity of every component of MAPs is regulated by phosphorylation. We therefore
examined whether phosphorylation of MAP2 or tau affects the
ability of each protein to induce tubulin polymerization.
   As shown in Fig. 6a, the unphosphorylated tau induced
tubulin polymerization efficiently when mixed with phosphocellulose-purified tubulin, while the same concentration of
the phosphorylated tau promoted less polymerization. In the
same way, phosphorylation of MAP2 resulted in marked loss of
the ability of the protein to promote tubulin polymerization,
as is illustrated in Fig. 6b. These results indicate clearly
that phosphorylation of MAP2 and tau inhibits the ability of
each protein to induce tubulin polymerization.

VIII. CONCLUSIONS

   We have demonstrated here that phosphorylation of MAP2
and tau proteins, both of which constitute the major part of
brain MAPs, regulates the abilities of these proteins to
interact with both actin and tubulin. Phosphorylation of
these proteins inhibits their tubulin-polymerizing activity
and modulates their actin-crosslinking ability. In addition,
it was found that MAPs-actin interaction *in vitro* is controlled by subtle changes in the physiological range of pH.
Because MAPs are considered to play a key role in regulating
microtubule assembly and mediating the interaction of microtubules with actin filaments, it is highly probable that
phosphorylation and dephosphorylation of MAPs and change in
pH regulate microtubule-actin interaction as well as microtubule assembly, and therefore control the cytoskeletal
networks in cells.

REFERENCES

1. Kirschner, M. W. (1979) *Inter. Rev. Cytol. 54*, 1-71
2. Raff, E. C. (1979) *Inter. Rev. Cytol. 59*, 1-96
3. Griffith, L. M. and Pollard, T. D. (1978) *J. Cell Biol. 78*,
   958-965
4. Sloboda, R. D., Rudolph, S. A., Rosenbaum, J. L. and
   Greengard, P. (1975) *Proc. Natl. Acad. Sci. U.S. 72*,
   177-181

5. Rappaport, L., Leterrier, J. F., Virion, A. and Nunez, J. (1976) *Eur. J. Biochem.* *62,* 539–549
6. Cleveland, D. W., Hwo, S-Y. and Kirschner, M. W. (1977) *J. Mol. Biol.* *116,* 227–247
7. Jameson, L., Frey, T., Zeeberg, B., Dalldorf, F. and Caplow, M. (1980) *Biochemistry* *19,* 2472–2479
8. Islam, K. and Burns, R. (1981) *FEBS Lett.* *123,* 181–185
9. Jameson, L. and Caplow, M. (1981) *Proc. Natl. Acad. Sci. U.S.* *78,* 3413–3417
10. Shelanski, M. L., Gaskin, F. and Cantor, C. R. (1973) *Proc. Natl. Acad. Sci. U.S.* *70,* 765–768
11. Nishida, E., Kumagai, H., Ohtsuki, I. and Sakai, H. (1979) *J. Biochem.* *85,* 1257–1266
12. Weingarten, M. D. , Lockwood, A. H., Hwo, S-Y. and Kirschner, M. W. (1975) *Proc. Natl. Acad. Sci. U.S.* *72,* 1858–1862
13. Williams, R. C., Jr. and Detrich, H. W. III. (1979) *Biochemistry* *18,* 2499–2503
14. Nishida, E., Kuwaki, T. and Sakai, H. (1981) *J. Biochem.* *90,* 575–578
15. Spudich, J. A. and Watt, S. (1971) *J. Biol. Chem.* *246,* 4866–4871
16. Nishida, E. (1981) *J. Biochem.* *89,* 1197–1203
17. MacLean-Fletcher, S. D. and Pollard, T. D. (1980) *J. Cell Biol.* *85,* 414–428
18. Lowry, O. H., Rosebrough, N. J., Farr, A. L. and Randall, R. J. (1951) *J. Biol. Chem.* *193,* 265–275
19. Bradford, M. M. (1976) *Anal. Biochem.* *72,* 248–253
20. Laemmli, U. K. (1970) *Nature* *227,* 680–685

CHAPTER 27

THE ASSOCIATION OF MAP-2 WITH MICROTUBULES, ACTIN FILAMENTS, AND COATED VESICLES

Richard F. Sattilaro[1]
William L. Dentler[1]

Department of Physiology
and Cell Biology
Center for Biomedical Research
McCollum Laboratories
University of Kansas
Lawrence, Kansas

I. INTRODUCTION

Although cytoplasmic microtubules appear to be necessary for the positioning and movements of a variety of organelles, the manner by which they function is not understood. Morphological studies have revealed filamentous bridge structures that link microtubules to ciliary and plasma membranes, to the membranes of mitochondria, pigment and secretory granules, and synapatic vesicles, as well as to actin and inter-mediate-sized filaments. Evidence for strong physical attachments between microtubules and membranes has come from studies in which microtubule-membrane complexes have been isolated from cilia (1) and from the endocrine pancreas (2,3) The microtubules and membranes were linked by filamentous proteins and the complexes were stable to shearing during homogenization, centrifugation, and resuspension. The roles of these bridges in the positioning and movements of organelles is unknown at this time. They may serve to propel structures along microtubules in a manner similar to that of the dynein-based sliding of microtubules during ciliary and flagellar movements. Alternatively, the bridges may merely

[1]Supported by NIH grants AM 21672 and GM 24583

serve to anchor organelles to microtubules in order to position them within the cytoplasm or to orient them with respect
to other motile machinery. It is certain that the bridges
link organelles to microtubules, and it is likely that the
binding and release of the organelles from microtubules is
basic to their function. Therefore, it was considered essential to characterize the proteins that comprise the bridge,
to determine conditions that regulate the reversible association of microtubules with organelles, and to identify proteins or lipids associated with the organelles that regulate
the specificity of interaction with microtubues.

What proteins compose the bridge structures? To date,
the only microtubule-membrane bridge that has been identified
has been the high molecular weight dynein-like bridge that
links ciliary membranes to ciliary outer doublet microtubules
(1). Cytoplasmic microtubules with attached secretory granules have been fractionated from the endocrine pancreas but
these fractions were not sufficiently homogeneous to permit
biochemical analysis (3). Since microtubule-organelle complexes cannot as yet be dissected in vitro, we have examined
proteins that co-purify with microtubules to determine if the
microtubule-associated proteins may serve to anchor organelles to microtubules. Part of the rationale for studying
these proteins, and particular the major high molecular
weight microtubule-associated protein, MAP-2, is that this
protein can be purified in quantities sufficient for biochemical analysis. Furthermore, MAP-2 is a likely candidate
for a microtubule-membrane bridge protein since (A) MAP-2
attaches to the walls of microtubules assembled in vitro and
forms filamentous projections similar in appearance to the
microtubule-membrane bridge filament observed in situ (4, 5,
6), (B) MAP-2 is associated with microtubules that are cross-
linked by filamentous bridges in vivo, as determined by
immunofluorescence staining in nucleated erythrocyte marginal
bands (7), (C) MAP-2 has been shown to bind to pituitary
secretory granule membranes (8) and has been shown to link
endocrine pancreatic secretory granules to microtubules in
vitro (2), and (D) the cyclic-AMP stimulated phosphorylation
of MAP-2 (9) suggests a number of possible mechanisms that
may regulate the association of MAP-2 with membranes of
microtubules.

Direct evidence that MAP-2 can bind organelles to microtubules was obtained by use of dark field light microscopy
and electron microscopy. Suprenant and Dentler (2) isolated
insulin-containing secretory granules from the endocrine
pancreas and found that the granules would bind to MAP-
containing but not to MAP-free microtubules. Examination of
the secretory granules and MAP-containing microtubules by
electron microscopy revealed filamentous MAP-bridges linking

the secretory granules to microtubules. In addition, these
studies showed that cyclic AMP stimulated the binding of the
granules to microtubules, presumably by stimulating MAP-2
phosphorylation (9), and that ATP, but not other nucleotides
or ADP, released the granules from the microtubules. Similar
results were simultaneously and independently determined by
electron microscopic analysis of mixtures of purified
clathrin coated vesicles and microtubules (10). The binding
and release of secretory granules from microtubules was
assayed using dark field light microscopy, which permitted
the direct observation of individual microtubules and secretory granules in solution. Coated vesicles were too small to
be unambiguously detected by this technique. These results,
therefore, provided direct evidence for the MAP-2 dependent
association of microtubules with membrane bound organelles
and indicated that these interactions may be regulated by the
phosphorylation of MAPs.

II.   ATTACHMENT OF COATED VESICLES TO MAP-MICROTUBULES

        In addition to secretory ganules, MAPs also bind clathrin
coated vesicles to microtubules in vitro. Initial observations showed that microtubules polymerized from brain
homogenates often contained coated vesicles attached by
filamentous bridges (Suprenant and Dentler, unpub. obs.).
Recent data indicates that coated vesicles can co-purify,
under specific conditions, with microtubules prepared by
cycles of temperature dependent polymerization (10). In this
study, vesicles only attached to MAP-containing microtubules
and not to MAP-free microtubules. Thin sections revealed
that MAP filaments linked the clathrin coats to the microtubules (Fig. 1). The association of coated vesicles with
microtubules was stimulated in a concentration dependent
manner by cyclic AMP (Fig. 2). The increase in the number of
associations as the cyclic AMP concentration was varied from
1 µM to 1 mM (the ATP concentration was maintained at 0.01
mM) was quite similar to the cyclic AMP concentration-
dependent phosphorylation of MAP-2 reported by Sloboda et.
al. (9). If the concentration of ATP was raised to 2 mM, the
coated vesicles were dissociated from the microtubules.
        To better quantify the association of clathrin with
MAP-2, affinity chromatographic techniques were employed.
Protein was solubilized from purified coated vesicles and was
bound to CN-Br activated Sepharose beads (10). The beads
were then mixed with microtubule protein or MAP-2 fractions,
were washed with buffer, and were then prepared for electrophoresis. In most experiments, protein was not eluted from

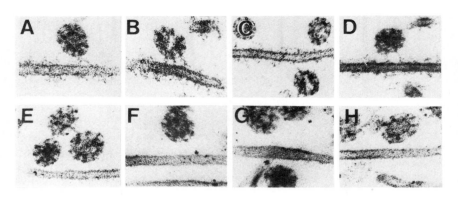

FIGURE 1.   Attachment of coated vesicles to microtubules.
Microtubules were assembled with (A-D) or without (E-H) MAPs
in the presence of coated vesicles and the mixtures were
fixed by perfusion with 1% glutaraldehyde followed by osmium
postfixation.   Thin sections of these mixtures showed that
coated vesicles attached to the MAP-filaments (A-D).   Even
though coated vesicles lay adjacent to MAP-free microtubules,
direct associations between the vesicles and microtubules
were never observed.   X120,000.

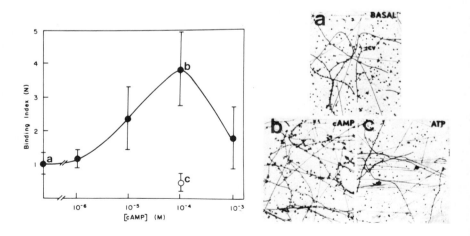

FIGURE 2.   The effect of cAMP on the association of
coated vesicles with MAP-containing microtubules.   Maximum
binding occurred in the presence of 0.1 mM cAMP (b).   The
addition of 2 mM ATP released most of the bound coated vesicles (c).   The binding index (N) is the relative frequency
of bound vesicles per unit length of microtubules (10).

the beads prior to electrophoresis due to the small quantities of protein that were used. As a control for non-
specific associations of protein with clathrin or other
coated vesicle proteins, bovine serum albumin was bound to
another set of beads and these beads were mixed with microtubule protein or MAP-2 in parallel with the coated vesicle
protein beads. Preliminary results from these experiments
are shown in Figure 3. Tubulin readily bound to both the
coated vesicle beads and to the BSA-coated beads. Thus,
although MAP-2 binds to the coated vesicle proteins immobilized on the beads, the affinity of the coated vesicle proteins for tubulin makes it impossible to mix the beads with
cell homogenates or with relatively crude microtubule fractions in order to determine MAP binding. It is important to
note that although immobilized coated vesicle protein could
bind tubulin, coated vesicles did not bind to MAP-free microtubules.

The results presented below show that clathrin coated
vesicles also bind to the region of the MAP-2 molecule that
projects from MAP-actin bundles, and that this region is
likely to be that portion of the MAP-2 molecule that projects
from the microtubule surface.

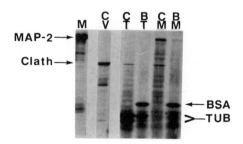

FIGURE 3. The attachment of microtubule proteins to coated
vesicle proteins. Coated vesicle proteins (CV) were solubilized from purified coated vesicles and were coupled to CNBr-
activated Sepharose 4B (10). Bovine serum albumin was bound
to a second aliquot of Sepharose. Immobilized coated vesicle
proteins and albumin were mixed with tubulin or MAP-2
enriched preparations (M), and the protein-coated beads were
washed with 50 mM PIPES, 1 mM EGTA, 0.5 mM $MgSO_4$, and 1 mM
DTT (pH 6.5). Electrophoretic analysis of the proteins bound
to the beads revealed that tubulin bound to both coated
vesicle protein (CT) and BSA (BT) whereas MAP-2 bound to
coated vesicle protein (CM) but not to BSA (BM). BSA preferentially bound the small amount of contaminating tubulin in
the MAP preparation (BM).

III. THE FORMATION OF MAP-ACTIN BUNDLES

MAP-2 also binds to actin filaments in vitro. Griffith and Pollard(11) initially reported that MAPs could cross-link actin filaments and assayed this cross-linking by low shear viscometry. The cross-linking was reversed by ATP, ITP, pyrophosphate, and, to a lesser extent, by other nucleotides. Thin sections revealed that actin filaments appeared to associate with the filamentous MAPs that projected from the microtubule surface.

Our initial studies of MAP-2 were designed to identify potential actin-binding sites on the molecule. When purified skeletal muscle actin was polymerized in the presence of MAP-2, the actin filaments were organized into discrete bundles (12) of 5 - 7 filaments and had an average diameter of 26 nm (Fig. 4). The bundles were composed of 20% MAP-2 and 80% actin, as determined by sedimentation of the MAP-actin bundles and analysis of the proteins by SDS-poly-acrylamide gel electrophoresis (Fig. 5). This corresponded to a molar ratio of approximately 1 mole MAP-2 : 28 moles of actin.

MAP-2 wrapped about the actin bundles and formed discrete striations spaced at 28 nm intervals along the bundles (Fig. 4). Heavy metal shadowing revealed that a portion of the MAP-2 molecule extended for a distance of 12 nm from the MAP-actin bundle (Fig. 4). These appeared similar to the MAP-2 projections attached to the microtubule surface. To determine the relationship between these 12 nm projections and those associated with microtubules, the MAP-actin bundles were mixed with coated vesicles. Coated vesicles readily attached to the 12 nm projections (Fig. 4). Assuming that a specific region of MAP-2 binds to coated vesicles, these results suggest the end of the MAP-2 molecule that lies farthest from the microtubule wall is that region that pro-jects from the MAP-actin bundle. This region may bind coated vesicles and, possibly, other organelles, but does not appear to bind to actin.

ATP both inhibited the formation of MAP-actin bundles and dissociated intact bundles. When intact MAP-actin bundles were incubated in the presence of 1 mM ATP, the bundles began to unravel and individual actin filaments were observed to be studded with periodic arrays of MAP-2 molecules (Fig. 4G). Similar results were observed if AMP-PCP or pyrophosphate were substituted for ATP, but no inhibition of bundle forma-tion occurred if AMP, GTP, or 2 mM EDTA were added to the MAP-actin mixtures. These results, together with the fact that MAP-2 cross-links actin filaments, suggests that there are at least two actin-binding sites on MAP-2 and that one of

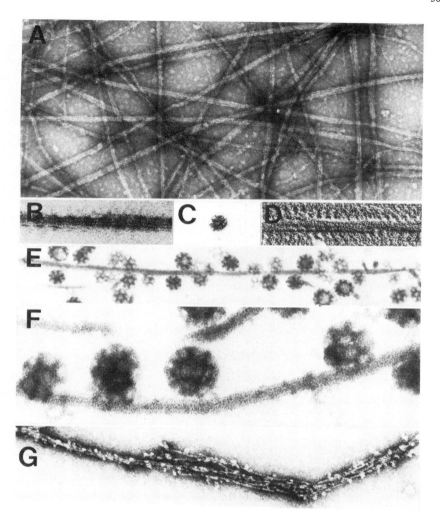

FIGURE 4. Morphology of MAP-actin bundles. Negatively
stained bundles (A,B) revealed the tight packing of actin
filaments into uniform bundles and the periodic MAP striations on the bundle surface (B). Hexagonal packing of actin
filaments is seen in a cross-section (C). Periodic 12 nm
MAP-2 projections from the bundles are revealed by metal
shadowing (D). Coated vesicles appear to bind to these
projections, which are similar to the coated vesicle-binding
sites attached to microtubules (E, F). MAP-2 attachments to
individual actin filaments are revealed when the bundles are
dissociated with ATP (G). A,x100,000; B,C,D,F,x200,000;
E,x60,000; G,x130,000.

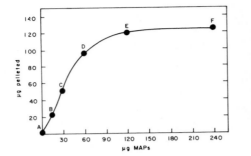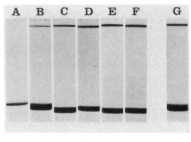

FIGURE 5. Sedimentation and electrophoretic analysis of MAP-actin bundles. Actin (140 µg) was polymerized in the presence of 0 - 240 µ of a heat-stable MAP-2 fraction in a final volume of 200 µl PEMD (see 12). The mixtures were centrifuged for 15 min at 20,000 x g to sediment MAP-actin bundles but not individual actin filaments. Pelleted protein was analyzed by SDS-polyacrylamide gel electrophoresis. Gel A shows the sedimentation of pure actin and Gels B-F contain aliquots of the MAP-actin mixtures shown on the graph. Gel G shows the sedimentation of MAP-actin bundles through a cushion of 10% sucrose in PEMD buffer (12).

these is sensitive to ATP. These results also indicate that MAP-2 contains a coated vesicle binding site.

Since MAP-2 is known to be a microtubule associated protein, it was of interest to determine if tubulin binds to any of the above mentioned sites. When purified tubulin was mixed with MAP-actin bundles, the bundles rapidly dissociated. If the tubulin concentration was lower than that necessary for microtubule assembly (0.1 mg/ml or less) nearly all of the bundles dissociated or were partially splayed. If the tubulin concentration was raised to 0.5 mg/ml, the bundles completely dissociated and microtubules assembled (Fig. 6). Although actin filaments were frequently observed to be wrapped around the microtubules, it is not certain that the actin filaments were attached to microtubules by the MAPs. These results show that at least one of the actin-binding sites on MAP-2 binds to tubulin and that this site(s) has a higher affinity for tubulin than for actin. In addition, it would appear that the portion of the MAP-2 molecule that is distal to its microtubule binding site contains at least one other actin binding segment. This latter segment is likely to be distinct from the coated vesicle binding site of MAP-2.

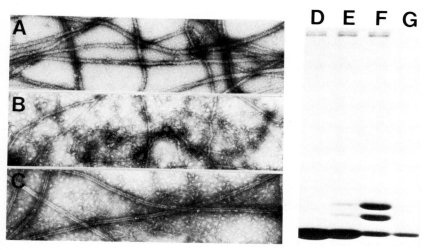

FIGURE 6. Competition for MAP 2 by actin and tubulin. MAP-actin bundles were formed by 0.8 mg/ml actin and 0.2 mg/ml MAP-2 and were incubated with 0, 0.1 mg/ml, or 0.5 mg/ml purified tubulin prior to examination by electron microscopy or by electrophoresis. MAP-actin bundles (A,D) were dissociated by 0.1 mg/ml tubulin, although a mixture of partially dissociated bundles remained (B) that trapped individual actin filaments upon sedimentation (E). Tubulin apparently bound to the MAPs as they were dissociating from the actin filaments, as judged from the tubulin sedimented in E, even though no microtubules were formed. Higher concentrations of tubulin (C,F) resulted in the formation of microtubules which were wrapped with actin filaments (C). Gel G shows the amount of pure actin that sedimented under these conditions. A,B,C, x 50,000.

III.   IDENTIFICATION OF MAP-2 FRAGMENTS
THAT BIND ACTIN OR TUBULIN

In order to identify microtubule-binding domains on MAP-2, purified bovine brain microtubules were digested with α-chymotrypsin for 1-5 minutes. The digestions were terminated by the additon of 2 mM phenylmethylsulfonylfluoride (PMSF), and the microtubules were sedimented through 40% sucrose containing PMSF. The pellets and supernatants were analyzed by SDS-polyacrylamide gel electrophoresis (Fig. 7A). Chymotrypsin cleaved MAP-2 into at least 4 major fragments of Mr 250,000, 185,000, 145,000 and 35,000 and at least one minor fragment of Mr 72,000. The Mr 35,000 fragment and a

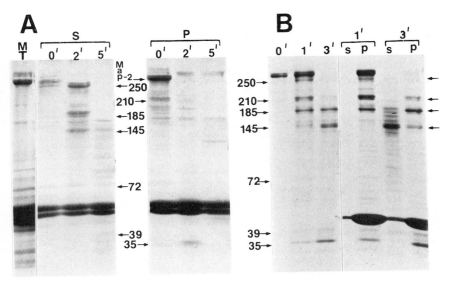

FIGURE 7. Determination of MAP-2 fragments that bind to tubulin and actin. A. Microtubules were assembled with MAP-2 and were digested for 2 and 5 minutes with α-chymotrypsin (see text). Microtubules were pelleted and the pellets (P) and supernants (S) were analyzed by electrophoresis to identify microtubule-binding domains. B. Purified MAP-2 (0) was digested with α-chymotrypsin for 1 and 3 minutes and was then copolymerized with actin. Supernatants (S) and pellets (P) were analyzed to determine actin-binding sites on MAP-2. See text for description of the fragments.

FIGURE 8. Identification of domains on MAP-2 that bind actin (A) and tubulin (T). This model is based on that of Vallee (14) and shows the microtubule-binding site as a Mr 35,000 site to the left of the diagram. The remainder of the molecule is the projection portion that forms the MAP-2 filaments (4,5,6). Actin binds to both the Mr 35,000 and Mr 39,000 fragments while tubulin only binds to the Mr 35,000 assembly-promoting fragment. A small amount of actin may bind to the Mr 72,000 fragment, as shown in Figure 7. The Mr 72,000 fragment is the portion of MAP-2 that projects from the MAP-actin bundle and is proposed to be that which binds to coated vesicles in vitro.

210,000 component sedimented with the microtubules, while the other fragments remained in the supernatant. The 210,000 fragment could be produced by brief enzyme digestions (1 min) and clearly pelleted with microtubules (not shown). These results are consistent with those reported by Vallee (13, 14). The Mr 35,000 fragment is the microtubule assembly-promoting region, which binds to the microtubule wall, while the high molecular weight fragments that do not bind to the microtubules are portions of the filamentous projections distal to the microtubule wall (13, 14). This region of the MAP-2 molecule is called the projection fragment (13).

To identify actin-binding domains of MAP-2, purified MAP-2 fractions were subjected to limited proteolysis by α-chymotrypsin (15). The major fragments produced included those of Mr 250,000, 210,000, 185,000, 145,000, 35,000, and smaller amounts of Mr 72,000 and 39,000 polypeptides were also produced (Fig. 7B). When actin monomers were polymerized in the presence of these fragments and the filaments sedimented and analyzed by SDS-polyacrylamide gel electrophoresis, it was found that the Mr 250,000, 210,000, 185,000, 39,000, and 35,000 fragments co-sedimented with the actin filaments but that very little of the Mr 145,000 fragment bound to the actin. Although a small quantity of the 72,000 fragment did sediment with actin, we could not determine if it was a true actin binding fragment since it was produced in limited quantities.

The results of these fragmentation experiments reveal that there are specific regions of MAP-2 that can bind to tubulin or actin and that other regions bind to neither of these proteins. The Mr 35,000 microtubule assembly-promoting fragment can bind to either tubulin or actin but clearly has a higher affinity for tubulin than for actin. The dissociation of MAP-actin bundles by tubulin probably occurs as a result of the displacement of actin by tubulin at this site. The Mr 145,000 fragment, which is a portion of the projection fragment, has no apparent affinity for tubulin and little for actin. Based on these results we have modified the model presented by Vallee (14) to include the actin-binding sites revealed in our fragmentation experiments (Fig. 8). We propose that the Mr 39,000 actin-binding site is the ATP-sensitive site that is released from actin upon addition of ATP with the accompanying dissocaition of the MAP-actin bundles. The Mr 72,000 fragment is likely to be the site that projects from the MAP-actin bundles (Fig. 4) and is that fragment that binds coated vesicles to both microtubules and to MAP-actin bundles. This may be the site on MAP-2 that mediates the attachment of secretory granules or other organelles with microtubules.

IV. CONCLUSION

These results show that MAP-2 can mediate the association of coated vesicles, secretory granules, and actin filaments with microtubules and that MAP-2 can induce the formation of discrete actin filament bundles in vitro. In addition, studies of MAP-2 fragments produced by limited digestion revealed sites on MAP-2 that bound tubulin or actin or neither of these proteins. It is interesting that the sites to which actin and tubulin bind are also those that are phosphorylated in vitro (14). Moreover, Vallee (14) has shown that a Mr 72,000 fragment is also phosphorylated in vitro and we have tentatively identified a Mr 72,000 fragment as a coated vesicle binding site on MAP-2. Since phosphorylation of MAP-2 appears to be important in the regulation of the association of endocrine pancreatic secretory granules (2) and coated vesicles (10) with microtubules in vitro, these results suggest that specific fragments of MAP-2 may be responsible for regulating the interactions between organelles and microtubules.

Do these properties of MAP-2 reveal a function of MAP-2 in vivo or do they simply reveal that MAP-2 can nonspecifically associate with a variety of different proteins? The identification of specific regions of MAP-2 that can bind to different proteins certainly suggests that specific interactions can occur in vivo. The functions of these interactions are less certain. A probable role for MAP-2 is that of attaching organelles or actin filaments to microtubules in order to position them within the cytoplasm, or to set up a motile machinery by which they are moved along the microtubules. Another role for MAP-2 may be to position microtubules within the cytoskeleton. Matus and Bernhardt (16) have recently shown that MAPs precede the appearance of microtubules in developing dendrites. It is possible that the MAPs are bound to actin filaments or, possibly, clathrin associated with the plasma membrane, and that the microtubules associate with the MAPs as they grow into the dendrite. It is becoming clear that MAPs will be important in the role of cytoplasmic microtubules both with regard to their assembly and their ability to interact with organelles or cytoskeletal filament systems. Further analysis of the domains on MAP-2 that regulate these interactions will aid in determining the nature of the interactions as well as the mechanisms responsible for determining their specificity.

REFERENCES

1. Dentler, W. L., Pratt, M. M., and Stephens, R. E., J. Cell Biol. 84:831 (1980).
2. Suprenant, K. A., and Dentler, W. L., J. Cell Biol. in press.
3. Dentler, W. L., Biol. Bull. in press.
4. Dentler, W. L., Granett, S., and Rosenbaum, J. L., J. Cell Biol. 65:237 (1975).
5. Murphy, D. B. and Borisy, G. G., Proc. Natl. Acad. Sci. USA. 72:2696 (1975).
6. Kim, H., Binder, L. I., and Rosenbaum, J. L., J. Cell Biol. 80:266 (1979).
7. Sloboda, R. D. and Dickerson, K., J. Cell Biol. 87:170 (1980).
8. Sherline, P., Lee, Y. C. and Jacobs, L. S., J. Cell Biol. 72:380 (1977).
9. Sloboda, R. D., Rudolph, S. A., Rosenbaum, J. L., and Greengard P., Proc. Natl. Acad. Sci. USA 72:177 (1975).
10. Sattilaro, R. F. Dentler, W. L., and LeCluyse, E. L., Ms. in preparation.
11. Griffith, L. and Pollard, T. D., J. Cell Biol 78:958 (1978).
12. Sattilaro, R. F., Dentler, W. L., and LeCluyse, E. L., J. Cell Biol. 90:467 (1981).
13. Vallee, R. B. and Borisy, G. G., J. Biol. Chem. 254:2834 (1978).
14. Vallee, R. B., Proc. Natl. Acad. Sci. USA 77:3206 (1980).
15. Sattilaro, R. F., Dentler, W. L., and LeCluyse, E. L., Ms. in preparation.
16. Matus, A. and Bernhardt, R., J. Cell Biol. 91:323a (1981).

CHAPTER 28

ACTIN-MICROTUBULE INTERACTIONS

Thomas D. Pollard[1]
S. Charles Selden

Department of Cell Biology and Anatomy
Johns Hopkins University School of Medicine
Baltimore, Maryland

Linda M. Griffith

Department of Structural Biology
Sherman Fairchild Center
Stanford University School of Medicine
Stanford, California

We will critically review the original evidence that
actin filaments can interact with microtubules (1) and present
our new evidence that the actin cross-linking activity of the
microtubule-associated proteins (MAPs) is (a) contained in
both the tau and MAP-2 fractions and (b) is modified by the
extent of MAP phosphorylation. The work has been carried out
principally by Dr. Griffith in the period 1977 to 1980 (1,2)
and Dr. Selden since 1980.

I. ORIGINAL EVIDENCE FOR ACTIN-MICROTUBULE INTERACTION

The original test for actin-microtubule interaction (1)
was to mix purified muscle actin monomers with depolymerized
brain microtubule protein, induce polymer formation, and then
measure the viscosity. By high shear Otstwald-type capillary
viscometry, both actin and microtubule protein polymerized in
less than 5 min. The mixture had a viscosity more than 2 times
the viscosity of the sum of the components suggesting
that a complex formed. At low shear rates in a simple falling
ball viscometer (3), the two separate polymer types maintained

[1]Supported by NIH Research Grant GM 26132.

Biological Functions of Microtubules
and Related Structures
311

a constant apparent viscosity after 10 min. In contrast, the
mixture had an apparent viscosity 10 times higher than the
components. In addition, the apparent viscosity of the
mixture continued to increase for at least 2 hours. The apparent viscosity of these mixtures depended on the concentration of both polymers. At suitable concentrations (actually
far below those found in cells) the mixtures formed a gel.
Electron micrographs (Figure 1) of these gels showed random
mixtures of actin filaments and microtubules, with some of the
actin filaments closely associated with microtubules. The
actin filaments were bare and the microtubules had numerous
thread-like lateral projections, suggesting that the bulk of
the high molecular weight MAPs were bound to the tubules.

     Formation of a high viscosity complex required MAPs
because mixtures of actin filaments with pure tubulin polymers
(made in two different ways) did not have a high apparent
viscosity. Furthermore, MAPs separated from tubulin (Figure
2A) are actin gelation factors. At about 100 µg/ml they cause
1 mg/ml actin to form a gel. This ability of MAPs alone to
cross-link actin filaments was unexpected and suggested that
MAPs either have multiple actin-binding sites or that aggregated MAPs crosslink actin filaments.

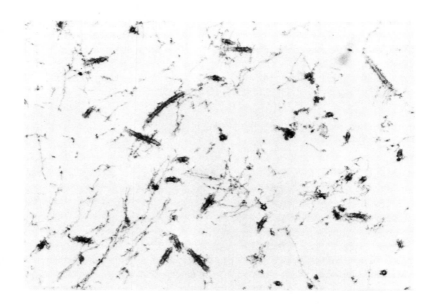

*FIGURE 1. An electron micrograph of a thin section of an
actin filament-microtubule gel.*

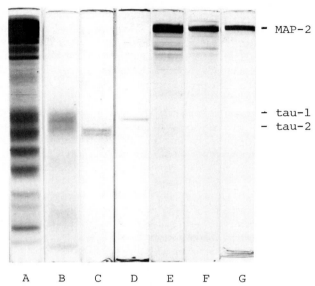

FIGURE 2.   Polyacrylamide gel electrophoresis in SDS of
microtubule associated protein (MAP) fractions. A. Crude
MAPs eluted from phosphocellulose with 0.8 M KCl. B. Crude
tau prepared by DEAE-Sephadex chromatography of crude MAPs.
C. Tau-2 doublet purified from crude tau by hydrocyapatite
chromatography. D. Tau-1 purified from crude tau by hydroxyapatite chromatography. E. Leading half of the high molecular
weight MAP peak (primarily MAP-2) from DEAE-Sephadex chromatography of crude MAPs. F. Trailing half of the high molecular
weight MAPs peak from DEAE-Sephadex. G. MAP-2 purified by the
method of Kim et al. (4). All of these preparations except
(F) have actin gelation activity.

     One striking feature of the actin-microtubule and actin-
MAPs interaction was the ability of nucleotides, especially
ATP and ITP to inhibit the apparent viscosity. This inhibition occurred without substantially changing the viscosity of
either the actin or the microtubules, indicating that the
nucleotides act on cross-links between the polymers.

II. DIFFICULTIES WITH INTERPRETING THE ACTIN-MICROTUBULE
    EXPERIMENTS

     We recognize that there are a number of difficulties with
interpreting these experiments:

1.   The falling ball assay is not quantitative. Because the
polymer samples are not Newtonian fluids, the observed viscosity is highly non-linear. Small differences between samples
can be amplified because the shear rate is lower in the sample
with the higher viscosity. Consequently one cannot simply
add the apparent viscosities of components to estimate the
viscosity of a non-interacting mixture.

2.   The high viscosity of mixtures of the two polymers could
be due to simple entanglement. We think that entanglement is
unlikely to account for the high viscosity of actin-
microtubule mixtures, because actin-pure tubulin polymer
mixtures and actin-microtubule-ATP mixtures have low viscosities (1).

3.   In our initial studies, the apparent viscosity of actin-
microtubule mixtures was quite low at physiological ionic
strength, raising questions about the physiological relevance
of the interaction. We have now found that actin and microtubule protein can form a gel at physiological ionic strength
providing that the anion is glutamate rather than chloride,
which was used in the original studies. Potassium glutamate
inhibits MAPs-actin interaction, but this is compensated for
because microtubules grow longer in the presence of glutamate.
The long tubules form a gel with actin filaments at lower
concentrations than short tubules. On the other hand,
chloride inhibits both MAPs-actin interaction and microtubule
polymerization (2).

4.   Because MAPs are polycations (in the sense that they bind
to phosphocellulose) and many polycations, including polylysine, histones, ribonuclease, lysozyme and aldolase, will
nonspecifically cross-link actin filaments (5), it is possible
that the interaction of MAPs with actin is nonspecific as well.
This type of artifact is hard to rule out, but the different
potencies of various nucleotides in inhibiting actin gelation
by MAPs provided some evidence for specificity (1).

5.   MAPs doe not bind tightly to actin filaments. Using
pelleting assays, we have not detected any binding of MAPs to
actin. Given the concentrations of the proteins tested, we
calculated that the K_d is greater than 10^{-3} M. On the other
hand, Sattilaro *et al*.(6) have detected some binding of heat
stable MAP-2 to actin, although the affinity did not seem to
be high.

III. ISOLATION OF MAPS SUBFRACTIONS CAPABLE OF CROSS-LINKING
     ACTIN FILAMENTS

To deal with these concerns about the specificity of the
presumed interaction of actin filaments with microtubules via
MAPs and to establish the mechanism of the interaction, we
have fractionated MAPs to identify the proteins active in
cross-linking actin filaments (2). By phosphocellulose, DEAE-
cellulose and hydroxyapatite chromatography, we purified two
tau fractions (Figure 2B,C,D) and high molecular weight
fractions (Figure 2E). By heating and gel filtration, we
also purified MAP-2 (Figure 2E). All of these fractions were
capable of cross-linking actin and all had about the same
specific activity as the crude MAPs fraction. The specific
activity was determined by measuring the MAPs concentration
dependence of the actin-MAPs apparent viscosity. At the point
of incipient gelation, there was a very sharp transition from
a low viscosity liquid to a gel (Figure 3). This critical
gelling concentration is a quantitative assay for cross-
linker specific activity.

During the fractionation of the MAPs we found that a
subset of the high molecular weight MAPs (Figure 2F) had
little or no actin crosslinking activity. The inactive MAPs
were in the trailing half of the broad peak of high molecular
weight MAPs which eluted from DEAE-cellulose with a KCl
gradient. Since the polypeptide composition was similar
across the peak (compare Figure 2E and F), we suspected that
some secondary modification of the protein was responsible
for both the wide range (50 to 400 mM) of KCl required to
elute the peak and the variation in activity across the peak.
Based on the recent work of Nishida et al. (7) and our
own experiments described in the next section, variable
phosphorylation of the high molecular weight MAPs is the
most likely explanation. The more anionic (presumably more
phosphorylated) fraction elutes later and has low actin
crosslinking activity.

In addition to testing the various purified MAP fractions
for actin gelation activity, we also examined samples by
electron microscopy. High concentrations of whole MAPs,
purified MAP-2 and tau all caused actin filaments to form
bundles. Independently, Sattilaro et al. (6) have made
the same observation with purified MAP-2. The bundles with
MAP-2 had lateral projections while those with tau did not.

IV. PHOSPHORYLATION OF MAPS REGULATES INTERACTION WITH
    ACTIN FILAMENTS

Recently, Nishida *et al.* (7) reported experiments
suggesting that phosphorylation of MAPs might regulate their
interaction with ectin. They preincubated microtubule
protein in $CaCl_2$ with or without ATP and cAMP prior to
boiling in high salt and mercaptoethanol to obtain a soluble
heat-stable fraction enriched in MAPs. They found that the
MAPs preincubated with ATP had less actin gelation activity
than the MAPs preincubated without ATP and concluded that
this was due to a change in the extent of phosphorylation.

We repeated and confirmed their observations (Figure 3).
Preincubation with ATP gives MAPs with a considerably higher
critical gelling concentration than preincubation without ATP.
Because this treatment could have changed MAPs in ways other
than the extent of phosphorylation, we measured the MAPs
phosphate content after various pretreatments. We found
a correlation of actin crosslinking activity with extent
of phosphorylation (Figure 4). Under the conditions
used by Nishida *et al.* (7), ($CaCl_2$ $+$ ATP and cAMP), we
found a difference of only about 1.2 phosphates per MAP-2
and a small difference in critical gelling concentration
(Figure 4). However, using EGTA rather than $CaCl_2$ we were
able to obtain greater extremes in the phosphorylation:
down to 3.3 phosphates per MAP-2 without ATP and up to
7.7 with ATP and cAMP. The largely dephosphorylated MAPs
have the lowest critical gelling concentration (highest
specific activity) while the critical gelling concentration
of MAPs with 7.7 phosphates is greater than the highest MAP
concentration we tested. Perhaps the most interesting aspect
of the analysis is the striking difference between smples
preincubated with cAMP and ATP with or without Ca^{++}.
Although these samples differ by only 1.5 phosphates per
MAP, the samples incubated in EGTA (7.7 phosphates per MAP)
had a critical gelling concentration more than 4 times higher
than the samples incubated with Ca^{++} (6.3 phosphates per MAP).
This result suggests that there may be one or two specific
phosphorylation sites which strongly influence the ability
of MAPs to interact with actin and that a specific Ca^{++}
sensitive kinase or phosphatase may be involved.

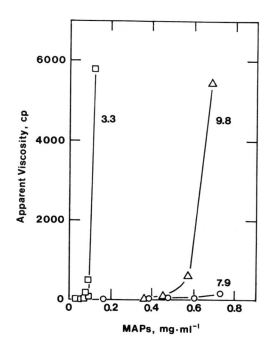

FIGURE 3.   Dependence of the low shear apparent viscosity
of actin filaments on the concentration of MAPs prepared in
different ways. Samples of twice cycled pig brain microtubule proteins (C_2S) at 12 mg/ml were preincubated in
24 mM PIPES, 17 mM MES, 5mM $MgCl_2$, 90 mM KCl, 0.15 mM GTP,
1.1 mM $CaCl_2$, pH 6.8 for 30 minutes at 37°C with or without
1 mM ATP and 0.05 mM cAMP, with $CaCl_2$ to 5.9 mM final or
EGTA to 2 mM. EGTA alone , □ ; EGTA plus ATP, △ ; EGTA plus
ATP plus cAMP, ○ . Then they were made 0.8 M in NaCl and
70 mM 2-mercaptoethanol and boiled for 5 minutes. Precipitated proteins were removed by centrifugation and the supernatants consisting primarily of MAP-2 were tested for actin
gelation activity. The test samples consisted of 0.3 mg/ml
muscle actin and variable concentrations of MAPs in
100 mM KCl, 2.5 mM EGTA, 1 mM $MgCl_2$, 0.6 mM MES, 2 mM HEPES,
0.4 mM TRIS-HCl, 0.1 mM DTT, 0.06 mM ATP, 0.04 mM $CaCl_2$,
pH 7.5. The viscosity was measured after 10 minutes
incubation at 22°C. The protein bound phosphate content of
each fraction was measured by the method of Stull and Buss
(8) and expressed in the figure as moles of phosphate
per 300,000 g protein.

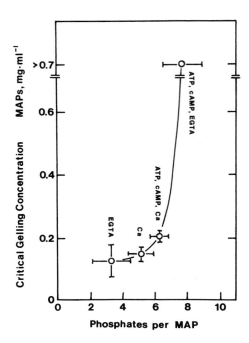

FIGURE 4. *Dependence of the critical gelling concentration on MAP phosphate content. The critical gelling concentration was measured as described in Fig. 3 as the MAPs concentration necessary to give a viscosity of 1000 cp. The microtubule protein samples were preincubated in 2 mM EGTA or 5.9 mM CaCl$_2$ with or without 1 mM ATP and 0.05 mM cAMP before boiling and clarification. The error bars are one standard deviation of three separate experiments.*

Together with the work reported by Nishida et al. (7) at this Symposium, our results on MAP phosphorylation provide a new line of evidence that the MAPs-actin interaction is meaningful. The work also suggests that actin-microtubule interaction might be regulated by specific kinases and/or phosphatases acting on the MAPs.

V. CONCLUSIONS

1. Both tau and high molecular weight MAPs have a high affinity for microtubules and a low affinity for actin.

2. Actin filaments can bind to microtubules via one or more of the different MAPs. Given the low affinity of MAPs for actin, this occurs largely because of the high local

concentration of the MAPs projecting from the surface of the microtubules.   In other words, this is probably a high valency, low affinity association which might be easily made or broken in the cell.

3.   The association of MAPs with actin is probably regulated, at least in part, by phosphorylation of specific sites on the MAPs.   Phosphates at these sites strongly inhibit interaction with actin filaments.

4.   It is impossible to know whether the interaction between actin and microtubules observed in the test tube is used by the cell, but we have established that it is possible at physiological polymer and salt concentrations.   If actin filaments and microtubules do interact in cells they may simply play a structural role.   Furthermore, the ability of agents which depolymerize either actin filaments (9,10) or microtubules (11,12) to inhibit fast axonal transport, provides some evidence that the two polymers may cooperate in some forms of intracellular particle movement.

REFERENCES

1.   Griffith, L. M., and Pollard, T. D., *J. Cell Biol. 78,* 958 (1978).
2.   Griffith, L. M., and Pollard, T. D., *J. Biol. Chem.* (1982b) (submitted for publication).
3.   MacLean-Fletcher, S., and Pollard, T. D., *J. Cell Biol. 85,* 414 (1980).
4.   Kim, H., Binder, L. I., and Rosenbaum, J. L., *J. Cell Biol. 80,* 266 (1979).
5.   Griffith, L. M., and Pollard, T. D., *J. Biol. Chem.* (1982a) (submitted for publication).
6.   Sattilaro, R. F., Dentler, W. L., and LeCluyse, E. L., *J. Cell Biol. 90,* 467 (1981).
7.   Nishida, E., Kuwaki, T., and Sakai, H., *J. Biochem. 90,* 575 (1981).
8.   Stull, J. T., and Buss, J. E., *J. Biol. Chem. 252,* 851 (1977).
9.   Isenberg, G., Schubert, R., and Kreutzberg, G. W., *Brain Res. 194,* 588 (1980).
10. Schwartz, J. H., *Sci. Amer. 242,* 152 (1980).
11. Banks, P., and Till, R., *J. Physiol. (London) 252,* 283 (1975).
12. Dahlström, A., Heiwall, P. O., and Larsson, P. A., *J. Neurol. Transm. 37,* 305 (1975).

CHAPTER 29

ON THE COEXISTENCE OF GLIAL FIBRILLARY ACIDIC (GFA) PROTEIN AND VIMENTIN IN ASTROGLIAL FILAMENTS

Kazuhito Yokoyama
Hiroshi Mori
Masanori Kurokawa

Department of Biochemistry
Institute of Brain Research
Tokyo University Faculty of Medicine
Hongo, Bunkyoku
Tokyo, Japan

The astroglial 10 nm filaments isolated from the degenerated mouse optic nerve are composed of two major polypeptides with molecular weights of 56,000 and 45,000 (1). Considering that glial fibrillary acidic (GFA) protein has so far been thought to comprise a single protein species (2, 3), and also that astroglial filaments isolated from the human gliosed brain were mainly composed of a single species of polypeptide of 49,000 molecular weight (4), the simultaneous occurrence of two polypeptides in our purified filaments requires explanation.

An attempt to characterise these two polypeptides in the present study indicates that the larger polypeptide is homologous to vimentin, the protein subunit of a type of intermediate filaments in many cells, especially in mesenchymal cells. The molecular weight of the smaller polypeptide varies with species, being 51,000 in rats and 45,000 in mice. Despite the interspecies variation in the molecular weight and also in one-dimensional peptide maps, the polypeptide cross-reacts with anti-GFA protein antiserum raised against human spinal antigen. In the spinal cord cytoskeleton in newborn animals, vimentin prevails, while the level of GFA protein is still very low. Vimentin rapidly declines during the subsequent maturational stage, and becomes almost undetectable in the adult spinal cord, where GFA protein is virtually the only component constituting non-neuronal intermediate filaments. In contrast to the spinal cord, vimentin coexists with GFA protein in astrocytes from the adult optic nerve.

Biological Functions of Microtubules
and Related Structures

321

CYTOSKELETAL PROTEINS FROM THE SPINAL CORD AND OPTIC NERVE —
INTERSPECIES VARIATION

    The cytoskeletons were prepared from the spinal cord and op-
tic nerve of male Wistar rats and male *dd* mice as described
(5). On electron microscopy, cytoskeletal preparations from
the spinal cord and optic nerve of the adult animals consist
mainly of 10 nm filaments of smooth contour, either loosely re-
ticulated or tightly packed (Fig. 1, B and C). Electrophoretic banding patterns of cytoskeletons from the adult rat spinal cord are shown in Fig. 1A and Fig. 2.

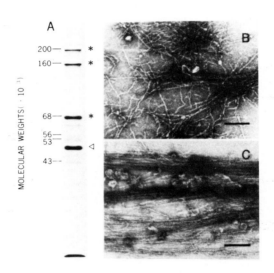

*FIGURE 1. Cytoskeletal preparations isolated from the spinal
cord of the adult Wistar rat. [A] Electrophoretic banding pattern; (*) neurofilament polypeptide triplet; (◁) P51. [B and
C] Negatively stained electron micrographs. Loosely reticu-
lated [B] and tightly packed [C] filaments are shown.   Bar,
0.2 µm. Reproduced from reference 5.*

    The optic nerve cytoskeleton contains, in addition to the
neurofilament triplet, two major polypeptides, i.e. P56 and P51
in rats, and P56 and P45 in mice (Fig. 3A). P56 and P45 in

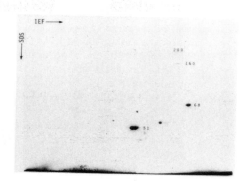

FIGURE 2. *Two-dimensional electrophoretic banding pattern of cytoskeletons isolated from the adult rat spinal cord*

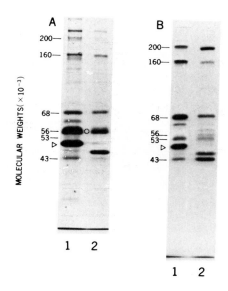

FIGURE 3. *Comparison of cytoskeletal proteins between the optic nerve and spinal cord, and between rats and mice: [A] optic nerve; [B] spinal cord; (1) rats; (2) mice. In both species P56 (○) is seen in the optic nerve, but not in the spinal cord. In both the optic nerve and spinal cord, P51 (▷) is present in rats; this is replaced by P45 (◄) in mice. Reproduced from reference 5.*

the mouse optic nerve were previously shown to be copurified
with the astroglial 10 nm filaments (1). In contrast with the
optic cytoskeletons, those prepared from the spinal cord of the
adult animal lack P56; P51 and P45 thus represent the major
non-neuronal cytoskeletal protein in the spinal cord of the
adult rat and mouse, respectively.

PRESENCE OF VIMENTIN IN THE OPTIC NERVE CYTOSKELETON

   Because of the absence of P56 in the spinal cord cytoskeleton, we have explored the possibility that this polypeptide
represents the subunit of intermediate filaments other than
neurofilaments and astroglial filaments. One dimensional pep-

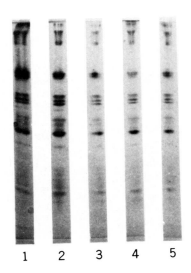

FIGURE 4. *One-dimensional peptide maps of the spinal and optic P56, and of the fibroblast intermediate filament protein
(vimentin): (1) P56 from the newborn rat spinal cord; (2) P56
from the adult rat optic nerve; (3) P56 from the adult mouse
optic nerve; (4) vimentin from Don-6 cells; and (5) vimentin
from L·P3 cells. Each lane contained approximately 20 μg of
the indicated protein, which was subjected to a limited proteolysis with 0.5 ng of* Staphylococcus aureus *V8 protease. Reproduced from reference 5.*

tide maps of the optic nerve P56 from rats and mice are very
similar to one another (Fig. 4, lanes 2 and 3). The homology
in peptide maps was ascertained also between the rodent P56
and the intermediate filament protein from two different lines
of fibroblasts (Fig. 4, lanes 2–5), indicating that P56 is
vimentin.

IDENTIFICATION OF GFA PROTEIN IN CYTOSKELETONS ISOLATED FROM THE OPTIC NERVE AND SPINAL CORD

Attempts have been made to characterise the smaller component of astroglial filament protein that shows the interspe-

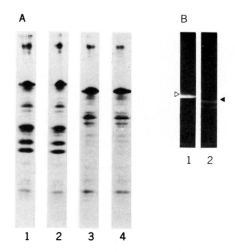

*FIGURE 5. Species difference in GFA protein between rats and
mice. [A] One-dimensional peptide maps of rat P51; (1) spinal;
(2) optic: and of mouse P45; (3) spinal; (4) optic. [B] Identification of the rat P51 and mouse P45 as GFA protein. The
spinal cytoskeletal preparation from the rat (1) and mouse (2)
was electrophoresed in a small gel, and the gel was treated
with rabbit anti-GFA protein antiserum raised against the human antigen, and further with fluorescein isothiocyanate-con-
jugated goat IgG directed against rabbit IgG. Electrophoretic mobility of P51 in rats (▷) and P45 in mice (◄) was determined by comparison with the equivalent lanes electrophoresed
in the same gel and stained with Coomassie brilliant blue.
Reproduced from reference 5.*

cies variation in its molecular weight (Fig. 3). One-dimensional peptide mapping indicates the homology between the optic nerve polypeptide and the spinal cord polypeptide in each species, i.e. P51 in rats (Fig. 5, lanes 1 and 2), and P45 in mice (Fig. 5, lanes 3 and 4). However, the peptide maps exhibit many dissimilarities between the rat and mouse.

In the next step, immunoreactivity of P51 and P45 with anti-GFA protein antiserum was examined. Significant fluorescence was produced by P51 in rats and P45 in mice (Fig. 5B), indicating that these are the GFA protein despite their interspecies variation in the molecular weight and in one-dimensional peptide maps.

POSTNATAL CHANGES IN THE COMPOSITION OF CYTOSKELETAL PROTEINS

P56 is present in the optic nerve cytoskeleton of the adult animal, but is almost undetectable in the spinal cord cytoskeleton from the adult animal (Fig. 3). However, P56 does predominate in the spinal cord cytoskeleton isolated from 2 day-old rats, where the level of GFA protein is still very low (Fig. 6, lane 1). The homology between the spinal P56 and optic P56 is established by one-dimensional peptide mapping (Fig. 4, lanes 1 and 2). The spinal P56 rapidly declines during the subsequent maturational stage, and becomes almost undetectable in the adult spinal cord, while the adult level of GFA protein seems to be attained within 2 weeks at latest (Fig. 6).

COMMENTS

Our biochemical studies, in accord with the recent immunocytochemical studies (6), clearly demonstrate an earlier expression of vimentin and a later expression of GFA protein in astrocytes, which indicates that the relative dominance of vimentin versus GFA protein is important in investigating biochemical events occurring in the course of differentiation and maturation of astrocytes. In the spinal cord, vimentin declines with age. However, in some astrocytes such as those in the optic nerve, vimentin does not undergo such a maturational decline, indicating that the relative dominance of vimentin versus GFA protein is also important to distinguish various subclasses of astrocytes at a given stage of maturation. Whether the vimentin filaments and GFA protein filaments are independently formed and then coexist, or whether vimentin and GFA protein can copolymerise to form 10 nm filaments in astrocytes, remains to be elucidated.

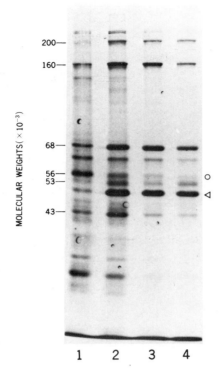

FIGURE 6. Maturational changes in the composition of cytoskeletal proteins in the rat spinal cord: (o), vimentin; (◁), GFA
protein (P51); 1, 2, 3 and 4 correspond to 2 day-, 2 week-, 4
week- and 10 week-old rats, respectively. Reproduced from
reference 5.

Despite a common antigenic cross-reactivity, the GFA protein in rats and mice differs to each other in the molecular
weight and in the one-dimensional peptide maps. In guinea
pigs, the molecular weight of GFA protein was estimated to be
around 52,000 (data not shown). Such an interspecies variation
may explain, at least in part, the disagreement in the apparent molecular weight of the GFA protein estimated by different
workers (2,3). Molecular basis for this interspecies variation would be an important issue to be solved.

ACKNOWLEDGEMENTS

We thank Drs H. Ishikawa, S. Tsukita, N. Inoue and M. Schachner for their valuable discussions and suggestions. Generous gifts of anti-GFA protein antiserum from Drs A. Bignami and D. Dahl, and of fibroblasts from Drs T. Marunouchi and Y. Yagisawa are gratefully acknowledged.

REFERENCES

1. Tsukita S., Ishikawa H. and Kurokawa M. (1981) *J. Cell Biol.* *88,* 245-250

2. Eng L. (1980) *In* "Proteins of the Nervous System" (R.A. Bradshaw and D.M. Schneider, ed.) pp. 85-117, Raven Press, New York

3. Bignami A. and Dahl D. (1980) *Adv. Cell. Neurobiol. 1,* 285-310

4. Goldman J.E., Schaumburg H.H. and Norton W.T. (1978) *J. Cell Biol. 78,* 426-440

5. Yokoyama K., Mori H. and Kurokawa M. (1981) *FEBS Letters 135,* 25-30

6. Schnitzer J., Franke W.W. and Schachner M. (1981) *J. Cell Biol. 90,* 435-447

CHAPTER 30

THE MICROTUBULE-NEUROFILAMENT NETWORK:
THE BALANCE BETWEEN PLASTICITY AND STABILITY
IN THE NERVOUS SYSTEM[1]

R. J. Lasek
J. R. Morris[2]

Neurobiology Center and
Department of Anatomy
Case Western Reserve University
Cleveland, Ohio

I. INTRODUCTION

Microtubules (MT) are important structures in the development and maintenance of neuronal form. In neurons, as in most
other cell types, the MT are structurally associated with
intermediate filaments to form a network (1,2). In neurons,
this network can be called the microtubule-neurofilament network (MT-NF network). These structures are important during
axonal development and regeneration, when they act to consolidate the progress which is made by the growth cone (3,4). The
extension of a new axon or axonal branch involves two processes. The first is the extension of microspikes and the formation of the growth cone (4). The microspikes contain actin
microfilaments but no MT or NF. The second step is the elongation of the axon behind the growth cone. Elongation involves
the MT and NF which form the backbone of the axon proper. The
extension of the MT-NF network into a new axonal branch during
development apparently stabilizes that structure and reduces
the possibility that it will be retracted.
The MT-NF network provides the cytoskeletal backbone for
neuronal processes. If the MT are destabilized by colchicine,

[1]*Supported by NIH-Ag 00795*
[2]*Present address: Cancer Research Center, Massachusetts
Institute of Technology, Cambridge, Massachusetts.*

this can lead to both an increase in the activity of filopodia along the lateral surface of the neurite and ultimately to the retraction of the neurite (3,5). These pharmacological experiments illustrate the importance of the dynamics of MT in stabilizing neuronal shape. The dynamic properties of MT or any other polymer are determined by the equilibrium between monomer and polymer. If monomer can be added or subtracted from the ends of the polymer then the length of the polymer will be susceptable to changes in the equilibrium conditions for polymerization (6). However, if the off reaction is very slow or blocked, then the polymer will be extremely stable and require some reaction which alters the rate of the off reaction (7). Two kinds of polymer can be recognized: soluble polymer, which depolymerizes when the equilibrium conditions are changed by reducing the monomer concentration, and stable polymer, which does not depolymerize when the monomer concentration is reduced. NF are examples of stable polymers because they remain assembled if the plasma membrane is disrupted or removed and monomer is permitted to diffuse into the surrounding solution (7). By contrast, MT apparently exist both as stable polymers and soluble polymers in cells (6). Soluble MT are probably more dynamic structures than their stable counterparts because the soluble MT are in equilibrium with monomeric tubulin. Theoretically, the soluble polymers will undergo changes in length, if the equilibrium conditions are altered so as to change the rate of monomer addition or subtraction from the ends of the soluble polymers. Studies of MT *in vitro* have demonstrated that a number of factors may contribute to the equilibrium between monomer and polymer (6). However, very little is known about the factors which regulate these equilibria *in vivo*. Measurements of the fraction of tubulin which is present in monomer, soluble polymer and stable polymer in cells would provide basic information for understanding the equilibrium conditions *in vivo*. Studies of this type on neuroblastoma and PC 12 cells in tissue culture indicate that the ratio of polymerized tubulin to monomer increases during neurite extension (8,9).

What fraction of tubulin is present in the form of monomer and polymer in adult neurons which have achieved a stable morphology? Information which is relevant to this question can be obtained by analyzing the squid giant axon, which offers several advantages. The most important of these is that axoplasm can be obtained from the giant axon by extrusion without contamination by materials from other cells. More than 95 % of the axoplasm is removed from the giant axon by extrusion, and the structural integrity of the extruded axoplasm does not appear to be substantially altered (7). When the axoplasm is extruded it apparently yields under pressure at a plane which is a few microns deep to the plasma membrane.

This explains why extruded axoplasm emerges as a cylinder and a thin layer of axoplasm remains associated with the plasma membrane.

Axoplasmic cylinders extruded from giant axons are stable and retain the original shape of the axon for 24 hours or longer after extrusion into a physiological buffer (7). Using extruded axoplasm, we have developed a method for measuring the amount of monomer and polymer in the squid giant axon. The rate of protein elution from axoplasm into an external buffer is measured and compared with the theoretical rate for diffusion of monomer from a cylinder of comparable size (10). Physical theory predicts that free monomer will elute from the axoplasmic cylinder at a rate equivalent to diffusion. However, protein which is present in the form of soluble polymer should elute more slowly because of the delay inherent in the depolymerization reaction. The following paper presents the results of these experiments with an evaluation of the role of MT and NF in the plasticity and stability of neruronal structure.

II. METHODS

Axoplasm was extruded directly into a buffered solution by the method described previously (7). The buffer employed in these experimetns simulates the internal conditions of axoplasm in terms of osmolarity, ionic strength and pH. The buffer contained 165 mM isethionic acid, 132 mM taurine, 74 mM betaine, 16 mM alanine, 18 mM glycine, 100 mM aspartic acid, 6 mM arginine, 4 mM glycerol, 1 mM glucose, 344 mM potassium, 147 mM chloride, 65 mM sodium, 18 mM phosphate,10 mM magnesium, 1.0 mM ATP, 0.5 mM GTP, 1.0 mM EGTA and 0.5 mM PMSF. The elution of proteins from the axoplasmic cylinders was analyzed by extruding the axoplasm into 0.5 ml of buffer in a glass dish. The dish had nine depressions each which contained 0.5 ml of buffer. The axoplasm was then repeatedly transfered from one depression to the next at intervals of 1,2,3,4, 8,16,30,50,80, and 120 min. after extrusion. Each bath contained the proteins which eluted from the axoplasm during the period when the axoplasm was immersed. Thus the first bath contained the proteins eluting between 0 and 1 min., the second between 1 and 2, and so on to the last bath which contained the proteins released between 80 and 120 min. The proteins in each bath were concentrated by TCA precipitation. The samples including the remaining axoplasm were prepared for SDS-PAGE and analyzed on slab gradient (5-16 %) polyacrylamide gel electrophoresis using the buffer method of Laemmli (11). The gels were stained for quantitative analysis with Coomassie

blue and protein concentration in individual bands was measured
by densitometric scanning (at 600 nm wavelength).

III. RESULTS

When axoplasm is extracted with buffer P for 120 min, most
of the actin and tubulin are soluble. However, 27 % of the
total actin and 15 % of the total tubulin remain stably associated with the extracted axoplasmic cylinder which we have
called the axoplasmic ghost (7). Essentially all of the NF
protein remains in the ghost. Electron microscopic investigations demonstrate that the NF remain polymerized under these
conditions and that a fraction of the MT also remain in the
ghost. Actin microfilament were not readily detected with the
electron microscope, but this may be due to their short length.

The soluble proteins exit from the axoplasmic cylinder at
different rates which can be observed by examining the electrophoretic profiles in Fig. 1.

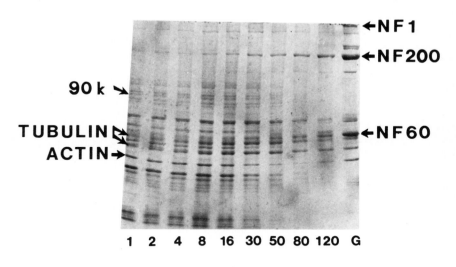

*FIGURE 1. shows a gel used in the analysis of the elution
of proteins from a cylinder of axoplasm. It is clear that
protein elution is not simply related to monomeric molecular
weight because all proteins of similar molecular weight do not
elute at similar rates.*

The protein designated 90K was almost completely extracted from the axoplasm at 16 min and is not detectable in the residual ghost. On the other hand, tubulin is extracted much more slowly from the axoplasm and most of it leaves the cylinder after 8 min. In Fig. 1 some NF proteins were present in the later extraction baths. However, the actual proportion of NF protein which is released from the cylinder is much smaller than that indicated by inspection of Fig. 1 because the ghost proteins are quantitatively underrepresented on the gel in order to avoid overloading. Only a small fraction of the total ghost protein was loaded on the gel. Furthermore, the small amount of NF protein which appeared in the buffer arose from small pieces of axoplasm that broke off of the ghost during the later transfers of the cylinder from one bath to another. We found that as the number of transfers increased the fragility of the cylinder also increased. This experimental artifact had very little effect on the quantitative results, because similar rates of tubulin and actin elution were obtained by another method in which the axoplasm was transferred only once and no neurofilament protein appeared in the bath (Morris and Lasek, unpublished observation).

Quantitative analysis of the rate of elution of tubulin and actin are illustrated in Fig. 2. These rates are compared with the theoretical rate of elution for globular proteins with molecular weights of tubulin or actin (110,000 and 45,000, respectively). Both tubulin and actin diffuse from the cylinder more slowly than diffusion theory predicts suggesting that a significant fraction of the soluble tubulin and actin in the axon are bound in a nondiffusable form. If the eluted proteins were freely diffusable in the axoplasm the elution kinetics should fit the model of a single exponential when the data are plotted on semi-log axes (Fig. 3). However, Fig. 3 indicates that the elution of tubulin and actin can be modeled by the sum of two exponential components. That is, there is an early rapid efflux lasting 10-20 min and a more slowly eluting form. This type of curve is typical of two component systems in compartmentation analysis (12).

One likely explanation for the presence of two distinct rates of efflux is that the faster component represents monomeric protein which is diffusable and that the slower component is protein bound as soluble polymer which depolymerizes when the monomer concentration falls. If this is the case, then the rate of the faster component should be similar to that predicted for simple diffusion. The faster component was extracted from the data by the method of exponential peeling (12). Fig. 3 illustrates an example of the linear regression lines generated by this method. We found that in each case the faster component was linear (correlation coefficient = 0.999) and that the rate of the faster component agreed with

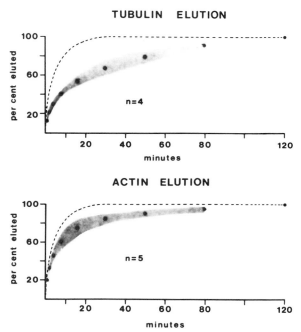

FIGURE 2. shows the average data (filled circles) for the elution of actin or tubulin from axoplasm. The shaded regions represent the 95 % confidence envelopes for the data and "n" indicates the number of axons used. The dashed line represents the elution profile predicted by theory for the diffusion of monomeric tubulin or actin.

the rate predicted by diffusion theory. Thus, the faster component appears to be a measure of the amount of monomer in axoplasm, and it is possible to use these data to estimate the amount of monomer and soluble polymer in the axon. This is done by extrapolating the linear regression for the slower component to zero time where it intersects the ordinate. This point should be equivalent to the amount of soluble polymers which are in the axoplasm at the beginning of the experiment. The results of this analysis are shown in Table I.

If the slower component is related to the amount of soluble polymer in the axoplasm, then it should be possible to block the efflux of the slower eluting proteins with drugs such as taxol or phalloidin which selectively stabilize microtubules or microfilaments respectively (13,14). The following experiment was done to test this prediction. Axoplasm was extruded directly into buffer P containing either 10 µM taxol or 10 µM phalloidin and the amounts of tubulin and actin which

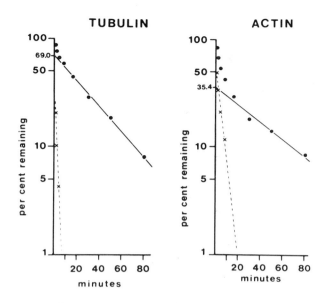

FIGURE 3. illustrates elution data obtained from a single representative sample of axoplasm. A model function consisting of a sum of two exponential peeling. The crosses represent the fraction of protein remaining after the slower exponential term was subtracted. The two lines (solid and dashed) represent the linear regression lines fit to the "peeled" first component or to the later time points respectively.

TABLE I. The Concentration of Various Forms of Tubulin and Actin Determined by the Kinetic Elution Paradigm

	Stable Polymer	Soluble Polymer	Monomer	
Tubulin	15 ± 1.3	58.8 ± 2.9	26.2 ± 2.9	n = 4
	(0.84)	(3.29)	(1.47)	
Actin	27 ± 2.4	26.2 ± 3.5	46.8 ± 3.5	n = 5
	(0.38)	(0.37)	(0.66)	

Data is expressed as per cent ± standard error.
(mg/ml)

eluted from the axoplasm were measured. The effects of taxol
were specific. It reduced the amount of tubulin which eluted
but did not reduce the elution of other proteins except for
two high molecular weight proteins which may be microtubule
associated proteins. Similarly, phalloidin was very specific
for actin. Estimates of the amount of actin monomer obtained
with phalloidin are virtually identical to those obtained by
the kinetic elution method. In the presence of phalloidin,
42.4 % (standard error ± 2.2, n = 2) of the actin eluted from
the axoplasm which is similar to the value obtained for monomeric actin by the kinetic elution method (Table I). However,
the estimates of monomeric tubulin obtained with taxol were
significantly greater than those obtained by the kinetic
elution method. In the presence of taxol 36.5 % (standard
error ± 3.5 %, n = 3) of the tubulin was extracted which is
somewhat larger than the value predicted by the kinetic elution experiments (Table I). The large value for monomer which
was obtained in the taxol experiment may be explained by the
fact the taxol was introduced into the axoplasm through the
buffer. It is possible that some of the microtubules begin to
depolymerize before the taxol completely stabilizes the microtubules. In any case, the results of the taxol and phalloidin
experiments support our interpretation of the kinetic elution
experiments that the faster and slower components of efflux
are a measure of the amount of monomer and soluble polymer
respectively.

IV. DISCUSSION

We have observed two rates of elution for actin and tubulin
from axoplasmic cylinders. The first component of elution is
similar to that of a freely diffusing protein. This result
suggests that the internal networks within the axoplasm do not
significantly impede the mobility of proteins such as tubulin
or actin and that these proteins can be relatively free to
diffuse in the neuron. However, 75 % of the tubulin and 50 %
of the actin are not freely diffusable. We propose that this
tubulin and actin are in the form of polymerized MT or microfilaments respectively. In the case of tubulin, this proposal
is supported by the results with taxol. We have also obtained
similar values for the amounts of monomeric and polymerized
tubulin by extracting axoplasm with microtubule stabilizing
solutions which include DMSO, glycerol, and GTP to stabilize
the soluble microtubules (Morris and Lasek, unpublished).
Furthermore, electron microscopic studies demonstrate that the
number of microtubules in the axoplasm declines dramatically
as the second component of tubulin elutes from the axoplasm

(Morris, Lasek and Hodge, unpublished).

Actin microfilaments are not prominent structures in differentiated axons. However, microfilaments have been demonstrated in axons by employing heavy meromyosin to decorate them (15,16). Our results suggest that approximately one half of the actin in the squid giant axon is polymerized (Table I). Evidence that most and possibly all of this actin is actually in the form of microfilaments is provided by the results with phalloidin. Microfilaments are selectively stabilized by phalloidin (14), and we observed that phalloidin blocked the slower eluting actin. Thus, it is likely that this fraction of actin is in the form of soluble microfilaments which are in equilibrium with the monomeric pool. If, as seems likely, the remaining stably polymerized actin is also in the form of microfilaments, then the axon contains 0.75 mg/ml of actin in the form of microfilaments. These biochemical results indicate that microfilaments are far more abundant in the axoplasm than morphological studies have suggested. By comparison, MT which are prominent structure in the axon contain 4.1 mg of tubulin (Table I). This is about five times the amount of protein which appears to be present in microfilaments (Table I). In fact, since microfilaments consist of only two protofilaments and MT consist of 13 protofilaments, the total length of all the microfilaments in the axon may exceed that of MT. Although the aggregate length of microfilaments is relatively great, the individual microfilaments may be very short. This could explain why microfilaments are so difficult to detect in neurons with the electron microscope though the biochemical results indicate that they may be extremely abundant.

The values we have obtained for the concentration of monomer in axoplasm are a measure of the equilibrium conditions for polymerization in the axon. In the case of tubulin the monomer concentration at equilibrium is approximately 3 fold greater than that determined *in vitro* under conditions which are employed to polymerize tubulin. When mammalian tubulin is polymerized *in vitro* the monomer concentration at equilibrium is 0.4 mg/ml (17), as compared to 1.5 mg/ml in the squid giant axon. Olmsted (9) has also found that the monomer concentration in neuroblastoma cells with neurites (0.8 mg/ml) is also higher than expected from *in vitro* experiments. Neurons must contain factors other than tubulin which regulate the monomer-polymer equilibrium. One likely set of regulatory factors are the microtubule associated proteins (MAPs) (6). The MAPs in the squid giant axon have not been identified, however, we have noted a high molecular weight protein which is probably a MAP because it is not extracted from axoplasm when taxol is present but elutes when taxol is removed.

The presence of a large fraction of monomeric actin and tubulin in the axon may provide a reserve of these proteins

for the formation of additional polymer. One situation in
which these reserves may be important is during the regenera-
tive response to axonal injury. When the axon is severed it
sprouts a growth cone and a new axon elengates. Generation of
a growth cone apparently requires the formation of actin microfilaments which comprise the cytoskeleton of the microspikes.
The large fraction of actin monomer in the axon may provide
for a rapid response to such an injury. Similarly the availa-
bility of tubulin monomer may provide for the initial formation of MT in newly regenerated axons.

A. *Soluble Polymers - Stable Polymers and the Dynamics of the Cytoskeleton*

The polymers in the axon can be subdivided into a soluble
class and stable class. The fact that soluble polymer depolymerizes when the monomer concentration is diluted indicates
that at least one end of these polymers is free to exchange
with the monomer pool. However, both ends of stable polymers
must be blocked because these polymers do not release monomer
into solution even when the axoplasm is extracted for 24 hours
(7). Approximately one fourth of the total polymerized tubulin
in the axon is stably polymerized (Table I). The figure for
actin is even higher. About one half of the microfilaments
appear to be stable polymer. It is not surprising to find
that a large proportion of the cytoskeletal elements are stably
polymerized in a mature neuron. Cellular differentiation involves the consolidation of cellular morphology and a reduction of the variety of forms that a cell can take. The
presence of stable polymers which do not readily depolymerize
will tend to preserve the architecture of the cytoskeleton and
resist change.
Many unanswered questions remain about the soluble and
stable polymers of the neuron. For example, what factors are
responsible for the stabilization of these polymers? Are the
soluble polymers and stable polymers located in different
regions of the cytoplasm or are they uniformly distributed?
This latter question can be asked another way. Are the soluble polymers and stable polymers two completely different sets
of structures or are they coextensive? Individual axonal MT
may contain regions which are stable and regions which are
soluble. If this is the case, the stable segment of the MT
would be located at the slow growing or minus end, because of
the unequal rates of monomer addition and subtraction at the
two ends of the MT (18). The differences in the rates of
reaction at the plus and minus ends of MT results in treadmilling. That is, tubulin tends to move from the plus end
toward the minus end of the MT (19). However, if an MT
contains a stable segment, treadmilling will be blocked when

the stable segment reaches the minus end of the MT because the tubulin will not be able to leave the stable end of the MT. As Kirschner (20) has proposed, MT which have a stable minus end will compete with those MT which can treadmill. Eventually the MT which contain stable segments will extinguish those which do not contain a stable segment. Thus, the presence of stably polymerized tubulin in the axon can substantially influence the dynamics of the axonal cytoskeleton.

B. Cytoskeletal Polymers Influence the Relative Plasticity or Stability of Neuronal Morphology

During development, neurons undergo a continuing series of morphological changes which result in the circuitry of the adult nervous system. This capacity for change which characterizes developing organisms declines as the animal matures. The capacity to undergo orderly change in response to external stimuli is termed plasticity by neurobiologists. Plasticity is often measured by damaging a region of the nervous system and measuring the capacity of the system to compensate for the injury. For example, axonal sprouting or regeneration are examples of neural plasticity. Plasticity declines as an animal matures and it continues to decline after maturation. Thus, the oldest animals show the least plasticity in their nervous systems. This loss of plasticity during development and aging must result from some alteration in the properties of the structure determining elements in the nervous system such as the cytoskeletal polymers.

Our observations demonstrate that the cytoskeletal polymers in the squid giant axon vary in their dynamic properties. NF represent one extreme end of the spectrum because they exist exclusively in the form of stable polymers and do not appear capable of undergoing reorganization by simply changing the monomer-polymer equilibrium. This contrasts with the MT and microfilaments which consist in great part of soluble polymer. Each of these polymers contributes to the overall properties of the axonal cytoskeleton. Thus, axons or regions of an axon may vary in their properties depending on the relative proportion of stable polymers or soluble polymers. For example, cytoskeletons with a large number of NF should tend to be very stable and resist change. In this regard, it is interesting that NF are present in large numbers in axons and dendrites with the greatest diameters (21). Analyses of the ratio of NF to MT in axons during development demonstrate that the NF/MT ratio increases as axons increase in diameter (22). Initially in development axons have cytoskeletons which consist principally of MT and only few NF. As the neurons mature, both MT and NF increase but the NF increase more rapidly than the MT. If, as we propose, NF tend to stabilize the cytoskeleton, this

developmental progression will result in a reduction of the
capacity of the cytoskeleton to undergo change and consequently
to a reduction in plasticity. The regulation of these processes may involve changes in transcription or translation in
the neuron cell body. That is, increases in the fraction of NF
entering the axon will probably require increased synthesis of
the NF subunits. Although there isn't data concerning the
regulation of the synthesis of NF subunits in embryonic neurons, studies on regenerating neurons suggest that NF export
can be regulated by the neuron. During the regeneration of rat
motor axons, the ratio of NF/MT exported into the axon declines
(23). This apparently results from a reduction in the synthesis of NF subunits by the cell body. This reduction in the
amount of NF in the axonal cytoskeleton may represent a tendency of the regenerating neuron to return to a more plastic
state resembling the embryonic neuron.

At early stages of development when the axonal cytoskeleton
consists principally of MT, the neuron should be more plastic
because the MT can exist both as soluble polymers or stable
polymers. This capacity of tubulin to form either soluble
polymers or stable polymers raises the possibility that changes
in the plasticity of neurons may also result from alterations
in the proportion of soluble MT and stable MT in the cytoskeleton. In support of this hypothesis, Black and Green (8)
have noted that the fraction of colchicine sensitive MT
decreases when neurons in culture extend neurites. This
observation indicates that as neurons extend neurites the
ratio of stable MT to soluble MT increases. Changes in the MT
could be produced by altering the synthesis of regulatory
factors such as MAPs or by the post-translational modification of tubulin and MAPs (24). More information will be
required to determine which of the many possible regulatory
mechanisms is involved in changing the stability of the cytoskeleton during development and regeneration.

The proposal that the properties of the cytoskeleton are
related to the plasticity of neurons can be formulated more
generally as "the stabilization hypothesis of development and
aging". This hypothesis holds that during development there is
an increasing expression of mechanisms which stabilize the
morphology of neurons. These stabilizing mechanisms serve to
consolidate previous changes which have occurred during development. This tendency to reduce the amount of change in the
nervous system is one of the hallmarks of maturation. However,
a price must be paid for this increased stability and that is
a reduction in plasticity. As the organism continues to
develop, these stabilizing mechanisms may become increasingly
dominant so that plasticity continues to decline as the animal
ages after maturation.

Efforts to understand the changing plasticity of the

nervous system during development and aging naturally have tended to focus on the loss of certain mechanisms or the reduction of the efficiency of certain processes because plasticity generally declines as development proceeds. On the other hand, the stabilization hypothesis balances this perspective by drawing attention to structures or molecules which actively reduce plasticity as they fulfill their appropriate and natural role in the consolidation or maturation of the nervous system. Clearly, an encompassing theory of plasticity and stability must take both of these processes into account. That is, the plasticity of the nervous system or any part of the nervous system represents a balance between those mechanism which provide dynamic potential and those which stabilize structure. By continuing to elucidate the relative proportion of stable and dynamic structures in neurons, such as those provided by measurements of soluble and stable polymers in the giant axon, it should be possible to define the important developmental mechanisms that lead to changes in plasticity during maturation and aging.

REFERENCES

1. Lasek, R. J., *Neurosci. Res. Prog. Bull. 19,* 7 (1981).
2. Shelanski, M. L., Leterrier, J. F., and Liem, R. K. H., *Neurosci. Res. Prog. Bull. 19,* 32 (1981)
3. Bray, D., Thomas, C., and Shaw, G., *Proc. Natl. Acad. Sci. USA 75,* 5226 (1978).
4. Yamada,K. M., Spooner, B. S., and Wessells, N. K., *J. Cell Biol. 49,* 614 (1971).
5. Daniels, M. P., *J. Cell Biol. 53,* 164 (1972).
6. Kirschner, M. W., *Internatl. Rev. Cytol. 54,* 1 (1978).
7. Morris, J. R., and Lasek, R. J., *J. Cell Biol. 92,* 192 (1982).
8. Black, M. M., and Greene, L. A., *J. Cell Biol. 87,* 81a (1980).
9. Olmsted, J. B., *J. Cell Biol. 89,* 418 (1981).
10. Morris, J. R., and Lasek, R. J., *Biol. Bull. 257,* 384a (1979).
11. Laemmli, U. K., *Nature 227,* 680 (1970).
12. Rubinow, S., "Introduction to Mathematical Biology", John Wiley and Sons, New York, (1975).
13. Schiff, P. B., and Horwitz, S. B., *in* "Molecular Action and Target for Cancer Chemotherapeutic Agents", p. 483, Academic Press, New York, (1975).
14. Wieland, T., *Naturewissenschaften 64,* 303 (1977).
15. LeBeux, Y. J., and Willemot, J. W., *Cell Tiss. Res. 106,* 1 (1975).

16. Metuzals, J., and Tasaki, I., *J. Cell Biol. 78,* 597 (1978).
17. Olmsted, J. B., Johnson, K. A., Marcum, J. M., Allen, C., and Borisy, G. G., *J. Supramol. Struct. 2,* 429 (1974).
18. Bergen, L. G., and Borisy, G. G., *J. Cell Biol. 84, 141* (1980).
19. Margolis, R. L., and Wilson, L., *Nature 293,* 705 (1981).
20. Kirschner, M. W., *J. Cell Biol. 86,* 330 (1980).
21. Peters, A., Palay, S. L., and Webster, H. deF., *in* "The Fine Structure of the Nervous System. The Neurons and Supporting Cells", W. B. Saunders Co., Philadelphia, (1976).
22. Friede, R. L., and Samorojski, T., *Anat. Rec. 167,* 379 (1970).
23. Hoffman, P. N., and Lasek, R. J., *Brain Res. 202,* 333 (1980).
24. Brady, S. T., *J. Cell Biol. 91,* 333a (1981).

CHAPTER 31

THE CYTOSKELETON IN MYELINATED AXONS

Shoichiro Tsukita

Harunori Ishikawa

Department of Anatomy
Faculty of Medicine
University of Tokyo
Tokyo

The cytoskeleton in myelinated axons is mainly composed of longitudinally oriented microtubules and neurofilaments which are connected with each other by wispy filamentous structures (1). Recently, it has been clearly shown that these cytoskeletal components themselves are transported anterogradely through the axon at the rate of 1-4 mm/day (2,3). Interacting with the slowly moving cytoskeleton, the smooth endoplasmic reticulum and multivesicular bodies are rapidly transported at the rate of 100-500 mm/day in anterograde and retrograde directions, respectively (4-6). Hence, the morphological analyses on the organization of the axoplasm, especially of the cytoskeleton is prerequisite to better understand the molecular mechanism of the fast and slow axonal transport. In this article is given a brief survey of our recent studies on the three-dimensional organization of the cytoskeleton in myelinated axons.

I. SERIAL SECTION STUDY

A. *Arrangement and Continuity of Microtubules*

The microtubules in myelinated axons are not randomly mixed with neurofilaments in the axoplasm (7). They tend to be arranged in groups or bundles among the neurofilaments, forming "microtubule territories". Interestingly, within such microtubule territories are found almost all the membranous organelles that are transported rapidly through the axon. It is

Biological Functions of Microtubules
and Related Structures

343

also well known that the fast axonal transport of these membranous organelles are blocked by colchicine (8,9).   Hence, microtubules have been repeatedly considered to be structures important to the phenomenon of fast axonal transport.   On the other hand, recent biochemical analyses have made it clear that microtubules are transported slowly through the axon in the anterograde direction (2,3).   When any involvement of microtubules in the axonal transport is considered, we face an important but unsettled question whether or not microtubules extend the entire length of the axon.   The arguement about the continuity of microtubules could be settled only by an elaborate work of serial thin sectioning and reconstruction.

It is generally difficult to fix the myelinated axons in mammalian peripheral nerves well enough for the electron microscopic observation.   The mouse saphenous nerve which runs just beneath the skin, however, can be fixed well by dropping the fixative *in situ* (5).   Taking advantage of this nerve, we analyzed the arrangement and continuity of microtubules in the myelinated axons by the serial section method (10).   For this study, approximately 20 nerves were fixed and one of the best preserved nerves were used.   About 100 serial transverse sections (100 nm in thickness) were cut, and the selected axons in each section were photographed.   After all microtubules were numbered, the course of individual microtubules was followed in two sets of transversely cut serial thin sections for the internodal region and in one set for the nodal region, and approximately 97%, 96% and 94% of all microtubules, respectively, were continuous within the examined 10μm-long segments (Fig.1). For example, in one of the internodal segments, 223 microtubules extended the entire length of the examined segment, and only 2 microtubules terminated and 4 started.   The terminating or starting points of such discontinuous microtubules were found to be independent of each other in their distribution within the segment, indicating that the discontinuity of the microtubules was representative of the *in vivo* situation, not an artifact due to fixation.   Assuming that all microtubules are homogeneous in length, their average length was calculated to be about 370-760 μm based on the formula described by Hardham and Gunning (11)   (L=2Na/T,   L; average length of microtubules, N; average number of microtubules per section,   a; length of the segment,   T; number of free ends of microtubules in the segment) (Table I).   Recently, Chalfie and Thomson (12) successfully have followed individual microtubules in the specifically differentiated unmyelinated axons of the nematode.   According to their analysis, most, if not all, microtubules extend less than the entire length of the axon;   their average lengths were calculated to be about 5-25 μm.   Since the axon length is about 40-80 mm in the mouse saphenous nerve and about 0.2-0.4 mm in the nematode nerve, it may be safe to conclude that in both

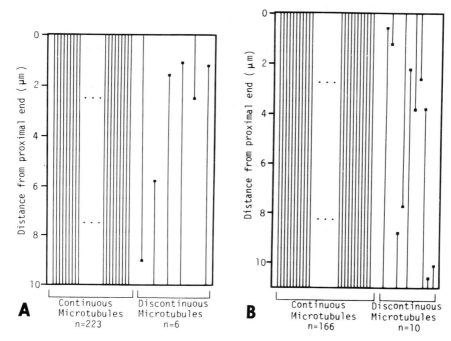

FIGURE 1.   Schematic diagram of microtubules along the internodal part (A) and the nodal part (B) of myelinated axons. (A): At the internodal part, 223 microtubules extend the entire length of the 10 μm-long segment, while 2 terminate and 4 start.   (B): At the nodal part, 166 microtubules extend the entire length of the 11 μm-long segment, while 4 terminate and 6 start.   (from ref. 10)

TABLE I.   Average Length of Microtubules   (from ref. 10)

Series [a]	Length of series	Avg. No. of microtubules	No. of termination	Calculated Avg. microtubule length
1	10 μm	227.2	6	757.3 μm
2	10 μm	256.8	10	513.6 μm
3	11 μm	170.1	10	374.2 μm

[a] Series 1 and 2 are from the internodal part of the myelinated axon, and Series 3 is from the nodal part.

types of nerves the microtubules are shorter than the axon. This conclusion is further supported by Bray and Bunge (13) in the similar analysis in the cultured sensory nerve axons, in which the average microtubule length is calculated to be 108 μm.

In the internodal region, each microtubule ran parallel to the axon without taking any spiral course. In the reconstruction series for the nodal region, the axon was shown to be extremely constricted at the node of Ranvier, especially at the paranode. Although the diameter of the axon strikingly decreased at the paranodal regions, there was no indication that microtubules frequently terminated or started in these constricted regions, resulting in a significant packing of microtubules. Interestingly, such a packing pattern of microtubules at the node was still maintained as a central core at the level of about 5 μm from the middle of the node. Such an arrangement of microtubules seems to make it easier for membranous organelles to pass through the constricted axon at the nodal region. In fact, Cooper and Smith (14) observed under the light microscope that round particles (membranous organelles) often passed through the node of Ranvier with no apprecialble change in velocity.

B. Arrangement and Continuity of Neurofilaments

Few observations have been reported on the arrangement and continuity of neurofilaments, although it has been proposed that neurofilaments might be discontinuous linear structures (15,16). Our serial section study also provided some informations of the continuity of neurofilaments (10). In the internodal parts of myelinated axons, neurofilaments tended to follow a spiral course to some extent. Therefore, it was very difficult not only to trace individual neurofilaments in serial transverse sections, but also to count the number of neurofilaments accurately. On the other hand, the number of neurofilaments sharply decreased toward the node of Ranvier, indicating that most of neurofilaments were discontinuous at the levels just proximal and distal to the node. This observation may point out the important issue that any model for the slow axonal transport should explain the discontinuity of most neurofilaments at the node. Interestingly, tubulin is shown to be transported at a rate similar to the triplet proteins of neurofilaments (2), though microtubules were estimated to be continuous at the node (Fig. 1).

II. FREEZE-ETCH REPLICA STUDY

A. *Non-Treated Axonal Cytoskeleton*

The question naturally arises how the motive force is generated to transport the membranous organelles at a fast   rate
and the cytoskeleton at a slow rate.   In this   connection,
conventional thin sections reveal that microtubules, neurofilaments, and membranous organelles are interconnected through
wispy filamentous structures (WFS) (17-20).   Therefore,   the
possible roles of such WFS in axonal transport have often been
discussed.   We thought detailed analyses of the three-
dimensional organization of the axonal cytoskeleton without
using chemical fixation were required for better understanding
of the physiological roles of the cytoskeleton in axonal transport.

The recent technical improvement in deep-etch, rotary-
shadowing raplica technique aided by rapid freezing has made
possible a direct electron microscopic analysis of fresh biological specimens without chemical fixation (21-23).   Using
similar techniques, we were able to visualize the organization
of the cytoskeleton in myelinated axons without   chemical
fixation (24).   Trigeminal nerves of the male albino rat were
dissected out and cut longitudinally into two pieces by a razor
blade.   The nerve tissue was rapidly frozen by slamming its
newly exposed surface against a pure copper block cooled to 4°K
by liquid helium, basically according to the procedure of
Heuser *et al.* (21).   The frozen tissue was brought into a cryo-
kit equipped on a Sorvall MT-2 ultramicrotome and cut with glass
knives at −120°C to remove the most superficial 10μm-thick layer
of the metal-contacted tissue surface.   The tissue was then
transferred into a conventional freeze-etching apparatus, in
which the exposed tissue surface was deeply etched and rotary-
shadowed with platinum to obtain replicas.

Freeze-etch replicas of fresh, unfixed samples from rat
trigeminal nerves provided three-dimensional views of the well-
developed cytoskeleton inside myelinated axons (Fig. 2).   The
characteristic feature of such freeze-etch replicas was that
longitudinally oriented microtubules and neurofilaments were
linked with interconnecting strands to form a dense, three-
dimensional latticework.   These strands can be considered to
correspond to the wispy filamentous structures (WFS) (17-20) or
microtrabeculae (25) seen in chemically fixed axons.   The
interconnecting strands may measure from 5 nm to 10 nm in thickness and from 15 nm to 50 nm in length, though the thickness
of individual strands varied along their length.   There seemed
to be some differences in the arrangement of such interconnecting strands between microtubule territories and neurofilament

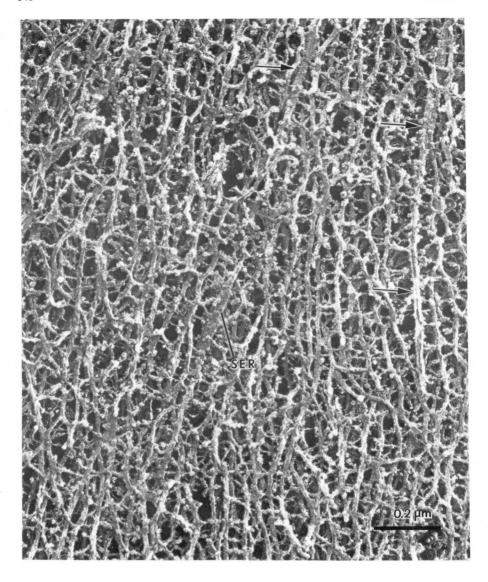

FIGURE 2. *Freeze-etch replica of longitudinally fractured myelinated axon in an unfixed trigeminal nerve. The axoplasm is seen to be densely packed fibrous networks, which mainly comprise microtubules (arrows), neurofilaments, and their interconnecting strands. SER: smooth endoplasmic reticulum. Note the granular structures which are closely associated with the axonal cytoskeleton.*

territories.   In the microtubule territories, the interconnecting strands also joined each other to form complicated networks, while in the neurofilament territories, the strands extended from a neurofilament to merely connect adjacent neurofilaments showing a ladder like appearance. The interconnecting strands appeared to be irregularly decorated with numerous granular structures about 10 nm in diameter, thus showing a bumpy or granular contour.

Membranous organelles such as smooth endoplasmic reticulum and mitochondria were also connected with microtubules and neurofilaments through the interconnecting strands (Fig. 2). Careful observations at higher magnification revealed that the interconnecting strands extending from neurofilaments and microtubules directly attached to the surface of the smooth endoplasmic reticulum and mitochondria, often in a periodic manner. Just beneath the axolemma were observed the well-developed meshworks composed of filamentous components, with wich microtubules and neurofilaments were connected through the interconnecting strands.   The   appearance   of   this subaxolemmal meshwork resembles the cytoskeletal network underlying the human erythrocyte membrane (26,27). The meshwork in the erythrocyte cytoskeleton is mainly composed of a high-molecular-weight, actin-binding protein, spectrin (220,000 and 240,000 M.W.), and actin (28).   Interestingly, from immunofluorescence study, Levine and Willard have reported that a high-molecular-weight, actin-binding protein, "fodrin" (240,000 and 250,000 M.W.), is localized at the periphery of the myelinated axon (29).   It is interesting to speculate that fodrin may be the major constituent of the subaxolemmal meshwork with which actin is closely associated, analogous to the spectrin-actin complex formed in the erythrocyte membrane.

B. Triton-Treated Axonal Cytoskeleton

The electron microscopic observation on replicas of freeze-etched, unfixed axons are very significant, since the possibility can now be ruled out that the interconnecting strands may be artifactual products of chemical fixation.   It is, however, still possible that with the freeze-etch replica technique deep-etching or freezing may cause an artifactual condensation of some soluble substances into strands between elements in the axoplasm.   To exclude this possibility, the nerve was pre-treated with Triton solution (1% Triton X-100,   6 mM Na^+-K^+ phosphate, 171 mM NaCl, 3 mM KCl, 0.5 mM EDTA, 0.1 mM PMSF, 1 mg/ml TAME, pH 7.0) and washed before freezing and replication. As a result, the neurofilament-associated interconnecting strands were found to be well preserved, whereas the microtubule-associated ones were removed together with microtubules

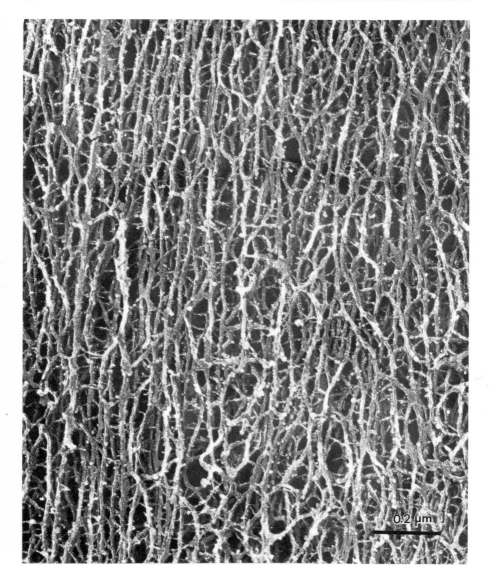

FIGURE 3. Freeze-etch replica of the axonal cytoskeleton of Triton-treated myelinated axon. Most of the neurofilament-associated interconnecting strands persist in Triton-treated axon, while the granular structures are completely removed. Most of microtubules disappeared during Triton-treatment, probably because of low temperature (4°C).

(Fig. 3). The reality of the interconnecting strands for neurofilaments was thus further substantiated and these strands are called here cross-linking filaments (CLF). The length of the CLF varied to a great extent, and interestingly the longer filaments tended to be thinner in diameter, suggesting their elastic nature. When the Triton-treated nerves were homogenized, such CLF were not found any longer, and instead, knoblike structures were seen on the neurofilaments. It is likely that these knob-like structures may represent the collapsed CLF. Combined biochemical and ultrastructural analyses suggest that the CLF may contain the high-molecular-weight subunits (200,000 M.W.) of the neurofilament triplets (2,30,31). It seems that this model is not incompatible with the antibody staining pattern described by Willard and Simon (32).

    After Triton treatment, the granular structures associated with the cytoskeleton disappeared completely. Recent biochemical analyses have clearly demonstrated that the slow transport system comprises two components, slow component a (SCa) and slow component b (SCb) (2,33), and that SCb comprises many kinds of soluble proteins such as enolase and creatine phosphokinase (34). Our fluorographic analysis revealed that many kinds of low-molecular-weight polypeptides generally associated with SCb were largely extracted by Triton treatment (24). Therefore, these SCb-associated polypeptides which may be soluble proteins represent the granular structures in freeze-etch replicas. The formation of granular structures and their association with the cytoskeleton may be explained in terms of artifactual condensation of such polypeptides during deep-etching. However, it is equally possible that these polypeptides may be transported slowly in aggregated forms and in labile and dynamic association with the cytoskeleton in living axons.

III. SUMMARY AND CONCLUSION

    Besides membranous organelles, the axoplasm in myelinated axons consists mainly of the following elements.
[1] *Microtubules*: Microtubules pursue a straight couse along the axis of the axon and tend to be arranged in groups to form "microtubule territories". Their average length is calculated to be about 370-760 μm.
[2] *Neurofilaments*: Neurofilaments tend to follow a spiral course to some extent. At the node of Ranvier, they decrease remarkably in number, indicating that most of them are discontinuous at the nodal region.
[3] *Microtubule-Associated Interconnecting Strands*: Microtubule-associated interconnecting strands join each other

to form complicated networks in "microtubule territories".
They are removed together with microtubules by Triton
treatment.

[4] *Neurofilament-Associated Interconnecting Strands (Cross
Linking Filaments; CLF)*: CLF extend laterally from a neurofilament to merely connect adjacent neurofilaments showing
a ladder-like appearance. Most of them persist even in
Triton treatment.

[5] *Granular Structures*: Many granular structures about 10 nm
in diameter are closely associated with the cytoskeletal
components.   They are completely removed by Triton treatment, suggesting that they are soluble proteins.

[6] *Subaxolemmal Meshwork*: Subaxolemmal meshwork underlies the
plasma membrane. This meshwork is composed of filamentous
structures, with which microtubules and neurofilaments were
connected through the interconnecting strands.

It is still premature to further discuss the physiological
roles of these elements in the axonal transport, but we believe
more detailed biochemical and morphological analyses on these
elements will lead to better understanding of the molecular
mechanism of fast and slow axonal transport.

ACKNOWLEDGMENTS

We wish to thank Drs. Jiro Usukura and Sachiko Tsukita,
Department of Anatomy, University of Tokyo, for their collaboration in the freeze-etch replica study. We also wish to thank
Prof. Eichi Yamada, Department of Anatomy, University of Tokyo,
and Prof. Masanori Kurokawa, Institute of Brain Research, University of Tokyo, for their helpful discussions throughout this
work.   Rapid freezing with liquid helium was carried out in
the Cryogenic Center, University of Tokyo.

REFERENCES

1. Wuerker, R. B., and Kirkpatrick, J. B., *Int. Rev. Cytol.*
    33, 45 (1972).

2. Lasek, R. J., and Hoffman, P. N., *in* "Cell Motility" (R.
    Goldman, T.Pollard, and J.Rosenbaum, eds.), p. 1021. Cold
    Spring Harbor Laboratory, New York, (1976).

3. Mori, H., Komiya, Y., and Kurokawa, M., *J. Cell Biol. 82*,
    174 (1979).

4. Tsukita, S., and Ishikawa, H., *Brain Res. 174*, 315 (1979).

5. Tsukita, S., and Ishikawa, H., *J. Cell Biol. 84*, 513 (1980).

6. Smith, R. S., *J. Neurocytol. 9*, 39 (1980).

7. Smith, D. S., Järlfors, U., and Cameron, B. F., *Ann. N.Y. Acad. Sci. 253*, 472 (1975).

8. Kreutzberg, G. W., *Proc. Natl. Acad. Sci. U.S.A. 62*, 722 (1969).

9. Dahlström, A., *Phil. Trans. R. Soc. London Ser. B 261*, 325 (1971).

10. Tsukita, S., and Ishikawa, H., *Biomed. Res. 2*, 424 (1981).

11. Hardham, A. R., and Gunning, B. E. S., *J. Cell Biol. 77*, 14 (1978).

12. Chalfie, M., and Thomson, J. N., *J. Cell Biol. 82*, 278 (1979).

13. Bray, D., and Bunge, M. B., *J. Neurocytol. 10*, 589 (1981).

14. Cooper, P. D., and Smith, R. S., *J. Physiol. 242*, 77 (1974).

15. Weiss, P. A., and Mayr, R., *Proc. Natl. Acad. Sci. U.S.A. 68*, 846 (1971).

16. Kreutzberg, G. W., and Gross, G. W., *Cell Tissue Res. 181*, 443 (1977).

17. Lane, N. J., and Treherne, J. E., *J. Cell Sci. 7*, 217 (1970).

18. Yamada, K. M., Spooner, B. S., and Wessells, N. K., *J. Cell Biol. 49*, 614 (1971).

19. Burton, P. R., *J. Cell Sci. 12*, 567 (1973).

20. Metuzals, J., Montpetit, V., and Clapin, D. F., *Cell Tissue Res. 214*, 455 (1981).

21. Heuser, J. E., Reese, T. S., Dennis, M. J., Jan, Y., Jan, L., and Evans, L., *J. Cell Biol. 81*, 275 (1979).

22. Heuser, J. E., and Salpeter, S. R., *J. Cell Biol. 82*, 150 (1979).

23. Usukura, J., and Yamada, E., *Biomed. Res. 2*, 177 (1981).

24. Tsukita, S., Usukura, J., Tsukita, S., and Ishikawa, H., *Neuroscience* (submitted).

25. Ellisman, M. H., and Porter, K.R., *J. Cell Biol. 87*, 464 (1980).

26. Tsukita, S., Tsukita, S., and Ishikawa, H., *J. Cell Biol. 85*, 567 (1980).

27. Tsukita, S., Tsukita, S., Ishikawa, H., Sato, S., and Nakao, M., *J. Cell Biol. 90*, 70 (1981).

28. Branton, D., Cohen, C. M., and Tyler, J., *Cell 24*, 24 (1981).

29. Levine, J., and Willard, M., *J. Cell Biol. 90*, 631 (1981).

30. Schlaepfer, W. W., and Freeman, L. A., *J. Cell Biol. 78*, 653 (1978).

31. Mori, H., and Kurokawa, M., *Cell Struct. Funct. 4*, 163 (1979).

32. Willard, M., and Simon, C., *J. Cell Biol. 89*, 198 (1981).

33. Black, M. M., and Lasek, R. J., *J. Cell Biol. 86*, 616 (1980).

34. Brady, S. T., and Lasek, R. J., *Cell 23*, 515 (1981).

CHAPTER 32

CELL CYCLE-DEPENDENT ALTERATION OF MICROTUBULE
ORGANIZATION OF A MOUSE CELL LINE, L5178Y[1]

Ichiro Yahara
Fumiko Harada

The Tokyo Metropolitan Institute of Medical Science
Tokyo, Japan

I. INTRODUCTION

Microtubules (MTs) are cellular structures that exist in a
wide variety of eukaryotic cells, and are involved in many
cellular events exhibited by these cells, such as mitosis,
humoral secretion, the movement of pigment granules, and
axonal transport (1). Since the spatial distribution and the
structural organization in cells of MTs could be implicated
in their functions, efforts has been made to elucidate them.
Recently, an immunofluorescence method originally devised by
Lazarides and Weber (2), made it possible to see the whole
organization of cytoplasmic MTs using specific antibodies
raised against tubulin (3). Using this method, we have found
an ordered array of cytoplasmic MTs in spherical cells including lymphocytes and thymocytes, only a part of which could be
observed by transmission electron microscipy (4)(Fig. 1).
However, other spherical cells, especially lymphoma cells, did
not appear to have a highly organized system of MTs because an
immunofluorescence staining pattern of these cells with anti-
tubulin antibody was diffuse (4)(Fig. 2). We describe in this
paper that mouse lymphoma L5178Y cells contain a MT organization which changes as the cell cycle progresses. In addition,
we show the evidence that unpolymerized tubulin exists in a
large amount in L5178Y cells, which interferes with the detection of tubular images of MTs by the immunofluorescence method.

[1]This work was supported in part by Grants-in-aid from
the Ministry of Education, Science and Culture of Japan.

II. IMMUNOFLUORESCENCE ANALYSIS

A. Tubular Image of Fluorescence Could Not Be Detected in L5178Y Cells

The indirect immunofluorescence method showed that the MT organization of mouse lymphocytes consists of 5-20 tubular structures, each of which is assocaited with the organization center-like structure at one end (Fig. 1a). Some tubular structures could be individual MTs, but others could be bundles of MTs. A model for the MT organization of lymphocytes is shown in Fig. 1b.

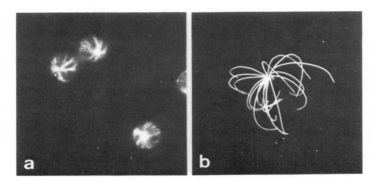

FIGURE 1. Microtubule organization of mouse lymphocytes. (a) Visualization by the indirect immunofluorescence method. (b) A model for MT organization of the lymphocyte.

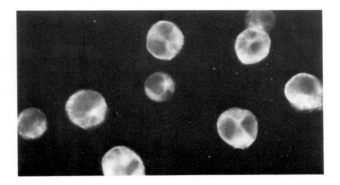

FIGURE 2. Visualization of MTs and related structures in L5178Y cells.

    The same method did not show clear images of MTs in cells
of a mouse lymphoma cell line, L5178Y (Fig. 2). Fluorescent
staining was very bright, suggesting that L5178Y cells are
rich in tubulin. The heterogeneity seen in the staining
patterns may result from difference in the stage of the cell
cycle among individual cells. We determined the cell cycle on
each cell whose MTs or related structures were visualized by
the immunofluorescence method.

B. *Identification of Phase of the Cell Cycle on Fluorescently
   Labeled Cells*

    Two conventional methods, double thymidine block and
hydroxyurea treatment, were used to synchronize lymphoma cells.
We have found, however, that cells incubated with hydroxyurea
or those treated by the double thymidine block method did not
reveal uniform staining patterns with anti-tubulin antibody
although these cells initiated synchronously DNA synthesis
after recovery from the blockade for synchronization. These
results suggest that both of the methods synchronized these
cells at the G1-S boundary with respect to the DNA cycle, but
neither did syncronize the cell cycle-dependent alteration of
cytoskeletal structures.

    We have devised a new tactic using exponentially growing
cells to investigate staining patterns specific for each phase
of the cell cycle. The method will be described in detail elsewhere[1]. Briefly, growing L5178Y cells were pulsed with [3]H-
thymidine for 30 min, and chased, if indicated, for 2, 4, 6,
or 8 hr. They were then fixed with formaldehyde and subjected

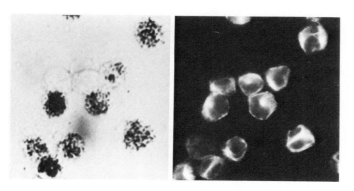

*FIGURE 3. Autoradiographic and immunofluorescence analysis
   of L5178Y cells.*

[1]*F. Harada, and I. Yahara, manuscript in preparation*

to cytocentrifugation. The cells collected on slide glasses
were treated with Triton X-100, and processed for the indirect
immunofluorescence staining with anti-tubulin antibody, after
which fluorescent images of arbitrarily selected cells were
photographed. The slide glasses were then dipped in nuclear
track emulsion, and processed for autoradiography. Finally,
we determined whether or not the cells photographed in immunofluorescent pictures contained silver grains in autoradiogram,
which indicated the incorporation of radioactive thymidine
into DNA. An example of radioactively and fluorescently
labeled cells is shown in Fig. 3.

If a cell is silver grain-positive, which had been pulse-
labeled with ^{3}H-thymidine and chased for 2 hr, the cell must
have existed in S phase 2 hr before the fixation for the
immunofluorescence staining. Similarly, we were able to identify immunofluorescently staining patterns of cells which had
been in S phase 0, 4, 6, or 8 hr before. According to analysis
of cell cycle periods (G1/S/G2+M was 6/10/4) under the culture
conditions, we identified the staining pattern specific for
each phase of the cell cycle. Typical staining patterns of
cells in different phases of the cell cycle are shown in
Fig. 4. We could classify more than 90% of examined cells into
either one of the 4 phases, G1, S, G2 and M, although variations in the staining pattern were detected to some extents
within cells at the same stage of the cell cycle.

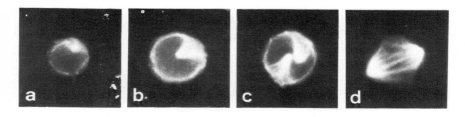

FIGURE 4. *Typical staining patterns with anti-tubulin
antibody of L5178Y cells at different cell cycle
(a) G1, (b) S, (c) G2, and (d) M cell.*

II. ELECTRON MICROSCOPIC ANALYSIS

Tubular structures could not be detected in L5178Y cells
by the fluorescence microscopic method while they were observed
in lymphocytes and fibroblasts. This fact led us to examine by
electron microscopy whether or not L5178Y cells contained MTs.
By refering to the fluorescent image of cells whose cell cycle

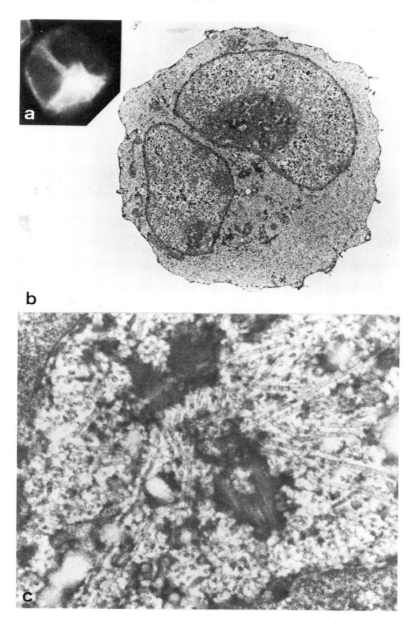

*FIGURE 5. MT organization of a L5178Y cell at G2. (a)
Immunofluorescent image (x 1980), (b) Electron microscopic observation of a G2 cell in ultrathin section
(x 7,200), (c) the centriole region of the cell shown
in (b) (x 43,200).*

phases were identified by autoradiography as shown above, we could determine the phase of the cell cycle also for cells observed in ultrathin sections by transmission electron microscopy (Fig. 5).

Electron microscopic results showed that:
—— G1 cell contains a single pair of centrioles with one of which a small number of MTs are associated.
—— S cell contains two pairs of centrioles which appear being duplicated. MTs are attached to only one of the two centriole pairs.
—— G2 cell contains two pairs of centrioles both of which are associated with MTs. Number of MTs increases as the cell cycle progresses from G1 to G2.
—— M cell contains mitotic spindles or related structures.

The evidence was given that MTs exist in L5178Y cells whereas they are hardly visible as fluorescent tubules by immunofluorescence microscopy. This led to the possibility that the cytoplasm of L5178Y cells might contain unpolymerized tubulins in a large amount in addition to MTs, which could interfere with the detection of MT structures by the immunofluorescence method.

III. EVIDENCE FOR THE PRESENCE OF A QUANTITY OF UNPOLYMERIZED
     TUBULIN IN L5178Y CELLS

A. *Treatment of Cells with Triton X-100 in MT-stabilizing*
   Buffer before Fixation for Immunofluorescence Staining

Osborn and Weber (5) have shown that extraction of soluble proteins from cells in MT-stabilizing medium made it possible

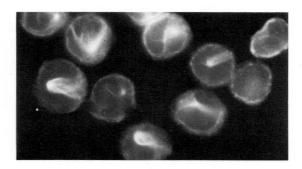

FIGURE 6. *Immunofluorescence visualization of detergent-*
   treated L5178Y cells.

to see a clear image of the MT organization even in round
cells. Before fixation with formaldehyde, L5178Y cells were
treated at room temperature for 5 min with 0.2% Triton X-100
in the MT-stabilizing buffer (5). Labeling of neutral detergent-treated and fixed cells with anti-tubulin antibody gave
much more sharp images of MT organization than that of non-
extracted and fixed cells (Fig. 6). Fluorescent tubules were
detected on some detergent-extracted cells. Determination of
phases of the cell cycle could be done also for detergent-
extracted cells (Fig. 6).

B. Radioimmunoassay

The method described by Eichhorn and Peterkofsky (6) was
followed for separation of polymerized and unpolymerized
tubulin in cell extracts. L5178Y cells were sonicated at 22°C
for 20 sec in a MT stabilizing medium (50%,w/w glycerol, 10%
DMSO, 5 mM Na-phosphate buffer, pH 7, 5 mM $MgCl_2$). The crude
lysates were centrifuged at 25,000 rpm in a Beckman type 40
rotor at room temperature for 1 hr so as to sediment MTs but
not unpolymerized tubulin. The precipitates were dissolved in
RIPA buffer (0.1M NaH_2PO_4, 10 mM EDTA, 10 mM EGTA, 10 mM NaF,
1% sodium deoxycholate, 1% Triton X-100, 0.1% SDS, pH 7.3)(7).
The amount of tubulin contained in the supernatants and precipitates was determined by the radioimmunoassay method described
by Van De Water, III and Olmsted (8). For one assay, 2 ng of

TABLE I.   Amounts of Tubulin Present as Sedimentable
and Nonsedimentable Forms

	Total Protein (μg)	Tubulin[a] (μg)	Tubulin Content (μg tubulin/mg protein)
C3H cells			
Whole cells	510	2.5	4.9
Supernatant	360	0.7	1.9
Pellets	220	1.5	6.8
L5178Y cells			
whole cells	73	2.2	30
Supernatant	23	0.82	36
Pellets	36	0.93	26

[a]Per 1 x 10^6 cells.

^{125}I-labeled mouse brain PC-tubulin (7,500-20,000 cpm/ng) was
used in 100 μl total reaction mixture containing diluted anti-
tubulin antibody and competitors. Immune complexes were precipitated after incubation with formaldehyde-fixed and SDS-
treated *Staphylococcus aureus*. Radioactivities assocaited with
immune complexes were then determined. Purified mouse brain
tubulin was used as the standard unlabeled antigens competing
with radio-iodinated tubulin for binding.

The results shown in Table I clearly indicate that the
content of tubulin in the total proteins was significantly
higher in L5178Y cells than that in C3H cells. It was also
shown that the percentage of unpolymerized tubulin in the total
tubulin was 47% for L5178Y cells and 32% for C3H cells. The
high content of unpolymerized tubulin in L5178Y cells would
account for the fact that fluorescent tubular structures could
not be detected by the immunofluorescence method whereas MTs
were found by electron microscopy.

C. *Two Dimensional Polyacrylamide Gel Electrophoresis*

According to the principle of the method, we measured by
radioimmunoassay the amount of tubulin antigens rather than
the amount of tubulin (9). To verify the results obtained by
the radioimmunoassay method, we have analyzed the total
extracts of L5178Y and C3H cells by two dimensional polyacrylamide gel electrophoresis (10). Supernatants and pellets fractionated by centrifugation in which unpolymerized and polymerized tubulin were contained, respectively, were also analyzed
by the same method. Results obtained using two dimensional
gels stained with Commassie brilliant blue confirmed the radioimmunoassay results shown in Table I (data not shown).

IV. DISCUSSION

Conventional synchronization methods such as hydroxyurea-
treatment or double thymidine block could not be used for ana-
lysing the cell cycle-dependent change of cytoskeletal structures because these methods were found to synchronize the DNA
cycle but not the cytoskeleton cycle. We developed, therefore,
a new method, a combination of autoradiography and immunofluorescence microscopy, for determination using exponentially
growing cells of alterations of cytoskeletal structures which
depend upon the progress of the cell cycle. This method would
be of generally useful for analysis on other cellular events

under the control of the cell cycle, particularly when synchronization method might affect these events.

By two dimensional PAGE, we have recently analyzed the
whole cell extracts from a variety of cultured mouse cells
including TCGF-dependent long term cultures of lymphocytes,
TCGF-dependent T cell clones, Con A-stimulated splenic lymphocytes, unstimulated lymphocytes, and other lymphoma cells such
as P388 with respect to the content of total, unpolymerized,
and polymerized tubulin. The results showed that tubulin
content of lymphoma cells was generally higher than that of
normal lymphocytes (unpublished observations). Furthermore,
as was shown for L5178Y cells, lymphoma cells were found to
contain a large amount of unpolymerized tubulin in addition to
MTs. This observation is compatible with the immunofluorescence microscopic result that tubular structures could not
be detected in lymphoma cells when they were stained with
anti-tubulin antibody (ref. 4 and this paper).

An abundance of unpolymerized tubulin in lymphoma cells,
when compared with normal fibroblasts and lymphpcytes, would
be explained in terms of either different expression of
tubulin multigenes or difference in regulation systems for MT
organization. The first explanation seems possible because
differences in the expression of tubulin genes were recently
found for different tissues in the same species (11) and in
different stages during development (12). The second is also
possible because there are many factors, such as calmodulin
and Ca^{2+} (13), which affect polymerization of tubulin or MT
organization. Such factors may differ in different cell types.
Whichever possibility is the case for lymphoma cells, it
would be related to a problem as to whether or not transformed
cells does essentially differ from normal counterparts in the
organization of MT (5,14,15). In addition, the existence of
unpolymerized tubulin in a large quantity may affect other
cellular functions, for it has been reported that treatment
of cells with colchicine induced DNA synthesis in arrested
cultures (16), and decreased the level of tubulin mRNA (17).

REFERENCES

1. Roberts, K., and Hyams, J. S., eds., "Microtubules".
   Academic Press, London, (1979).
2. Lazarides, E., and Weber, K., *Proc. Natl. Acad. Sci. U. S.
   A. 71*, 2268 (1974).
3. Weber, K., Pollack, R., and Bibring, T., *Proc. Natl. Acad.*

Sci. U. S. A. 72, 459 (1975).

4. Yahara, I., and Kakimoto-Sameshima, F., *Cell 15,* 251 (1978).
5. Osborn, M., and Weber, K., *Cell 12,* 561 (1977).
6. Eichhorn, J. H., and Peterkofsky, B., *J. Cell Biol. 82,* 572 (1979).
7. Burr, J. G., Dreyfuss, G., Penman, S., and Buchanan, J. M., *Proc. Natl. Acad. Sci. U. S. A. 77,* 3484 (1980).
8. Van De Water, III, L., and Olmsted, J. B., *J. Biol. Chem. 255,* 10744 (1980).
9. Fulton, C., and Simpson, P., *in* "Microtubules" (K. Roberts, and J. S. Hyams, eds.), p. 117. Academic Press, London, (1979).
10. O'Farrell, P. Z., Goodman, H. M., and O'Farrell, P. H., *Cell 12,* 1133 (1977).
11. Kemphues, K. J., Raff, R. A., Kaufman, T. C., and Raff, E. C. *Proc. Natl. Acad. Sci. U. S. A. 76,* 3991 (1979).
12. Gozes, I., De Baetselier, A., and Littauer, U. Z., *Eur. J. Biochem. 103,* 13 (1980).
13. Chafouleas, J. G., Pardue, R. L., Brinkley, B. R., Dedman, J. R., and Means, A. R., *Proc. Natl. Acad. Sci. U. S. A. 78,* 996 (1981).
14. Brinkley, B. R., Fuller, G. M., and Highfield, D. P., *Proc. Natl. Acad. Sci. U. S. A. 72,* 4981 (1975).
15. Edelman, G. M., and Yahara, I., *Proc. Natl. Acad. Sci. U. S. A. 73,* 2047 (1976).
16. Crossin, K. L., and Carney, D. H., *Cell 23,* 61 (1981).
17. Ben-Ze'ev, A., Farmer, S. R., and Penman, S., *Cell 17,* 319 (1979).

CHAPTER 33

DIRECT VISUALIZATION OF FLUORESCEIN-
LABELED MICROTUBULES IN LIVING CELLS

Charles H. Keith[1]
Michael L. Shelanski

Department of Pharmacology
New York University School of Medicine
New York, New York

INTRODUCTION

Research on mitosis and cell motility has expanded over
recent years from a descriptive morphological viewpoint to
one encompassing a full range of biochemical, cell
biological and molecular biological approaches. The power
of this multidisciplinary approach is clear from the papers
which have been presented in the course of this symposium.
Nonetheless, the progression from optical studies on living
cells to biochemical investigations on purified proteins
has revealed, not surprisingly, the workings of a
biological "principle of uncertainty". In its simplest
statement, this principle argues that greater biochemical
precision is obtained at the expense of temporal and
spatial resolution.

Recent advances in immunocytochemical techniques have
enabled us to diminish the spatial uncertainties while
sequential assays of clonal cell populations enable us to
obtain some temporal data. Thus, by careful collection of
a set of "snapshots" over time, a dynamic process can be
investigated.

In recent years, attempts have been made to partially
circumvent these problems by direct observation of labeled
molecules in living cells. In these procedures the
appropriate protein is purified, labeled with a fluorescent

[1] Supported by NIH grant NS -15076 and the McKnight
Foundation. CHK is a fellow of the MDAA.

dye and reintroduced into the cell by pressure microinjection. If the procedure is successful, the labeled protein will then be incorporated into the appropriate subcellular structure. To date methods have been developed which enable actin (1,2), α-actininin (3), 130 kilodalton protein or vinculin (4) and tubulin (5) to be labeled in a manner appropriate for such studies.

In this paper, we will present a summary of recent findings from our laboratory on the behavior of microinjected, fluorescently labeled microtubule proteins and discuss some of the problems encountered and possible future developments.

MATERIALS AND METHODS

Microtubules were prepared from bovine brain by the method of Shelanski et al. (6) and labeled in the assembled state by the addition of 0.1 vol of dichlorotriazinyl amino fluorescein (DTAF) (5 mg/ml) in dimethylsulfoxide for ten minutes at 37° C. The tubules were then pelleted and washed by resuspension in reassembly buffer. After labeling, the microtubule proteins could be further fractionated into DTAF-tubulin and DTAF-MAPs by phosphocellulose chromatography (7). Both the labeled tubulin and the labeled MAPs show normal assembly and assembly promotion (5). This procedure results in the linkage of 0.3 M of DTAF per mole of tubulin dimer.

Pressure injection of labeled proteins into cultured cells was carried out as described by Feramisco (3). Between 10^{-13} and 10^{-14} liters of material were injected in each experiment. Tubulin concentrations were 10 mg/ml.

Cells were mounted in controlled environment chambers (8) and observed with a Leitz Ortholux epifluorescence microscope equipped with a Zeiss 63X 1.4 NA planapochromat objective. Illumination was with an HBO 100 source controlled by an electronic shutter synchronized to the exposure intervalometer. Images were recorded using an RCA 1030H or DAGE 65 silicon image intensifier camera. Display was either recorded photographically from the television monitor, or on a Panasonic 1/2" time lapse video tape recorder.

Taxol was obtained from Dr. John Douros, National Cancer Institute, Silver Spring, Maryland, USA.

RESULTS

Mobilization of DTAF-Tubulin into Microtubules

When either DTAF-tubulin or DTAF-microtubule proteins (tubulin plus MAPs) is injected into cultured cells, it is rapidly mobilized into fibrillar structures. This has occurred in all mammalian cell types tested, but is seen most clearly in large flat cells such as gerbil fibroma (CCL-146), PtK2 and 3T3. Immediately after injection, there is a diffuse cloud of fluorescence which is more intense near the center of the cell than at the periphery. This signal is so intense as to obscure any organization into organelles which might occur during this period. After 10 to 20 minutes the beginnings of fibrillar organization can be seen within this cloud and by one hour labeled fibres extend to the periphery of the cell. Four hours after injection, a highly complex fibrillar network is seen which is very similar in appearance to the microtubular network observed by immunohistological staining with anti-tubulin (9). Exposure to colcemid completely inhibits evolution of the fibrillar pattern.

While either labeled tubulin or microtubule proteins yield fibrillar labeling, DTAF-MAPs are not incorporated into or along linear structures in any of the cultured cell lines referred to in the previous paragraph. Instead, the labeled material is rapidly localized to small dots diffusely scattered throughout the cytoplasm (5). Injection of denatured labeled tubulin shows both diffuse and punctate labeling in 1-4 hours. DTAF alone or DTAF-bovine serum albumin show a diffuse distribution over the same time period.

The labeled microtubular arrays are prominent in the cell for at least 24 hours after injection. If the cells are not obviously damaged on injection, their viability seems unimpaired compared to non-injected cells. Frequent or prolonged observation under fluorescence illumination will cause cell damage with any injected fluorochrome. This damage can be minimized by the utilization of brief illuminating pulses (1/30th second) coupled with observation with a silicon image intensifier television camera.

Injection of DTAF-tubulin into cultured rat sympathetic neurons also results in labeling of fibrillar structures though the rounded shape of the cell body and the narrow diameter of the neurite make them difficult to resolve. However, it is clear that initial mobilization occurs in

the cell body and that label migrates from the cell body
into the neurite at a slow and apparently uniform rate.
The velocity of migration is on the order of several tenths
of a millimeter per day or roughly the equivalent of the
rate of slow axoplasmic transport (10) When the process
is fully labeled, fibrillar elements can be resolved at the
base of the growing tip. Injected TRITC-ovalbumin labels
only the cell body and the proximal portion of the neurite
and is not transported to the ending. In contrast to the
inability of fibroblastic cells to mobilize labeled brain
microtubule accessory protein into microtubules, brain MAPs
are distributed in sympathetic neurons in precisely the
same manner as DTAF-tubulin. These proteins include
MAP_1, MAP_2 and tau leaving open the possibility that
any one or all of the MAPs are so distributed.

Other labeled cytoskeletal components such as actin,
α-actinin and 130K protein (vinculin) show patterns which
differ markedly from that of labeled tubulin (2,3,4).
Recently, we have labeled clathrin with DTAF. This
material is distributed in fine points throughout the cell
(Fig. 1). These points correspond very well with loci of
TRITC-α_2 macroglobulin applied from the cell exterior.
Thus, at least in the case of gerbil fibroma cells, the

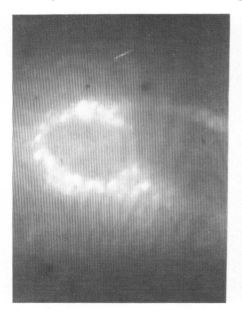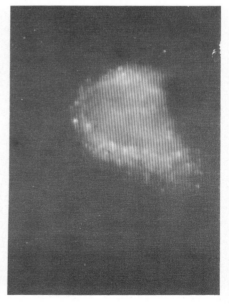

FIGURE 1. Gerbil fibroma cell injected with DTAF-clathrin
and exposed to rhodamine $\alpha_2 M$; DTAF image on right
rhodamine on left.

co-distribution of clathrin and $\alpha_2 M$ suggests that this material is organized in coated pits. As in the case of the aforementioned proteins, its distribution is completely distinct from tubulin.

Observation of Tubulin Cytoskeletons in Living Cells

When cells injected with DTAF-tubulin are observed repeatedly over time, changes are observed in the fibrillar pattern of their tubules. Existing tubules are seen to extend into newly forming regions of the cell (5) while other areas show apparent shortening of pre-existing microtubules. Preliminary experiments on mitotic plant cells (Haemanthus - in collaboration with Dr. A. Bajer) show selective incorporation into portions of the spindle apparatus with the precise locus of the label depending on the point in the cell cycle in which the cell is injected. Nonetheless, our attempts, to date, to utilize this technique to determine the loci of addition and removal of subunits from intact microtubules either in interphase or in mitosis have failed because of low resolution and high background levels.

The dynamics of large scale rearrangements of microtubules after treatment of cells with taxol, on the other hand, are readily observed with these techniques. Our experiments were carried out using two different protocols. In the first case, gerbil fibroma or PtK2 cells were injected with DTAF-tubulin, incubated for four hours to allow maximal incorporation of the label into microtubules, then taxol (6 µM) was added to the medium.

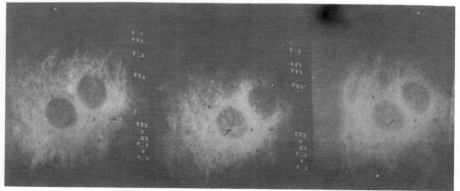

FIGURE 2. Gerbil fibroma cell injected with DTAF-tubulin and then exposed to taxol. Time series at 5 minutes (left), 30 minutes (center) and 90 minutes (right).

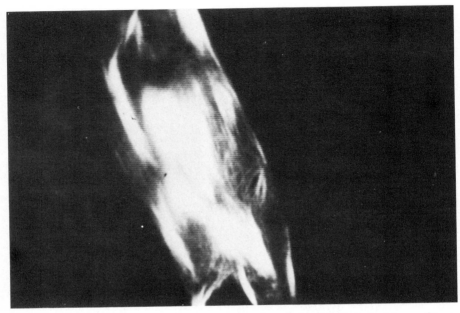

FIGURE 3.   PtK2 cell preinjected with DTAF-tubulin after
four hours exposure to taxol.

Within the first 30 minutes of exposure to taxol, existing
fibrillar structures were seen to thicken.   Over the next
hour, this coalescence of formed elements was seen to
progress in all parts of the cytoplasm (Fig. 2).   By four
hours, the bulk of the labeled microtubules was organized
into very thick parallel bundles (Fig. 3).

In the second case, the cells were exposed to taxol for
four hours prior to injection and then injected with
DTAF-tubulin in the continued presence of taxol.   In
contrast to the gradual coalescence of individual fibrils
seen in the former case, the labeled material was organized
directly into thick elements with fine linear striations.
In many cases, groups of these bundles radiated out from a
common center (Fig. 4).

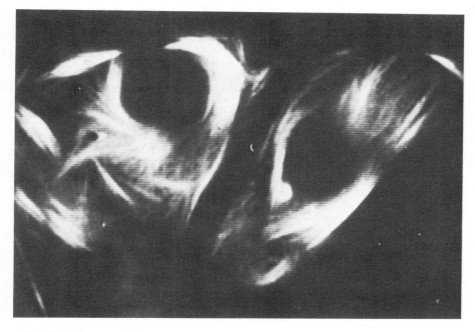

FIGURE 4.   PtK2 cells injected with DTAF-tubulin in the presence of taxol. Note radiation from "center" on left.

DISCUSSION

As we previously reported (5), tubulin and its accessory proteins can be labeled with dichlorotriazinyl amino fluorescein (DTAF) at or near neutrality without compromising the ability of these molecules to assemble into microtubules in vivo. When either DTAF-tubulin or DTAF-microtubule proteins are injected into cultured cells, there is a rapid evolution of a colchicine-sensitive, fluorescent fibrillar network. This labeling of microtubules could be due to either the incorporation of DTAF-tubulin into new and pre-existing cytoplasmic microtubules or the specific staining of pre-existing microtubules by the labeled subunits. Our data showing the time-dependent elongation of microtubules at the cell periphery and the selective labeling of only portions of the mitotic spindle in Haemanthus argue against a staining reaction. On careful examination of EM sections of injected cells, we have been unable to find double-walled microtubules which might suggest staining rather than incorporation. However, this approach would not detect a single DTAF-tubulin per turn of the microtubule helix, though the fluorescent signal from such a label would be detectable with our methods. A more definitive answer will require electron microscopic immunohistology using anti-fluorescein and colloidal gold. At the current time, our data clearly point to incorporation of DTAF-tubulin into microtubules, but do not rule out the possibility that selective staining also occurs.

A clear answer to the problem is required before the method can be used to study the processes of microtubule growth, shortening, polarity and treadmilling in vivo. However, questions pertaining to the movements of intact microtubules such as might be involved in microtubule sliding during anaphase or the taxol-induced rearrangements described above can be approached with the specific labeling obtained regardless of whether it is due to incorporation, staining or both.

Taxol is a plant alkaloid which has been demonstrated to favor microtubule assembly and inhibit disassembly (11). It is an effective antimitotic agent and has been shown to cause the bundling of microtubules in treated cells (12) with the development of prominent cross-bridging to intracellular membranes (12,13). The studies we report here attempt to look at the sequence of events involved in the formation of these bundles using two parallel approaches. In the first, the microtubular cytoskeleton is prelabeled with DTAF-tubulin and the effects of taxol on the intact microtubules of this network were followed. Within the first hour after exposure to the drug, the fine fibrillar pattern of the microtubular network was seen to progressively thicken while maintaining approximately the same position in the cell. At later times the cell shape alters markedly and the microtubules are organized into large bundles of long parallel fibers. If one assumes that the pool of free DTAF-tubulin is small at the time of addition of taxol and that taxol favors the assembly of the cytoplasmic microtubules, these results suggest that the bundles of microtubules seen on taxol treatment are due in large part to aggregation of pre-existing microtubules rather than recruitment of subunits from the cellular pool or disassembly and reordering of cytoplasmic microtubules.

In the second set of experiments the cells were exposed to taxol for four hours prior to injection to permit extensive formation of microtubular bundles and to favor the mobilization of free tubulin subunits into these arrays. Based on in vitro results, one would expect that injected DTAF-tubulin would add preferentially at microtubule ends or other assembly sites under these conditions. Alternatively, the injected material might self-assemble independently of other elements of the taxol stabilized cytoskeleton. The results obtained show relatively rapid organization of the DTAF-tubulin into thick bundles which are shorter in length than those seen in the cells used in the preceding experiments. Preferential labeling of linear elements either toward the cell periphery or near the center which would suggest addition to either the plus or minus ends of existing microtubules was not seen. However, the labeled bundles were often seen to radiate out from a common perinuclear center (Fig. 4) raising the possibility that the ability of the organizing center to nucleate new assembly of microtubules is neither saturated nor blocked by taxol treatment. The lack of addition to existing elements argues that cytoplasmic microtubules are stabilized and that each tubule has reached a limit to its growth. Factors limiting further microtubule growth in these cells remain to be identified as do the factors which cause the pre-existing microtubules to move together into bundles and the newly formed tubules to bundle on assembly.

The specificity of brain MAPs for neuronal cells raises many questions about the existence of highly specific intracellular recognition mechanisms. The data presented are clearly too limited to allow any general conclusion to be drawn. Experiments are underway in our laboratory to purify MAPs from non-neuronal cells using the approach of Bulinski and Borisy (14). These MAPs will then be fluorescently labeled and injected into homologous and heterologous cell types to see if they are mobilized into the cytoskeleton and whether specificity exists. An alternate explanation is that the labeling in nerve cells is not due to the neuronal nature of the MAPs, but to specific biochemical features of the neuron. Since the neuron has been found to distribute MAPs asymmetrically between its dendrites and axons, this possibility must be given serious consideration.

The images presented here were obtained with a standard RCA 1030H silicon image intensifier television camera and photographed from the television monitor. These images suffer from high background noise, non-linearity of the camera intensity response and bleaching of the fluorochrome. The number of observations which can be made is limited by both bleaching and cellular damage in the beam. Several approaches can be taken to minimize these problems and increase the information which can be obtained from the system. Clearly, a fluorescent label which bleached less rapidly than fluorescein would be desirable. While rhodamine isothiocyanate has been used successfully to label actin (2), we have failed to preserve assembly competence in tubulin preparations at high specific activity of the label though it is useful at lower specific activities for energy transfer studies in vitro (15). A second alternative is to decrease the intensity of illumination and amplify the signal by use of higher gain detectors or digital methods. In the former approach one might use a SIT camera with an additional stage of image intensification (ISIT, RCA 1040H) or a multichannel optical intensifier while in the latter the image would be digitally encoded and the gain increased in the computer. Use of one method does not exclude the other and, indeed, the rather high noise levels encountered in many of these devices at low signal strengths argues for use of digital methodology in conjunction with image intensifiers. The system currently in use in our laboratory uses a Digital Graphics CAT-800 digitizer with a resolution of 485 x 512 lines x 256 gray levels and is capable of digitizing a television frame in real time. Images are recorded in real time on video tape and then analyzed one frame at a time using a variety of programs. Of particular value in fluorescence studies are digital subtraction algorithms which allow a high background to be subtracted and the residual signal to be amplified to full scale and quantitative analysis programs allowing each pixel to serve as a fluorometer for energy transfer and other quantitative studies. It should be obvious that both the black level and the gain of the television camera must be under manual control for these studies and that the intensity response of the camera must either be linear or corrections made for non-linearity.

In summary, the methods of "molecular cytochemistry" (1) are a useful adjunct to more commonly used cell biological approaches. Important questions remain to be answered about the mechanisms of labeling in these experiments prior to their application to many of the questions of microtubular dynamics in vivo. However, with the clarification of these biochemical questions, the preparation of additional labeled components and the application of the very powerful digital analysis programs which have been developed for earth satellite images, this approach promises to provide a unique window into the cytoplasm of the living cell.

REFERENCES

1. Taylor, D.L. and Wang, Y.-L. (1978) Molecular
    cytochemistry: incorporation of fluorescent actin into
    living cells. Proc. Natl. Acad. Sci. USA 75: 857-861.
2. Kreis, J.E., Winterhalter, K.H. and Birchmeier, W.
    (1979) In vivo distribution of fluorescently labelled
    actin microinjected into human fibroblasts. Proc.
    Natl. Acad. Sci. USA 76: 3814-3818.
3. Feramisco, J.R. (1979) Microinjection of fluorescently
    labelled α-actinin into living fibroblasts. Proc.
    Natl. Acad. Sci. USA 76: 3967-3971.
4. Burridge, K. and Feramisco, J. (1980) Microinjection
    of a 130K protein into living fibroblasts: a
    relationship to actin and fibronectin. Cell 19:
    587-595.
5. Keith, C.H., Feramisco, J.R. and Shelanski, M. (1981)
    Direct visualization of fluorescein-labelled
    microtubules in vitro and in microinjected
    fibroblasts. J. Cell Biol. 88: 234-240.
6. Shelanski, M.L. Gaskin, F. and Cantor, C.R. (1973)
    Microtubule assembly in the absence of added
    nucleotides. Proc. Natl. Acad. Sci. U.S.A. 70:
    765-768.
7. Cleveland, D.W., Hwo, S-Y, and Kirschner, M.W. (1977)
    Purification of Tau, a Microtubule-Associated Protein
    that Induces Assembly of Microtubules from Purified
    Tubulin. J. Mol. Biol. 116: 207-225.
8. Dvorak, J.A. and Stotler, W.K. (1971) A controlled
    environment culture system for high resolution light
    microscopy. Exptl. Cell Res. 68: 144-148.
9. Osborn, M. and Weber, K. (1976) Cytoplasmic
    microtubules appear to grow from an organizing center
    towards plasma membrane. Proc. Natl. Acad. Sci.
    U.S.A. 73: 867-871.
10. Grafstein, B., McEwen, B.S. and M.L. Shelanski (1970)
    Axonal transport of neurotubule protein. Nature 227:
    289-290.
11. Schiff, P.B., Fant, J. and Horwitz, S.B. (1979)
    Promotion of microtubule assembly in vitro by taxol.
    Nature (London) 277: 665-667.
12. Schiff, P.B. and Horwitz, S.B. (1980) Taxol stabilizes
    microtubules in mouse fibroblast cells. Proc. Natl.
    Acad. Sci. U.S.A. 77: 1561-1565.

13. Masourovsky, E.B., Peterson, E.R., Crain, S.M. and
     Horwitz,   S.B.   (1981)   Microtubule   arrays   in
     taxol-treated   dorsal   root   ganglion-spinal   cord
     cultures. Brain Res. <u>217</u>: 393-398.
14. Bulinski, J.C. and Borisy, G.G. (1979) Self assembly
     of microtubules in extracts of cultured HeLa cells and
     the   identification   of   HeLa   microtubule   associated
     proteins. Proc. Natl. Acad. Sci. U.S.A. <u>76</u>: 293-297.
15. Becker, J.S., Oliver, J.M. and Berlin, R.D. (1975)
     Fluorescence techniques for following interactions of
     microtubule   subunits   and   membranes.   Nature   <u>254:</u>
     152-154.

CHAPTER 34

ASSOCIATION OF THE CYTOSKELETON WITH CELL MEMBRANES
- A CONCEPT OF PLASMALEMMAL UNDERCOAT -

Harunori Ishikawa
Sachiko Tsukita
Shoichiro Tsukita

Department of Anatomy
Faculty of Medicine
University of Tokyo
Tokyo

Microtubules, intermediate filaments and microfilaments (or actin filaments) constitute the major part of the cytoskeleton in many types of cells. These fibrous structures are often closely associated with the cell membrane to function not only as cytoskeletons but also as motile machineries. Thin-section electron microscopy shows that fibrous structures are usually attached to the membrane via certain specialized devices. Here, the problem how cytoskeletal components are associated with the cell membrane is discussed with a special emphasis on a concept of plasmalemmal undercoat (1).

I. ASSOCIATION BETWEEN FIBROUS STRUCTURE AND CELL MEMBRANE

A. *Microtubules*

In some types of cells microtubules are closely associated with the cell membrane (plasmalemma). The typical examples are found in various vertebrate spermatids (2), in insect "tendon" cells (3, 4), and in pilar cells of the inner ear (5), in which a large number of microtubules are arranged parallel to the cell axis and attached at their one or both ends to the plasmalemma in an end-on fashion. During spermiogenesis, a spermatid in mammals forms a parallel bundle of microtubules around the nucleus. Such a manchette of the microtubules is generally belie-

ved to play an important role in transformation of a round nucleus into a small elongate or flattened one (2). At the end
of the manchette, microtubules are attached to the membrane in
an end-on fashion via the electron-dense plaque, which appears
as a circular "thickening" of membrane in low-power electron
micrographs, thus called a *nuclear ring*. It should be recognized that there is no direct link between the microtubules and
the membrane proper (Fig. 1). The dense plaque is closely applied to the cytoplasmic surface of the plasmalemma, being considered to be included in the category of the plasmalemmal undercoat, which will be defined below.

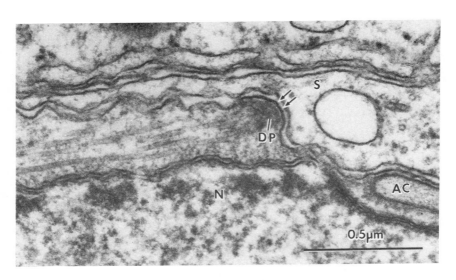

FIGURE 1. *Electron micrograph of part of a transforming*
spermatid. A manchette of microtubules around the nucleus (N)
is attached to the plasmalemma (arrows) via electron-dense pla-
que called nuclear ring (DP). S: Sertoli cell. AC: Acrosome.
Mouse testis.

The similar association of microtubules with the membrane
is found in epidermal *tendon cells* for striated muscles in some
species of insect. The tendon cells which connect muscle cells
with the cuticle contain a large number of parallel microtubules
along the cell axis. The strong links are formed between microtubules and the plasmalemma via electron-dense materials (3,4).
Similar situations are also found in other cell types in inver-
tebrates (6). Thus, microtubules not only play a role in formation of cell shape but also are used for transmitting the pull
through the firm attachment to the membrane.

Another type of the association of microtubules with the membrane is the lateral association between two systems. In developing neuromuscular junctions of chick embryo muscles, for example, dense materials which are called the postsynaptic density are applied to the cytoplasmic surface of the plasmalemma of a muscle cell. Closely associated with the post-synaptic density several micotubules run in a lateral association (1) (Fig. 2). The treatment with albumin before fixation may be used to well preserve microtubules for thin-section electron microscopy. By this treatment the association of microtubules with the membrane have been demonstrated in the postsynaptic regions in the rat (7).

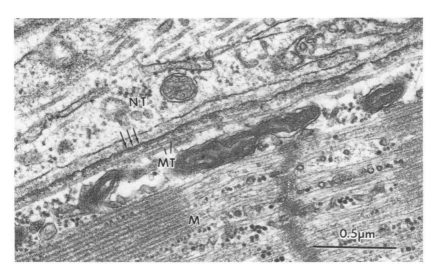

FIGURE 2. Developing neuromuscular junction in the posterior latissimus dorsi muscle from an 18-day chick embryo. The postsynaptic membrane (arrows) of a muscle fiber (M) is characterized by the dense plaque called postsynaptic density, with which microtubules (MT) are laterally associated.   NT: Nerve terminal.

B. Intermediate Filaments

Intermediate, 10 nm filaments form the firm attachment to the plasmalemma in several types of cells. We can see the typical examples for this attachment in the epithelial cells and cardiac muscle cells, in which the intermediate filaments are connected exclusively with the desmosomes. The mode of the

attachment was studied in some details; the filaments appear to
form hair-pin loops, which are embedded in the dense plaques of
the desmosome (8).   The desmosome,   a highly specialized
intercellular junction, thus provides the attachment site for
the filament by elaborating a multi-layered structure(9) on the
cytoplasmic surface of the apposed membranes. Similar situations
are observed in the vertebrate smooth muscle cells; the part of
dense plaques serves as the attachment site for the intermediate filaments (10).

C. *Microfilaments* and *Actin Filaments*

    Microfilaments tend to form parallel bundles in many types
of cells.   The bundles of microfilaments in cultured cells are
traditionally called as *stress fibers*, which usually run just
beneath the plasmalemma.   The lateral association may be involved along the bundle with the membrane(11). However, no further
detail can be resolved on the mode of the lateral association.
In certain cells, microfilaments also form a three-dimensional
network, which occurs as a gel being a cytoskeleton or motile
system.   Such microfilament networks also appear to be closely
associated with the plasmalemma (12, 13).

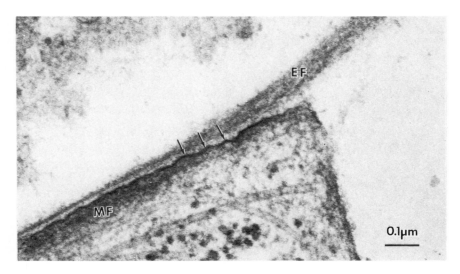

    *FIGURE 3.   Region of attachment of microfilaments in a
cultured cartilage cell. A microfilament bundle(MF) terminates
obliquely in the dense plaque applied to the inner surface of
the plasmalemma (arrows), from which an extracellular fibril
(EF) arises.*

The firm attachment of actin bundles with the membrane can be seen at their ends; the bundles anchor on the plasmalemma via electron-dense plaques (14, 15). On the opposite side of the membrane are attached some extracellular fibrous structures, probably *fibronectin* fibrils (Fig. 3). Through these adhesions some tension appears to be transmitted from the actin bundles eventually to the substratum.

One of the most typical examples of the association of the actin bundle to the membrane can be found in striated muscle cells (1).   At the ends of skeletal muscle fibers and at the intercalated disks of cardiac muscle cells, the myofibrils terminate on the plasmalemma (or *sarcolemma*) at the level of the Z disk. Again, actin filaments do anchor on the sarcolemma by way of the specialized dense plaques. At higher magnification, the dense plaques are seen to be layered structures: the sarco-lemma proper as a unit membrane, a less dense zone with verti-cal substructures, and a dense horizontal layer in thin-section electron microscopy (Fig. 4). Interestingly, actin filaments never penetrate the horizontal dense layer.

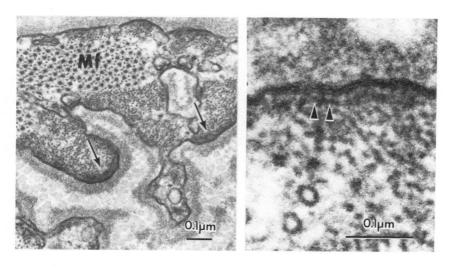

FIGURE 4. Transverse section of the myotendon junction of a snake tongue muscle. a, Low magnification. The myofibrils (Mf) attach to the sarcolemma via the dense plaque (arrows). The basal lamina overlying the attachment sites appears much thicker than that in other regions. b, Higher magnification. The dense plaque shows a layered organization with vertical and horizon-tal components (spearheads).

In smooth muscle cells, similar dense plaques can be found scattered along the sarcolemma as the anchoring sites for actin

filaments and intermediate filaments (Fig. 5). Higher magnification shows the layered construction of the dense plaque, to
which the filaments are attached (see Fig. 8*b*).. The idea that
the membrane itself in these regions is also specialized may be
supported by the observations on Triton X-100 treated muscle
cells. The unit membrane image of the sarcolemma was well preserved only in these regions, even though the dense plaques
tended to be detached from the membrane (Fig. 6). When smooth
muscle cells were treated with heavy meromyosin (HMM) S-1, arrowhead structures were formed along all actin filament in such
that the arrowheads pointed away from the membrane (16), indicating that the plaques determine the polarity of attached actin
filaments.

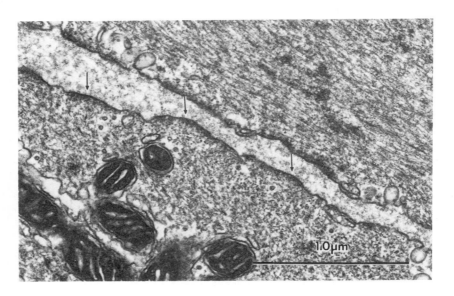

FIGURE 5. *Dense plaques in smooth muscle cells. The dense
plaques are seen scattered along the sarcolemma and provide the
attachment sites for actin filaments and intermediate filaments.
Guinea-pig ureter.*

II. A CONCEPT OF PLASMALEMMAL UNDERCOAT

It is well known that the plasmalemma at the initial segments and the nodes of Ranvier in myelinated axons is characterized by the membrane *thickening* (Fig. 6). At higher magnification, this membrane thickening is shown to be a complex of the

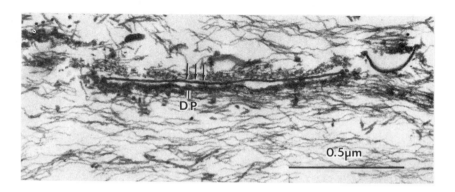

FIGURE 6. Dense plaque in a Triton-treated smooth muscle cells. The sarcolemma (arrows) is well preserved only at
the region of the dense plaques (DP). Chicken gizzard.

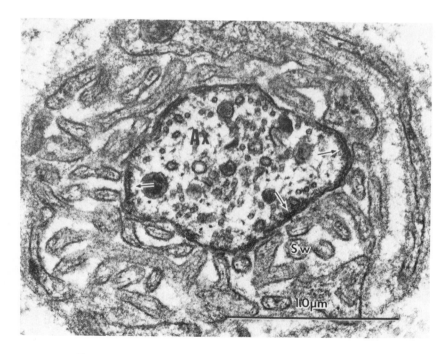

FIGURE 7.   Transverse section of the node of Ranvier of a
myelinated axon. At the node, the axon (Ax), which is surrounded
by many finger-like processes of Schwann cells (Sw), is
characterized by the membrane thickening (arrows). Rat spinal
nerve.

plasmalemma and the underlying dense plaque of a layered organization (17), but not the real thickening of the membrane proper.   Comparing the dense plaques in the myelinated axons
to those in smooth muscle cells, surprisingly they resemble
each other in appearance (Fig. 8). The dense plaques underlying
the plasmalemma are clearly resolved into the vertical and horizontal components.   Some types of such dense plaques are more
elaborate and thicker than the others.   All these observations
indicate that there are ultrastructural similarities among the
dense plaques in a wide variety of cell types (1).   Even
the desmosome may be included in the category.   This promted
us to call these structures collectively the   *plasmalemmal
undercoat* (1).

III. ERYTHROCYTE CYTOSKELETON AS A PLASMALEMMAL UNDERCOAT

In discussing the plasmalemmal undercoat, we have to refer
to the cytoskeletal network underlying the human erythrocyte
membrane.   The erythrocyte membrane has been shown to possess
the cytoskeleton closely applied to the membrane proper (18,19).
Using tannic acid for fixation, we could visualize such cytoskeletons in thin sections (20).   The erythrocyte membranes,
when cut transversely, show a layered structure which is resolved into vertical and horizontal components beneath the membrane proper (Fig. 9). Hence, the erythrocyte cytoskeleton can be
included within the category of the plasmalemmal undercoat.
When a section is cut tangential to the membrane, the cytoskeleton shows filamentous meshworks as an on-face view (Fig. 10).
Biochemically, the erythrocyte cytoskeleton is mainly composed
of spectrin and actin (21,22). When the ghost membranes were
treated with 0.1 mM EDTA, almost all of spectrin and actin are
removed from the membrane, leaving only granular components and
concomitantly the membrane becomes fragmented into inside-out
vesicles.   All the filamentous meshworks of the cytoskeletal
network disappear on such vesicles, suggesting that the filamentous meshworks are constructed of spectrin and actin.   Similar meshworks can be reconstructed by incubating these spectrin-
actin-depleted vesicles with the EDTA extract containing spectrin and actin (23).   It has been shown that the filamentous
components themselves are spectrin molecules.   Then, where and
in what form does actin exist?   So far, there is no actin filament detected within the meshworks.
When the original ghost membrane or spectrin-actin-reasso-
ciated vesicles are incubated with a low concentration of muscle G-actin under the condition which favors actin polymerization, actin filaments are formed, many of them sticking up from

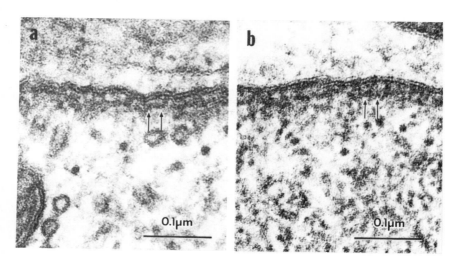

FIGURE 8.   Ultrastructural similarity in the plasmalemmal
undercoats.   a,   The node of Ranvier of the myelinated axon.
b,   The smooth muscle cell. Note the layered organization of
the plasmalemmal undercoats in both types of cells (see also
Figs. 4b and 9) (arrows).

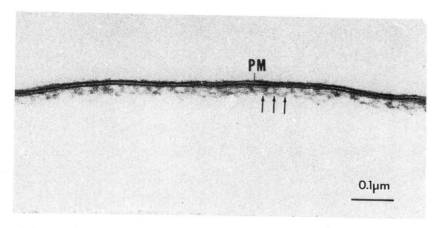

FIGURE 9.   Transverse section of the isolated human erythrocyte membrane. With tannic acid-glutaraldehyde fixation, the
cytoskeletal network underlying the membrane (PM) is clearly
visualized. The network is resolved into vertical and horizontal components (arrows).

the membrane (23). By the treatment with HMM S-1, such formed
actin filaments are shown to be polarized with arrowheads pointing toward the membrane. This polarity of actin filaments with
respect to the membrane is very unique to the erythrocyte membrane. The actin filaments are attached to the membrane via
spectrin molecules. Fig. 11 summarizes the experimental results obtained from this study(23). A possible model of the molecular organization of the erythrocyte cytoskeleton is deduced
from the biochemical and ultrastructural studies. The ends of
four to six spectrin tetramers join a junction point where
actin exists as a short, polarized filament, with which spectrin
molecules are connected to form a meshwork. Thus, actin filaments do provide the multiple attachment sites for spectrin.
Similar situations, though more or less modified, may be postulated in many other types of cells for the mode of attachment of
actin filaments to the membrane. It should be emphasized that
only in the erythrocyte the cytoskeleton-membrane association
can be discussed in terms of molecular organization.

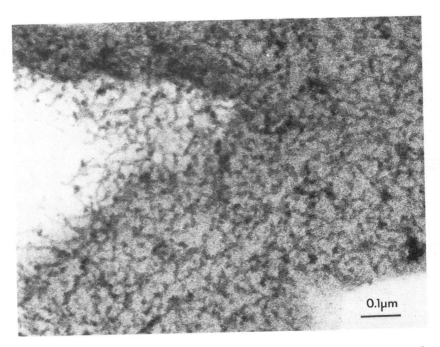

0.1μm

FIGURE 10. On-face view of the cytoskeletal network underlying the erythrocyte membrane. This section is cut tangen-
tially to the membrane. The horizontal components of the network are seen to form an anastomosing meshwork.

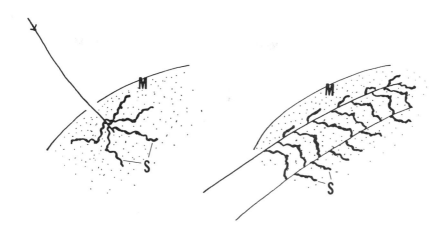

FIGURE 11.   Schematic illustration summarizing the experimental results for the attachment of muscle F-actin to the erythrocyte membrane.   Actin filaments with a definite polarity appear to be bound to membrane (M) via spectrin molecules (S). The filaments may provide the multiple attachment sites for spectrin molecules, serving as the junction points in the cytoskeletal network.

IV. SUMMARY AND CONCLUSION

Several examples of the so-called plasmalemmal undercoat are presented and discussed with special reference to their role in the cytoskeleton-membrane association. The structures which can be included within the category of the plasmalemmal undercoat are of a wide occurrence (1). The plasmalemmal undercoats vary in appearance among different membranes. However, there are some common features in various plasmalemmal undercoats; One of their ultrastructural characteristics is that the undercoats show a layered organization of dense material directly applied to the cytoplasmic surface of the plasmalemma as seen by thin-section electron microscopy. In many cases, these structures are found with close relation to the localization of certain specific functions of the plasmalemma. This undercoat itself is formed of peripheral proteins of membrane closely associated with the plasmalemma. The significance of the plasmalemmal undercoats may be primarily to provide a structur-

al support for the membrane, thus serving as "membrane skeleton" (21). With these supports the plasmalemma can perform
many important functions in its limited or whole surfaces. In
many cases, the undercoats are further sustained with the associated cytoskeleton or serve as the attachment and/or organizing sites for the cytoskeletal filamentous components.

REFERENCES

1. Ishikawa, H., Tsukita, S., and Tsukita, S., *in* " Nerve
    Membrane - Biochemistry and Function of Channel Proteins"
    (G. Matsumoto, and M. Kotani, eds.), p. 167. University
    of Tokyo Press, Tokyo, (1981).
2. Fawcett, D. W., Anderson, W. A., and Phillips, D. M.,
    Develop. Biol. 26, 220 (1971).
3. Auber, J., *J. Microscopie 2,* 325 (1963).
4. Caveney, S., *J. Cell Sci. 4,* 541 (1969).
5. Yamada, E., *Otologia Fukuoka 11,* 209 (1965).
6. Dustin, P., Microtubules. Springer-Verlag, Berlin-Heidel-
    berg-New York, (1978).
7. Gray, E. G., and Westrum, L. E., *Cell Tissue Res. 168,* 445
    (1976).
8. Kelly, D. E., *J. Cell Biol. 28,* 51 (1966).
9. Drochmans, P., Freudenstein, C., Wanson, J.-C., Laurent,L.,
    Keenan, T. W., Stadler, J., Leloup, R., and Franke, W. W.,
    J. Cell Biol. 79, 427 (1978).
10. Small, J. V., and Sobieszek, A., *Int. Rev. Cytol. 64,* 241
    (1980).
11. Ishikawa, H., *in* "Cell Motility: Molecules and Organization
    (S. Hatano, H. Ishikawa, and H. Sato, eds.), p. 417. Uni-
    versity of Tokyo Press, Tokyo, (1979).
12. Pollard, T. D., and Weihing, R. R., *CRC Critical Rev. of
    Biochem. 2,* 1 (1974).
13. Small, J. V., *J. Cell Biol. 91,* 695 (1981).
14. Mooseker, M. S. and Tilney, L. G., *J. Cell Biol. 67,* 725
    (1975).
15. Geiger, B., Dutton, A. H., Tokuyasu, K. T., and Singer, S.
    J., *J. Cell Biol. 91,* 614 (1981).
16. Tsukita, S., Tsukita, S., and Ishikawa, H., *Cell Struct.
    Funct. 6,* 422 (1981).
17. Chan-Palay, V., *Z. Anat. Entwickl.-Gesch. 139,* 1 (1972).
18. Lux, S. E., *Nature (Lond.) 281,* 426 (1979).
19. Branton, D., Cohen, C. M., and Tyler, J., *Cell 24,* 24
    (1981).
20. Tsukita, S., Tsukita, S., and Ishikawa, H., *J. Cell Biol.
    85,* 567 (1980).
21. Marchesi, V. T., *J. Membr. Biol. 51,* 101 (1979).

22. Steck, T. L., *J. Cell Biol.* *62,* 1 (1974).
23. Tsukita, S., Tsukita, S., and Ishikawa, H., *J. Cell Biol.* *90,*70 (1981).

CHAPTER 35

MICROTUBULES COMPOSED OF TYROSINATED TUBULIN ARE
REQUIRED FOR MEMBRANE EXCITABILITY
IN SQUID GIANT AXON

Gen Matsumoto

Electrotechnical Laboratory
Tsukuba Science City, Ibaraki

Hiromu Murofushi
Sachiko Endo
Hikoichi Sakai

Department of Biophysics and Biochemistry
Faculty of Science
The University of Tokyo, Tokyo

I. MORPHOLOGICAL STUDY OF THE DISTRIBUTION OF AXOPLASMIC
    MICROTUBULES

Electron microscopic observation of the giant axon of
squid *(Doryteuthis bleekeri)* shows that the axoplasmic microtubules (MTs) are densely distributed at the region underlying
the axolemma (Fig. 1A), and that they run parallel with the
longitudinal axis of the axon (Fig. 1B) forming networks with
neurofilaments, thin filaments and cross-bridges (1). The
characteristic distribution in the peripheral axoplasm, close
to the axolemma, becomes evident when the centripetal distribution of MTs is measured from the axolemma towards the center
of the axon (2) (Fig. 2). For the intact axon, the highest
density of about 100 MTs/μm^2 is observed at the region nearest
the axolemma, and the density decreases rapidly toward the
interior of the axon to 20 MTs/μm^2 at a distance 3 μm from the
axolemma. On the other hand, in colchicine-perfused axons
less than 10 MTs/μm^2 remain throughout the region observed
(Fig. 2). In these colchicine-perfused axons, it is consistently found that the electrical excitability is largely
lost.

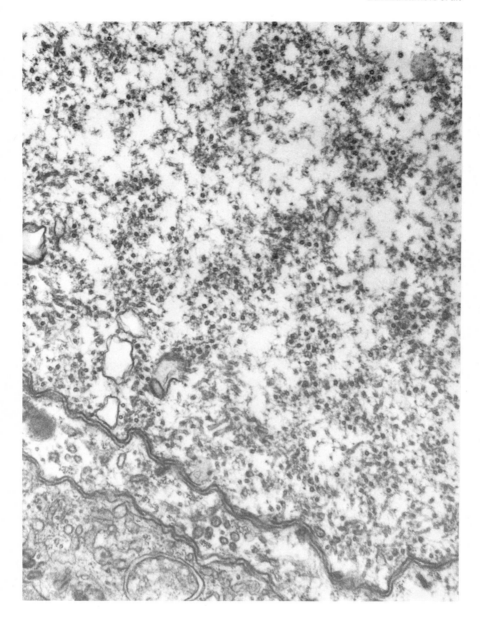

Fig. 1A. Cross section of the intact giant axon of squid (Doryteuthis bleekeri). *The number of MTs is more numerous in the peripheral axoplasm than in the interior. Note the network of interwoven thin filaments with MTs. AL, axolemma; SC, Schwann cell; MT, microtubules.*

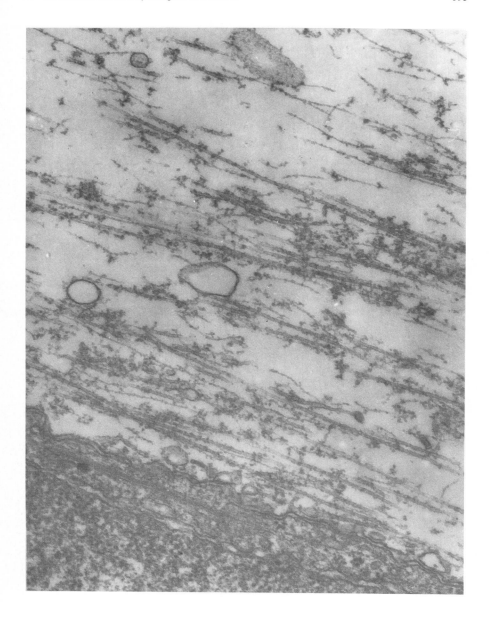

Fig. 1B. Longitudinal section of the intact giant axon of squid. MTs are more numerous close to the peripheral region of the axoplasm and run parallel to the longitudinal axis of the axon. AL, axolemma; SC, Schwann cell; MT, microtubules; NF, neurofilaments.

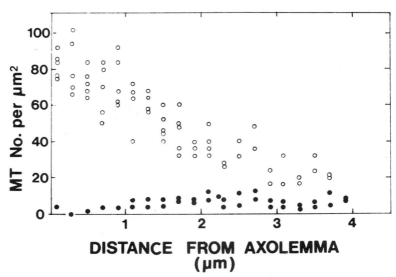

Fig. 2. Distribution of MTs in the peripheral region of the intact squid giant axon (open circles) and reduction in the number after internal perfusion of the axon with colchicine (closed circles). Before perfusing the axon, the axoplasm was extruded by the roller method (see section IV). The perfusion was carried out using 1 mM colchicine dissolved in the standard medium (380 mM KF and 25 mM K-HEPES at pH 7.25) for 6 min, and then the axon was fixed for electron microscopy.

The electron microscopic study has raised two questions concerning the physiological role of axoplasmic MTs; one is why MTs are densely distributed at the region just underlying the axolemma, and the other is why the axon loses its excitability when axoplasmic MTs are depolymerized by introducing colchicine inside the axon. The biochemical and electrophysiological studies described below intend to answer the above two questions.

II. SOME BIOCHEMICAL ASPECTS OF AXOPLASMIC PROTEINS OF THE SQUID GIANT AXON

Axoplasmic proteins of the squid giant axon were analyzed by SDS-polyacrylamide gel electrophoresis after the axoplasm was extruded by using a slender plastic tube (3). The extruded axoplasm was quickly suspended in a chilled glutamate-reassembly buffer (0.3 M K-glutamate, 10 mM K-MES, 1 mM GTP,

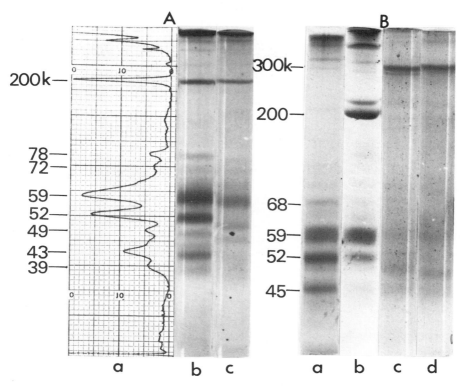

Fig. 3. (A) SDS-polyacrylamide gel (5 %) electrophoresis of axoplasmic proteins. a, Densitometric scan of gel b; b, 30,000 × g supernatant of axoplasm homogenate; c, 30,000 × g pellet of the homogenate. (B) SDS-polyacrylamide gel (4 %) electrophoresis of axoplasmic proteins and KI and Ca extracts of axoplasm-extruded axons. a, 100,000 × g supernatant of axoplasm homogenate; b, 100,000 × g pellet; c, KI extract; d, Ca extract.

1 mM EGTA, 0.5 mM $MgSO_4$ at pH 6.8), followed by centrifugation at 30,000 × g for 10 min. The electrophoregram of the
30,000 × g supernatant (Fig. 3A) shows that the major proteins
are α- and β-tubulin of 59,000 and 52,000 molecular weight
(27-29 % of the total proteins), and that the 43,000 and
200,000 molecular weight proteins are 7-10 and 5-6 % of the
total proteins, respectively. By purifying the axoplasmic
MT proteins by temperature-dependent assembly and disassembly
and analyzing them by SDS-polyacrylamide gel electrophresis,
it was found that α- and β-tubulin in the squid axoplasm are
similar to mammalian brain tubulin but the high molecular

weight MAPs (microtubule associated proteins) are different
from those of the mammalian brain (3, 4). Both soluble
(Fig. 3 A-a) and insoluble fractions (Fig. 3 B-b) of the extruded axoplasm contain a large amount of a polypeptide having
a molecular weight of 200,000. This can be regarded as a component of neurofilaments (5).

In order to analyze residual axoplasm (ectoplasm) associated with the axon whose axoplasm had been extruded, the axon
was suspended in an extraction medium of 0.6 M KI and 10 mM
HEPES (pH 7.3) or of 10 mM $CaCl_2$, 0.55 M NaCl and 10 mM HEPES
for 10-15 min at 0°C. The soluble axonal proteins were then
sedimented by centrifugation at 30,000 × g for 10 min. The
supernatant thus obtained was designated as KI (Fig. 3 B-c)
or Ca extract (B-d) (3). Fig. 3B shows that the main component in the KI or Ca extract has a molecular weight of 300K.
This polypeptide is not found in the extruded axoplasm.

Some characteristics concerning assembly and disassembly
of axoplasmic MT proteins were studied (3) after the MT proteins were purified by a cycle of temperature-dependent assembly
and disassembly. They polymerize with an optimum temperature
of about 25°C. The critical concentration of once-cycled MT
proteins for polymerization is 0.24 mg/ml. The polymerization-
supporting efficiency of anions is in the order glutamate~
F > Cl > Br > I.

III. SIMILARITY OF MEDIUM CONDITIONS FOR AXOPLASMIC
      MICROTUBULE ASSEMBLY *IN VITRO*, TO THOSE FOR MAXIMUM
      SODIUM AND POTASSIUM CONDUCTANCES

It has been elucidated that the squid axon, when internally perfused with medium unfavorable for axoplasmic MT assembly *in vitro*, eventually loses its excitability (6) as a
result of a decrease in both maximum sodium and potassium conductances (2) (Table 1). Colchicine, Ca ions, anions such as
Cl⁻, Br⁻, and I⁻, and cysteine or histidine modifying reagents
commonly destroy the maximum sodium and potassium conductances.
Furthermore, the effects of temperature and proteolytic
enzymes are interesting. The maximum sodium conductance of
the squid giant axon increases as the temperature increases
from 0 through 25°C, reaching a maximum at about 25°C. This
is followed by a loss of the membrane excitability at about
35°C (11). This behavior of the maximum sodium conductance as
a function of temperature resembles the temperature-dependence
of squid axoplasmic MT assembly (6) (see section II). Another
interesting point is the effect of proteolytic enzymes on the
Hodgkin-Huxley parameters (Table 1); enzymes other than
carboxypeptidase A suppress the sodium inactivation while

TABLE I. Effects of drugs and various internal perfusion media upon the maintenance of membrane excitability and parameters of the Hodgkin-Huxley model (@, very effective; o, effective; *, ineffective) CBP-A; carboxypeptidase A

Conditions		Effect on the membrane excitability	The H-H Parameters									Ref
			$\bar{g}_{Na}$	m	h	V_{Na}	$\bar{g}_K$	n	V_K	$\bar{g}_L$	V_L	
colchicine		10 µM colchicine is effective in destroying excitability	@	o	*		o	o		o	o	2, 6
ions	Ca^{2+}	100 µM Ca^{2+} ions suppress the excitability when applied internally for several min.	@	*			@			o		6, 7
	Mn^{2+}	1 mM Mn^{2+} ions have no effect on excitability	*	*	*	*	*	*	*			6
	anions	These ions maintain excitability in this decreasing order of effectiveness-glutamate~F>Cl>Br>I.	@				@					6, 8
protein modifying reagents	SH	These reagents block excitability at 1 mM within about 30 min.	@		*							6, 9
	histidine		o		*							
temperature		The excitability is lost above 35°C.	@	o	o		o	o	*	*	*	10, 11
proteolytic enzymes	CBP-A	0.1-1 mg/ml enzyme dissolved in a standard internal perfusion fluid suppresses the electrical excitability. The time required for the suppression depends on dose.	@	*	*		@	*	*	o	*	12
	pronase ficin trypsin chymotrypsin		o	*	@				o			12 13 14

carboxypeptidase A which removes the carboxy-terminal tyrosine
of α-tubulin always reduces the maximum sodium and potassium
conductances.

All of the effects of various MT assembly inhibitors upon
the Hodgkin-Huxley parameters, suggest that it is axoplasmic
MTs consisting of tyrosinated tubulin (2, 6) which may regulate the maximum sodium and potassium conductances (2).

IV. RESTORATION OF DETERIORATED MEMBRANE EXCITABILITY

Perfusion of the squid axon with purified MT proteins

The experimental results so far described lead to the
suggestion that axoplasmic MTs, consisting of tubulins retaining C-terminal tyrosine, may regulate the maximum sodium
and potassium conductances. In order to confirm this more
directly, we have investigated the experimental conditions
necessary to restore deteriorated excitability in squid giant
axons, and have been partially successful (2, 16, 17).

In the first restoration experiments, the excitability of
the axon was first deteriorated by a MT inhibitor using the
internal perfusion technique originally developed by Tasaki
(18). Then the axon was perfused with purified porcine brain
MT proteins dissolved in the standard perfusion medium.
However, this kind of experiments always failed to restore the
excitability. Even when a very fresh axon was perfused with
tubulin, the excitability was lost within a short period of
time.

*Requirement for factors other than purified tubulin for
restoration*

Addition of tyrosine, ATP and Mg^{2+} as well as K^+ to the
tubulin-containing perfusion medium favored restoration of the
excitability, which previously had been deteriorated by a MT
inhibitor (16). This mixture is similar to that used in the
in vitro reaction system for tubulin-tyrosine ligase. In
fact, further supplementation of the partially purified enzyme
has a pronounced effect on the restoration (2, 17). This
enzyme was highly purified from poricine brains by Murofushi
(19) using Sepharose-sebacic acid hydrazide-ATP affinity
chromatography and Sepharose-tubulin affinity chromatography.

Furthermore, we found another protein factor which could
be extracted from the axon whose axoplasm had been squeezed
out by a roller (3) (Fig. 3B, c and d). Because the
excitability is quickly lost by perfusing the axon with

Ca^{2+}-containing medium, the axoplasm-extruded axons (each cut longitudinally into two) were treated with a Ca^{2+}-containing buffer solution to extract the protein factor.

Finally, we identified the factors necessary for the restoration. First, porcine brain tubulin, tyrosinated by added tubulin-tyrosine ligase in the presence of ATP, Mg^{2+} and K^+, is required for this reaction. This tyrosinated tubulin will polymerize into MTs under these experimental conditions to about half the maximum level (3). Second, the 300 K dalton protein fraction is necessary for maintaining and stabilizing the restored excitability. If this protein fraction is omitted, the magnitude of the restoration is limited and soon declines (16). The possibility can be considered that this protein is located on the axoplasmic face of the axolemma. Third, the restoration is further supported by the addition of

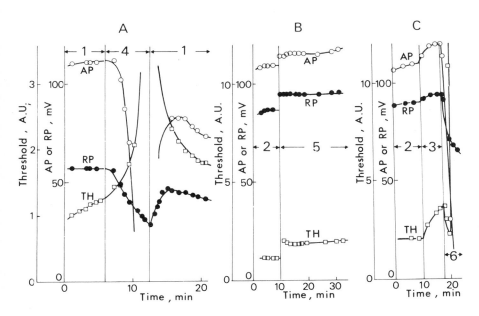

Fig. 4. Effect of phosphatases on the excitability of squid giant axons. In (A), perfusion medium contained 330 mM K-glutamate, 50 mM KF, 20 mM K-phosphate buffer at pH 7.2 (solution 1) and 1 unit/ml alkaline phosphatase (solution 4). In (B), perfusion medium consisted of 340 mM KF, 20 mM K-HEPES at pH 7.2 (solution 2) and 1 unit/ml phosphatase from E. coli. (solution 5). In (C), perfusion medium contained 340 mM K-glutamate and 20 mM K-HEPES at pH 7.2 (solution 3) and 1 unit/ml phosphatase (E. coli) (solution 6). AP, action potential; RP, resting potential; TH, threshold.

cAMP. This suggests that the phosphorylation of some components in the reaction system may be necessary for the restoration process.   In fact, perfusion of fresh axons with
K-glutamate perfusion medium containing phosphatases destroys
the excitability within a short period of time (Fig. 4).

    In the best recovery, the action potential is restored to
84 % after a lag phase (Fig. 5).   The threshold is recovered
to 1.85 times the control at 146 min, further decreasing

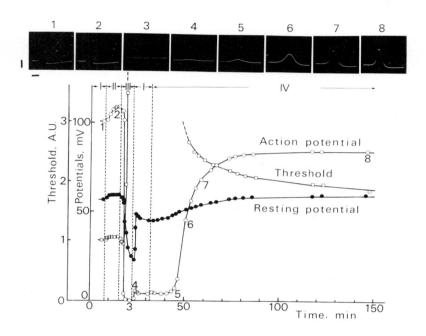

*Fig. 5.   Restoration of membrane excitability by tubulin-tyro-
sine ligase, MT proteins, 300 K protein fraction, reagents
which activate the ligase, and cAMP.   The axon was sequential-
ly perfused with solutions I through IV at 22°C.   Solution I,
360 mM KF and 40 mM K-phosphate buffer (pH 7.2); II, 360 mM
K-glutamate and 40 mM K-HEPES (pH 7.2); III, solution II
containing 0.2 mM CaCl$_2$ and 1 mM MgCl$_2$; IV, solution I containing 0.1 mg/ml 300 K protein fraction, 1 mg/ml 3-cycled
bovine brain MT proteins, 3 units/ml tubulin-tyrosine ligase,
0.5 mM ATP, 30 uM tyrosine, 0.5 mM MgCl$_2$ and 3 µM cAMP.   Eight
representative oscillograph records are shown on the top.   The
numbers refer to the respective positions in time along the
action potential records.   The horizontal and vertical bars
stand for 1 msec and 40 mV, respectively.*

thereafter. The resting potential recovers to a level even
lower than that of the control. The lag phase seems to be due
to the time necessary for microtubule assembly in the vicinity
of the axolemma.

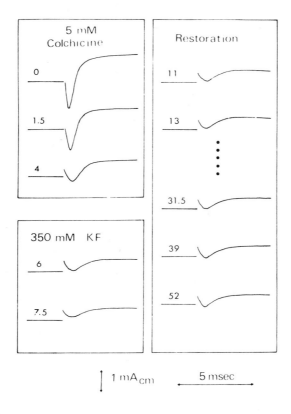

Fig. 6. Suppression and restoration of the membrane excitability of the voltage-clamped axon. The holding potential and
the amplitude of the clamping potentials were fixed at -50 and
45 mV, respectively. The Na and K currents were suppressed by
internally perfusing the axon with SIM containing 5 mM
colchicine (three records on the upper left). Then, the internal solution was switched to SIM at 4 min 45 sec after the
onset of the colchicine perfusion, causing no recovery (two
records on the lower left). At 8.5 min, the solution was
changed to SIM containing 1 mg/ml squid optic ganglia MT
proteins three-cycled, 40 µg/ml purified 300 K protein, 3 mM
$MgCl_2$ and 0.5 mM GTP (records on the right). Each figure represents time after the onset of the colchicine perfusion.
Temperature, 16°C.

Attempt to restore the Hodgkin-Huxley parameters, particularly
maximum sodium and potassium conductances

In the restoration experiments described above, the action
potential, resting potential, and threshold current have been
monitored as gross parameters of the excitability of squid
giant axons. In order to examine whether this recovery is due
to the restoration of the deteriorated Hodgkin-Huxley parameters, particularly maximum sodium and potassium conductances,
the restoration experiments were performed under voltage-clamp
(20). The internal perfusion technique (the roller method)
developed by Baker, Hodgkin and Shaw (21) was adopted to remove the axoplasm more completely.

Because the effect of Ca ions is not specific for MTs,
colchicine was used for deteriorating the excitability. When
the axon was perfused with standard internal medium (SIM,
370 mM KF, 30 mM K-HEPES and 4 % (v/v) glycerol at pH 7.2)
containing 5 mM colchicine for 5 min, both inward Na and outward K currents quickly decreased, and even after the perfusion medium was replaced with SIM, both currents continued to
decrease (Fig. 6). A restoration medium was then introduced
into the axon. The system consisted of 3×-cycled MT proteins
from squid optic ganglia, 300 K protein, GTP, Mg^{2+} and K^{+}.
Figure 6 shows a sign of recovery in the Na current after a
period of time. Using the same axon, the leak current was
found to increase with time during and after the colchicine
treatment (Fig. 7). Upon switching to the restoration medium.

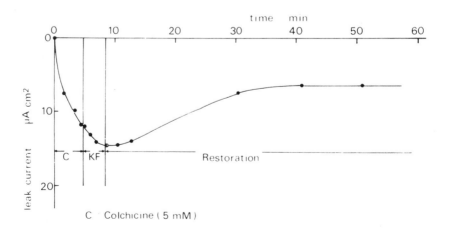

Fig. 7. Changes in the leak current at the holding potential
of -50 mV. The same experiment as in Fig. 6.

the leak current began to decrease corresponding to a partial
recovery of the resting potential. In this system, tubulin-
tyrosine ligase was not used, since more than 94 % of the
α-tubulin, which can be tyrosinated, possessed carboxyl terminal tyrosine when prepared from squid otpic ganglia or
axoplasm (15). In contrast, about half of the tubulin purified from porcine brain contained α-subunits which lacked the
tyrosine residue, and was incapable of restoring the action
potential when examined in the absence of tubulin-tyrosine
ligase. The absence of ligase in the restoration medium may
be one of the reasons why full restoration of the sodium
current was not attained. Full restoration of the sodium and
potassium currents is left for future study.

It should be pointed out here that if GTP was left out of
the restoration system, it failed to restore the membrane
excitability. This clearly indicates that a lack of MT assembly, by omitting GTP, suppresses the restoration, and
again leads us to the conclusion that the MTs underlying the
axolemma have an important role in maintaining membrane
excitability.

V.   CONCLUSION

A close correlation between the medium conditions necessary for axoplasmic MT assembly and those for the maintenance
of the membrane excitability of the squid giant axon was
described. When the axon is internally perfused with a medium
unfavorable for MT assembly, excitability is suppressed, as
measured by a decrease in the maximum sodium and potassium
conductances. Establishment of a restoration system for the
deteriorated giant axon provided evidence that tyrosinated
tubulin or MTs are necessary in order for the action potential
to be evoked, as well as for steady maintenance of the resting
potential. More stable restoration was attained by adding a
300 K protein fraction, isolated from axons, and cAMP to the
restoration system.

Voltage-clamping of the axon to quantify the recovery process of the Na, K and leak currents indicated that tyrosinated
squid tubulin with GTP and 300 K protein supported the return
of membrane excitability after the axon had been deteriorated
by 5 mM colchicine. These results strongly indicate that the
excitation of squid giant axons requires tyrosinated MTs and
probably 300 K protein, in close association with the
axolemma.

REFERENCES

1. Endo, S., Sakai, H., and Matsumoto, G., *Cell Struct. Funct.*
    4, 285 (1979).
2. Matsumoto, G., Murofushi, H., Endo, S., Kobayashi, T., and
    Sakai, H., *in* "Structure and Function of Excitable Cells"
    (D.C. Chang, I. Tasaki and W.J. Adelman, Jr. eds.), Plenum
    Press, New York (1982).
3. Sakai, H., and Matsumoto, G., *J. Biochem. 83*, 1413 (1978).
4. Matus, A., Bernhardt, R., and Hugh-Jones, T., *Proc. Natl.*
    Acad. Sci. USA. 78, 3010 (1981).
5. Pant, H. C., Shecket, G., Gainer, H., and Lasek, R. J.,
    J. Cell Biol. 78, 396 (1978).
6. Matsumoto, G., and Sakai, H., *J. Membrane Biol. 50*, 1
    (1979).
7. Gillespie, J. I., and Meves, H., *J. Physiol. (London) 315*,
    493 (1981).
8. Tasaki, I., Singer, I., and Takenaka, T., *J. Gen. Physiol.*
    48, 1095 (1965).
9. Baumgold, J., Matsumoto, G., and Tasaki, I., *J. Neurochem.*
    30, 91 (1978).
10. Kimura, J. E., and Meves, H., *J. Physiol. (London) 289*,
    479 (1979).
11. Matsumoto, G., and Ichikawa, M., to be submitted to
    Biophys. J.
12. Sevcik, C., and Narahashi, T., *J. Membrane Biol. 24*, 329
    (1975).
13. Armstrong, C. M., Bezanilla, F., and Rojas, E., *J. Gen.*
    Physiol. 62, 375 (1973).
14. Rojas, E., and Rudy, B., *J. Physiol. (London) 262*, 501
    (1976).
15. Kobayashi, T., and Matsumoto, G., to be submitted to *J.*
    Biochem.
16. Matsumoto, G., and Sakai, H., *J. Membrane Biol. 50*, 15
    (1979).
17. Matsumoto, G., Kobayashi, T., and Sakai, H., *J. Biochem.*
    86, 1155 (1979).
18. Tasaki, I., "Nerve Excitation: A Macromolecular Approach"
    Charles C. Thomas, Springfield, Ill. (1968).
19. Murofushi, H., *J. Biochem. 87*, 979 (1980).
20. Hodgkin, A. L., "The Conduction of the Nervous Impulse"
    Liverpool University Press, Liverpool (1964).
21. Baker, P. F., Hodgkin, A. L., and Shaw, T. I., *Nature*
    (London) 190, 885 (1961).

CHAPTER 36

NEW ROLES FOR TUBULIN IN MEMBRANE FUNCTION

R.D. Berlin, J.M. Caron, R. Huntley,
R.N. Melmed and J.M. Oliver

Department of Physiology
University of Connecticut Health Center
Farmington, Connecticut

It has been become clear that in many animal cells microtubule disassembly caused by drugs such as colchicine leads to marked changes in surface topography. For example, microtubule disassembly in macrophages and leukocytes leads to the formation of a bulge or protuberance: coated pits are localized (1), and pinocytosis and phagocytosis are restricted to this protuberance (2); various surface-bound ligands bind first to other regions but migrate to the protuberance (3). Elsewhere we have reviewed evidence for these topographical changes as well as the mechanism of ligand-receptor movement (4). More recently we have uncovered several physiological phenomena which indicate effects of microtubule disassembly on membrane transport processes responsible for regulating cell volume and intracellular pH (5). These suggest an even more direct effect of microtubules or tubulin on membrane structure. They may be particularly significant in the light of numerous reports of the presence of tubulin itself as a constituent of membranes (e.g.6). Recent studies of the interaction of tubulin with liposomes lends support to the notion that tubulin may directly influence membrane structure (7). In this chapter we will first outline the effects of microtubule disassembly on cell volume and intracellular pH. We will then describe recent experiments on the interaction of tubulin with neutral and acidic phospholipid liposomes with particular attention to the evidence for reversibility of binding and its implications for membrane organization.

Effect of Microtubule Disassembly on Cell Volume

We have explored the possibilities that cell volume is regulated by the status of microtubule assembly and cyclic AMP metabolism and may be coordinated with shape

Biological Functions of Microtubules
and Related Structures

changes (8). Treatment of J774.2 mouse macrophages with colchicine caused rapid microtubule disassembly and was associated with a striking increase (from 15-20% to approximately 90%) in the proportion of cells with a large protuberance at one pole. This provided a simple experimental system in which shape changes occurred in virtually an entire cell population in suspension. Parallel changes in cell volume could then be quantified by isotope dilution techniques. We found that the shape change caused by colchicine was accompanied by a decrease in cell volume of approximately 20% (Fig. 1) Nocodazole, but not lumicolchicine, caused identical changes in both cell shape and cell volume. The volume loss was not due to cell lysis nor to inhibition of pinocytosis.

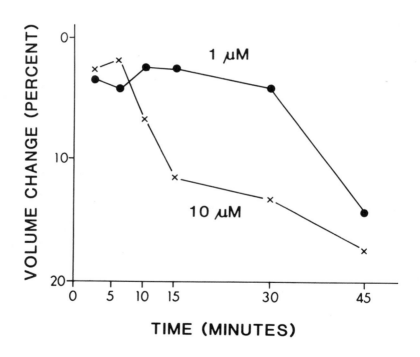

Figure 1. Effect of colchicine on cell volume. In these experiments, triplicate samples of cell suspension (4-8 x 10^5 cells/250 liter suspension) were layered over oil and centrifuged. ^{3}H-H$_2$O and

^{14}C-inulin were added to cell suspensions at the beginning of the incubation time, and the cell volume was determined at the times shown. Isotope equilibration is essentially instantaneous. The graph depicts the mean of seven experiments for each concentration (1 and 10 μM of colchicine). 5 μM colchicine (not shown here) gave an effect intermediate between those of the two concentrations shown. Control suspensions showed increases in volume of 0-4% over the course of the experiment as would be expected from the continued growth of cells in complex medium. Standard errors for all points are 2-3%. Volume changes were calculated relative to control cells at comparable times of incubation. (From 8).

The mechanism of volume loss was also examined. Colchicine induced a small (22%) but reproducible increase in activity of the ouabain-sensitive Na^+,K^+-dependent ATPase. However, inhibition of this enzyme/transport system by ouabain did not change cell volume nor did it block the colchicine-induced decrease in volume. On the other hand, SITS (4' acetamido, 4-isothiocyano-2,2'-disulfonic acid stilbene), an inhibitor of anion transport, inhibited the effects of colchicine, thus suggesting a role for an anion transport system in cell volume regulation (Fig. 2).

Because colchicine is known to activate adenylate cyclase in several systems and because cell shape changes are often induced by hormones that elevate cyclic AMP, we also examined the effects of cyclic AMP on cell volume. Agents that act to increase cyclic AMP (cholera toxin, which activates adenylate cyclase; IBMX, an inhibitor of phosphodiesterase; and dibutyryl cyclic AMP) all caused a volume decrease comparable to that of colchicine. To define the effective metabolic pathway, we studied two mutants of J774.2, one deficient in adenylate cyclase and the other exhibiting markedly reduced activity of cyclic AMP-dependent protein kinase (Table I). Colchicine reduced the volume of each mutant. In contrast, cholera toxin did not produce a volume change in either mutant. Cyclic AMP produced a decrease in the cyclase-deficient line comparable to that in the wild type, but did not cause a volume change in the kinase-deficient line. This analysis established separate roles for cyclic AMP and colchicine. The volume decrease induced by cyclic AMP requires the action of a cyclic AMP-dependent protein

Table I

Effect of Dibutyryl cAMP, IBMX, and Cholera Toxin on Macrophage Cell Volume

Mean percentage volume change*

Agent	Wild Type	n	Protein kinase deficient ($J7H_1$)	n	Adenylate cyclase deficient (CT_2)	n
Colchicine (10 M)	−20 (±1.75)	13	−17 (±2)	13	−13 (±1.25)	11
Dibutyryl cAMP (0.46 mM)	−18 (±6)	3	+1 (±3)	5	−12 (±4)	2
IBMX (1.2 mM)	−17 (±3.5)	3	−1 (±5)	4	+4 (±1.5)	2
Cholera toxin	−18 (±0.3)*	3	0 (±2)++	4	+1 (±4)++	3

In this table the volume change was determined at the end of 1 hr incubation
with dibutyryl cAMP, isobutylmethylxanthine (IBMX), and cholera toxin. 8 bromo-
cAMP (0.5 mM), not shown here, produced a volume reduction of 15% (n=4) in the
J774.2 cell line but was not tested in the mutant cells. Choleragenoid, the
inactive B subunit and competitive antagonist of cholera toxin (a gift of Professor
R. Finkelstein, University of Texas) did not produce a volume change in J774.2.
The volume decrease induced by dibutyryl cAMP was dose dependent; 0.11 mM, −6%;
0.23 mM, −12%; 0.46 mM, −18%. IBMX produced a 9% decrease at 0.15 mM (two experiments).
Mean (±SEM); n = number of separate experiments. From (8).
* 100 ng/ml.
++ 2.5 µg/ml.

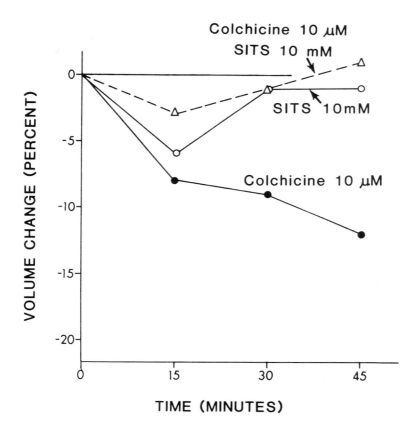

Figure 2.   Effect of SITS on the colchicine-induced
volume change. The reagents, 10 mM SITS (O), 10 μ M
colchicine (●), and 10 mM SITS + 10   μ M colchicine
(△), were added at time zero. The cell suspensions
were then placed in a 37°C water bath and agitated
frequently for approximately 10 min to ensure that
the SITS had properly dissolved. 0.25 mM HEPES
buffer (pH 7.4) was added to the medium before the
start of the experiment to minimize pH fluctuations.
The SITS was used in complete, protein-containing
medium. As noted in the text, a large part of the
SITS becomes tightly coupled to the protein under
these   conditions.   The   precise   SITS-free
concentration is thus unknown. The importance of
stabilizing cell volume through the use of the
normal   complex   medium   was   given   overriding
consideration. From (8).

kinase. The decrease caused by colchicine, on the other hand, does not require the kinase. In the context of this chapter, the most important concept to emerge from this study is that colchicine treatment appears to activate specific membrane functions: the ouabain-sensitive Na^+,K^+-ATPase and another (SITS-sensitive) system that leads to changes in volume. Of course, these findings do not allow one to conclude that there exist direct interactions between tubulin or microtubules with distinct membrane proteins. We shall see, however, that there are strong reasons to consider this possibility.

Effects of Microtubule Disassembly on Intracellular pH

A second example of changes in membrane function brought about by anti-microtubule agents is in the regulation of intracellular pH. In studies of the influence of pH <u>in vitro</u> (9), we had found that increasing pH causes rapid microtubule disassembly (Fig. 3) and that the action of podophyllotoxin is enhanced progressively with pH (Fig. 4).

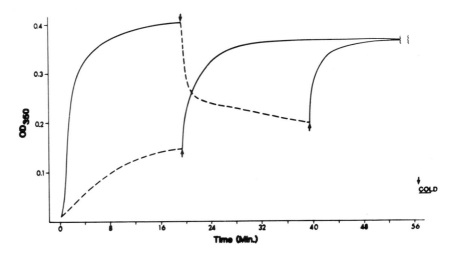

Figure 3. Microtubule assembly and disassembly at different pH values. The protein solution was incubated for 2 min at 30°C before initiation of assembly by Mg-GTP. The arrows indicate changes in pH induced by addition of Tris base (arrow down) or HCl (arrow up). Solid line indicates more acid condition, pH 6.65; broken line indicates more alkaline, pH 7.63. Total protein: 3mg/ml. From (9).

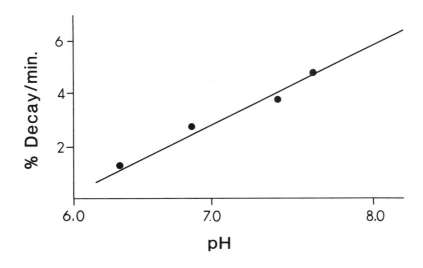

Figure 4. Summary graph of PLN disassembly as a function of pH. In these experiments, assembly was at pH 6.24. Aliquots were then alkalinized to varying pH by the addition of small volumes of concentrated Tris. After 6 min when a new steady state was nearly established, 50 μ M PLN was added as a X 100 concentrated solution in ethanol and the initial rate of disassembly measured. The results given as percent decrease of the initial OD are plotted as a function of the final measured pH. The line is drawn by least square fit: % decay/min = - 13.97 + 2.41 (pH), r^2 = 1.0. It should be emphasized that PLN blocks assembly at all pH. The experiment illustrated was performed with one tubulin preparation. Some variation between tubulin preparations was encountered, but the relative percent decay as a function of pH was approximately the same, though usually greater in magnitude than illustrated.

To test this as a possible regulatory mechanism in intact
cells, we sought to artificially manipulate intracellular
pH and examine the relative efficacy of anti-microtubule
drugs as a function of pH. To our surprise we found that
colchicine itself caused an increase in intracellular pH
(5).

We measured pH by the DMO equilibrium method. DMO
is a weak acid, 5,5 dimethyl-2,4-oxazolidinedione, whose
uncharged (undissociated) form freely crosses cell
membranes. It is therefore partitioned across membranes
according to the trans-membrane pH gradient, a more
alkaline pH tending to 'trap' the dissociated species.
It has been in widespread use since its introduction by
Waddell and Butler (10) and provides an estimate of mean
intracellular pH.

Triplicate parallel samples of cell suspensions were
incubated with either ^{14}C-inulin and ^{3}H-water or with
^{14}C-DMO. The cells were then pelleted by centrifugation
through a layer of silicone oil which removes the bulk of
the extracellular fluid from the cell pellet.
Radioactive counts in the pellet as ^{3}H-water provide a
measure of total pellet water; ^{14}C-inulin, of pellet
extracellular fluid. The concentration of DMO in the
cell water of parallel samples can then be calculated.
DMO concentration and pH of the medium (pH$_o$) are
determined directly on supernatants. Intracellular pH
(pH$_i$) was calculated according to the formula:

$$pH_i = pK_{A_{DMO}} + LOG\left(\left(\frac{(DMO)_i}{(DMO)_o}\right)\left(10^{(pH_o - pK_{A_{DMO}})} + 1\right) - 1\right)$$

As shown in Fig. 5, colchicine produced an elevation of
between 0.1 and 0.15 pH units. An increase was also
produced by podophyllotoxin or nocodazole but not by
lumicholicine indicating that the drug effects are
related to microtubule disassembly.

From Fig. 5 it is clear that intracellular pH (pH$_i$)
tends to parallel the extracellular pH (pH$_o$) but that
colchicine leads to a relative elevation of pH$_i$ as
compared to pH$_o$ throughout.

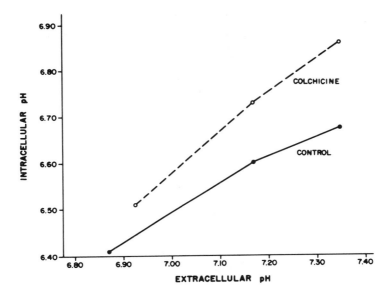

Figure 5. Effect of colchicine on intracellular pH
as a function of extracellular pH.   Suspensions of
cultured J774.2 mouse macrophages were incubated
with or without colchicine at 1 μ M in complete
medium adjusted to varying pH with MES or Tricine
buffers.   After 45 min.   Intracellular pH was
measured by the DMO method. Podophyllotoxin affects
intracellular pH similarly; lumicolchicine has no
effect.

This implies an active alkalinization by the cell, most
probably through proton extrusion. In animal cells this
is likely to occur by exchange of H^+ for Na^+ or by
exchange of HCO_3^- for Cl^- (11,12).   Mechanisms of the
Na^+/H^+ exchange may be sensitive to amiloride (11).
Indeed, we found that amiloride blocked the intracellular

alkalinization produced by colchicine (Fig. 6).

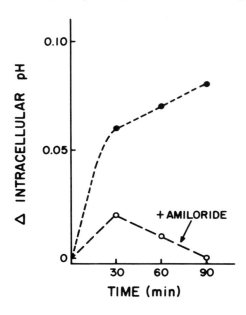

Figure 6. Shift in intracellular pH by colchicine
± amiloride, 3,5-diamino-N-(aminoiminomethyl)-6-
chloropyrazine-carboxamide. Intracellular pH was
measured by the DMO method after incubation for
varying intervals with 1 μM colchicine alone or in
the presence of 0.5 mM amiloride.

In other experiments we showed that amiloride did not
affect the uptake of colchicine, ruling out the trivial
possibility that the inhibition of colchicine-induced
alkalinization was due simply to a block of drug entry.
On the basis of these findings, a reasonable hypothesis
is that anti-microtubule agents activate an amiloride-
sensitive proton extrusion pump. Our preliminary
evidence suggests that the consequence of the resultant
elevation of pH$_i$ is to accelerate the action of
colchicine. We will not discuss this further here (See
Ref. 5). For our purposes the pump activation serves as
an   additional   example   of   the   effects   of
microtubules/tubulin on specific membrane functions.

How do such effects come about? One possibility is
that microtubule disassembly acts indirectly by setting
into motion a series of events that eventually affects
membrane function. For example, the marked cell shape
changes induced by microtubule disassembly lead to the
formation of a highly villous membrane (13). Regional
variations in membrane structure caused by extremes in
curvature could alter the orientation, function and even
distribution of embedded proteins. On the other hand,
tubulin may interact directly with membranes altering
their structure and function. The way in which anti-
microtubule agents could affect this process has recently
been clarified (14).

The Interaction of Tubulin with Membranes

The presence of tubulin in biological membranes has
now been reported by several investigators (e.g.
6,15,16). So far these studies have not allowed analysis
of the mechanism or consequence of tubulin interaction
with membranes. To isolate these processes, we began
studies of the interaction of purified microtubule
proteins with simple phospholipid bilayer vesicles (7).
These analyses demonstrated selective adsorption of
microtubule proteins from brain extracts or solutions of
tubulin purified by phosphocellulose chromatography.
Adsorption to vesicles has been confirmed (17) and
extended to natural membranes purified from mammalian
liver (18).

Two recent findings (14) are of particular potential
physiological significance: 1) binding of tubulin to
liposomes disrupts their structure, and 2) binding of
tubulin is reversible depending on the phospholipid
composition of the vesicle membrane.

We showed earlier that tubulin binding led to a
stacking and/or fusion of liposomes demonstrable by
negative stain (7). As a corollary it can be shown that
tubulin at very low concentrations can cause liposomes to
leak (14). Figure 7 shows the pattern of calcein release
from lipsomes caused by varying concentration of 6S,
phosphocellulose-purified   tubulin.   Corresponding
effects on liposome structure can be demonstrated by
negative stain with uranyl acetate. Unlike Klausner et
al (17) we find that binding at the critical transition
temperature of the lipid is not necessary for binding or
leakage to occur.

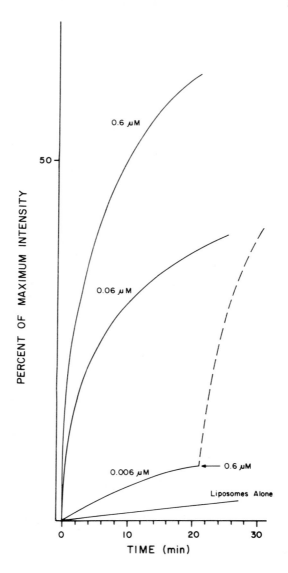

Figure 7

Figure 7.   MT protein-induced leakage of calcein
from phospholipid vesicles.   MT protein at final
concentrations of 0.6 µM, 0.06 µM, and 0.006 µM
was added to calcein-containing liposomes of egg
lecithin (10 µM) at 30°C. Leakage of calcein from
vesicles was measured as an increase in fluorescence
intensity. The rate and extent of dye release was
proportional to the protein concentration. Maximum
fluorescence intensity was determined by adding
deoxycholate to a final concentration of 0.025%
which completely disrupts the liposomes.

Most importantly, we have established that binding
of microtubule proteins to neutral phospholipid vesicles
can be reversed. The addition of GTP and magnesium
causes the assembly of microtubules in considerable
excess of what can be accounted for by the presence of
unadsorbed microtubule protein. In order to establish
this formation of microtubules from proteins previously
adsorbed to liposomes, we first established a
quantitative morphometric assay for microtubule
assembly.

The assay was based on determination of the combined
length of microtubules deposited on grids and negative
stained. To ensure accurate counting of microtubule
length, a 1/10 dilution of the microtubule sample in
microtubule stabilizing buffer was required before
negative staining. Because dilution is known to cause
disassembly of microtubules, it was necessary to show
that the microtubule stabilization buffers employed
were, indeed, preventing disassembly. Therefore,
microtubule protein, assembled at 1 mg/ml for 40 min at
30°C, was diluted 1/10, 1/20, and 1/50 in stabilizing
buffer, stained and processed according to a precise
protocol. A plot of total microtubule length per
photograph vs. assembled microtubules yields a straight
line with a correlation coefficient (r) of 1.00 (Fig.
8), thus indicating that dilution in the range tested
does not cause measurable disassembly of microtubules.
We then were able to establish the correlation between
total microtubule length measured on micrographs and the
mass of assembled microtubule in mg microtubule
protein/ml.

Microtubule protein, pre-incubated for 30 min. at
30°C was assembled at three known concentrations (1.00,

0.75 and 0.50 mg/ml) to steady state (40 min at 30°C), and processed for electron microscopy. Microtubule lengths were measured and the corresponding degree of microtubule assembly determined directly after pelleting by centrifugation. The linear relationship obtained is shown in Fig. 9.

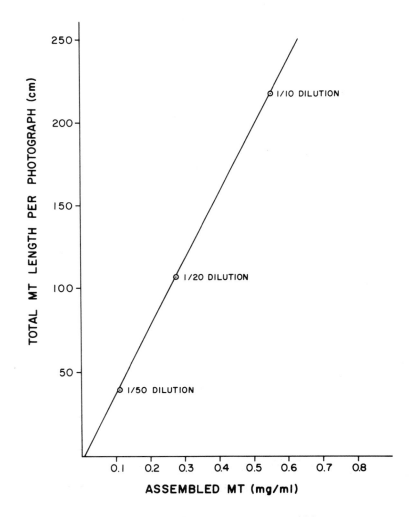

Figure 8. Legend on page 420.

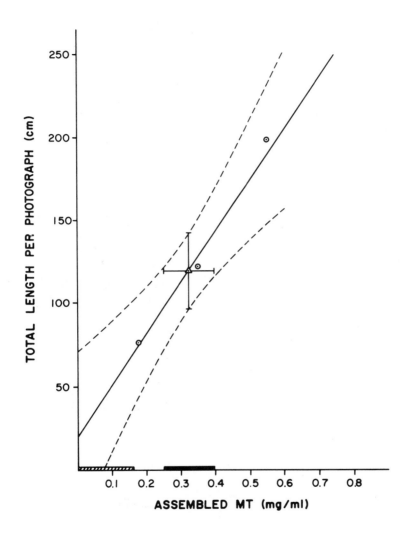

Figure 9. Legend on page 420.

(See Figures 8 and 9 on preceding pages)

Figure 8.   Stability of microtubules following dilution for negative staining.   MT protein, assembled at 1 mg/ml for 40 min. at 30°C, was diluted 1/10, 1/20, and 1/50 in stabilization buffer and immediately processed for negative staining. Pictures were photographed from 5 pre-determined positions per grid, enlarged to X 17,820, and MT assembly was quantitated.

Total MT length per photograph was plotted against assembled MT yielding a straight line with a correlation coefficient (r) of 1.00.

Figure 9.   Relation of microtubule length to assembled microtubules.   (0) MT protein, pre-incubated for 30 min. at 30°C, was assembled at 3 concentrations (1.00, 0.75, 0.50 mg/ml).   From the relationship of measured length to assembled protein at these concentrations a standard line was drawn. The mass of microtubules formed from liposomes, after introducing GTP to DMPC liposomes pre-adsorbed with MT protein, was calculated to be 0.25 -0.40 mg/ml with 95% confidence (▬▬).   This is significantly different from the zero point of assembled MT (▨▨▨).

Having this standard relationship in hand, microtubule protein at 1 mg/ml was first adsorbed onto DMPC vesicles for 30 min at 30°C, then GTP and Mg were introduced and the incubation continued for an additional 40 min at 30°C before negative staining. The microtubule lengths were then measured. From the correlation curve, Fig. 9, the quantity of microtubules assembled was calculated from this measurement to be between 0.25 and 0.40 mg/ml (values at the 95% confidence limits).   By contrast no microtubules are seen by negative stain when microtubule protein-liposome complexes are centrifuged out before assembly conditions are introduced.   This is consistent with measurements of the nonadsorbed protein after separation from liposomes on Ficoll gradients: approximately 0.23 mg/ml (roughly the critical concentration).   According to the statistical correlation, this zero point of assembled microtubules (amount of assembly in the absence of GTP) corresponds to 0.016 mg/ml (the range of the 95% confidence interval at this point).   By contrast the amount of microtubules

formed from liposome-protein mixtures shown by the point
△ in Fig. 9 is 0.25 - 0.40 mg/ml which is significantly
greater than expected from the amount of nonadsorbed
protein present initially.   Furthermore, 0.25 - 0.40
mg/ml assembled microtubule corresponds to 0.61 - 0.81
mg/ml total protein. Since the amount of nonadsorbed
protein present initially was 0.23 mg/ml, this implies
that the difference of 0.38 to 0.58 mg/ml, or 52 - 80% of
the adsorbed microtubule protein is desorbed from DMPC
vesicles after introduction of assembly conditions.
Blocking assembly by colchicine prevents desorption.

In contrast, microtubule protein binds irreversibly
to vesicles containing the acidic phospholipid,
phosphatidylserine (PS).   When microtubule protein
(1 mg/ml) is pre-incubated with PC/PS liposomes which
contain 50% (w/w) PS for 30 min at 30°C
(phospholipid:protein, 4000:1), no microtubule assembly
occurs after addition of GTP-Mg.   From SDS gel
electrophoretic analysis, this irreversible binding of
microtubule protein to PS-containing liposomes does not
appear to result from PS-containing vesicles binding
proteins which are different from those adsorbed onto
DMPC vesicles.

Implications

What do these observations tell us about the way in
which tubulin bound to biological membranes could affect
their structure and how could anti-microtubule agents
affect such mechanisms? Some possibilities are indicated
in the following diagram (Fig. 10). We suppose that
there is a steady-state of tubulin bound to membranes
reflecting both adsorption and desorption of tubulin.
Since microtubule assembly may be limited to certain
regions, for example, at the tips of cilia, it is
reasonable to suppose that the adsorption and desorption
of tubulin are spatially separate. This would provide
for a diffusion of tubulin on the membrane from
adsorption to desorption sites. Further, since tubulin
'bound' to phospholipids such as PS would not be
available for desorption, there would be no gradient for
diffusion between these sites and therefore a partial
separation of phoshatidyl choline and phosphatidyl serine
within the membrane could result. We do not believe that
these abundant phospholipid species would be partitioned
to any significant degree.   Rather we take them to
represent parts of a spectrum of membrane constituents

having various affinities or interactions with bound
tubulin and thus being transported or partitioned to
particular membrane regions.   The effect of anti-
microtubule   agents   would   be   to   effectively   block
desorption, arrest tubulin diffusion in the membrane and
thus the associated segregation of membrane constituents.
It follows from this hypothetical mechanism, applied to
the biogenesis of cilia, for example, that there are
regional differences in membrane lipid composition along
its length and that these differences would be eliminated
by agents that block microtubule assembly.   Further, the
cumulative evidence based on liposome leakage and
distortion by negative stain suggest direct effects of
tubulin binding on membrane mechanical stability.

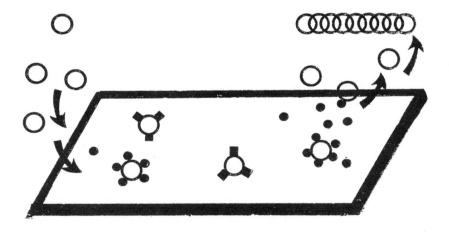

Figure 10.   Adsorption of microtubule proteins O
(left) onto membrane and desorption (right) onto
microtubules.   The protein may be adsorbed to
neutral ● or acidic ■ phospholipids but can only
be <u>de</u>sorbed from the former.   This leaves an excess
of neutral phospholipid at the desorption site which
is free to diffuse to the adsorption site and repeat
the cycle.   Tubulin bound to acidic phospholipids
will be dispersed more or less uniformly throughout.

In this way not only may tubulin binding to
membranes directly affect their structure and function,
but in addition a cycle of tubulin adsorption and
desorption driven by microtubule assembly may create
gradients of composition within the membrane. The
regulation of microtubule assembly/disassembly could
thus indirectly affect membrane function by altering the
segregation of constituents. This mechanism does not
eliminate others in which microtubules affect membrane
curvature or alter the interaction of other molecules
with the membrane. It does, however, suggest a testable
alternative and indicate further complexities in the
action of anti-microtubule agents.

REFERENCES

1. Pfeiffer, J.R., Oliver, J.M., and Berlin, R.D.
    Nature 286, 727 (1980).
2. Walter, R.W., Berlin, R.D., Pfeiffer, J.R. and
    Oliver, J.M. J. Cell Biol. 867, 199 (1980).
3. Berlin, R.D., and Oliver, J.M. J. Cell Biol. 77, 789
    (1978).
4. Oliver, J.M. and Berlin, R.D. Int. Rev. Cytol. In
    Press (1982).
5. Huntley, R., Oliver, J.M. and Berlin, R.D. J. Cell
    Biol. 91, 134a (1981).
6. Dentler, W.L. Int. Rev. Cytol. 72, 1 (1981).
7. Caron, J.M. and Berlin, R.D. J. Cell Biol. 81, 665
    (1979).
8. Melmed, R.N., Karanian, P.J. and Berlin, R.D. J.
    Cell Biol. 90, 761 (1981).
9. Regula, C.S., Pfeiffer, J.R., and Berlin, R.D. J.
    Cell Biol. 89, 45 (1981).
10. Waddell, W. and Butler, T. J. Clin. Invest. 38, 720-
    729 (1959).
11. Aicken, C.C. and Thomas, R.C. J. Physiol. 273, 295-
    316 (1977).
12. Russell, J.M. and Boron, W.F. Nature 264, 73-74
    (1976).
13. Albertini, D.F., Berlin, R.D. and Oliver, J.M. J.
    Cell Sci. 26, 57 (1977).
14. Caron, J.M. and Berlin, R.D. J. Cell Biol.
    Submitted.
15. Bhattacharyya, B. and Wolff, J. J. Biol. Chem. 250,
    7639-7646 (1979).
16. Stephens, R.E. Biochemistry 17, 2882-2891 (1978).

17. Klausner, R.D., Kumar, N. Weinstein, J.N., Blumenthal, R. and Flavin, M. J. Biol. Chem. <u>256</u>, 5879-5885 (1981).

18. Reaven, E. and Azhar, S. J. Cell Biol. <u>89</u>, 300-308 (1981).

CHAPTER 37

FAST AXOPLASMIC TRANSPORT OF
A CALMODULIN-RELATED POLYPEPTIDE

Tomoko Tashiro
Masanori Kurokawa

Department of Biochemistry
Institute for Brain Research
Tokyo University Faculty of Medicine
Hongo, Tokyo

INTRODUCTION

Neurons have developed a specialised system of axoplasmic
transport to convey macromolecules to their axons and terminals which lack the protein synthetic machinery. In most of the
vertebrate nerves studied, axoplasmic transport can be classi-
fied into two major velocity groups; the fast transport with
a speed of 240-400mm per day and the slow transport with a
speed of 1-4mm per day. The two groups differ totally in their
molecular properties as well. The slow transport consists of
a flow of cytoskeletal proteins (1-3), whereas the fast seems
to be a flow of membranous materials (4-10).

Identification of individual protein components in the
fast transport, however, has been hampered by the relative
complexity of molecular species and by the lower radioactive
labelling. We have chosen to investigate the rapidly transported proteins in the guinea pig vagal nerve where the protein
composition of the fast transport seemed simpler. This is
probably due to the relative homogeneity of this system with
respect to the extent of myelination and axon diameter.
Another advantage of the system is the possibility of recover-
ing nerve terminals from the myenteric plexus (11).

Analyses of the transported proteins by gel electrophoresis coupled to fluorography showed that the two proteins with
molecular weights of 26,000 (protein A) and 21,000 (protein B)
were the major components of fast transport in this system

Biological Functions of Microtubules
and Related Structures

(12). The former (protein A) was found to be chemically related to brain calmodulin. Calmodulin *per se* was not found in
the fast transport, but was detected among the slowly transported proteins.

MATERIALS AND METHODS

A. Axonal Transport in the Guinea Pig Vagal Nerve

Concentrated L-(^{35}S) methionine (40–50µCi in 0.25µl) was
injected into the dorsal motor nucleus of the vagus on each
side. At prescribed time intervals, animals were sacrificed.
The vagal nerve was dissected out and cut into 6mm consecutive
segments. Each segment was homogenised in SDS sample buffer
(2% in SDS) and aliquots were subjected to electrophoresis.

B. Gel Electrophoresis and Fluorography

SDS gel electrophoresis was carried out in Tris–glycine
system (13), and two–dimensional electrophoresis according to
O'Farrell (14). The gels were stained with Coomassie Brilliant
Blue R-250 (CBB) and further processed for fluorography (15).

C. Tryptic Peptide Mapping

Bovine brain calmodulin (1.5mg; purified according to ref.
16), or a mixture of calmodulin (1.5mg) and labelled protein A
(isolated by extraction from SDS gels; containing 5000cpm) was
digested with trypsin overnight at 37°C. The freeze–dried
digest was applied to a cellulose powder coated glass plate
(10x10cm) for two–dimensional peptide mapping (12). The plate
was first stained with ninhydrin, photographed and further
processed for fluorography.

D. Extraction of Labelled Proteins from the Nerve

Labelled nerves were frozen in liquid nitrogen, crushed
with air hammer, and homogenised in 0.32M sucrose containing
1mM EGTA and 10mM HEPES (pH 7.2). The homogenate was centrifuged at 1,000 g for 10 min. The sediment was washed , and the
combined supernatant was centrifuged at 100,000 g for 1 h to
yield the particulate and soluble fractions. Solubilisation
of proteins from the particulate fraction was achieved by
0.32% Lubrol PX.

E. *Fluphenazine Affinity Chromatography*

Immobilised Fluphenazine was prepared by coupling Fluphenazine maleate to epoxy-activated Sepharose 6B (17). A small column was packed and equilibrated with 10mM HEPES (pH 7.2) containing 0.2mM $CaCl_2$ and 0.1% Lubrol PX. After loading the sample, the column was washed with 10 volumes of equilibration buffer, then with 10 volumes of the same buffer containing 0.5M NaCl in addition, and finally with 10mM HEPES containing 2mM EGTA and 0.1% Lubrol PX.

RESULTS AND DISCUSSION

A. *Components of the Fast Transport in the Vagal Nerve*

Three hours after isotope injection, labelled proteins in the vagal nerve represent a distinct set of polypeptides conveyed by the fast transport (*Fig. 1a*). This polypeptide pattern is totally different from that of the slow transport seen at 7 days post injection (*Fig. 1b*), which consists mainly of cytoskeletal proteins (actin, α- and β-tubulin).

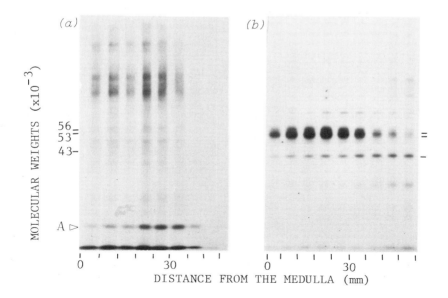

Fig. 1. Fluorographic visualisation of proteins migrating in the guinea pig vagal nerve 3 h (a) and 7 days (b) after labelling with L-(³⁵S)methionine. Gels were 10% in acrylamide.

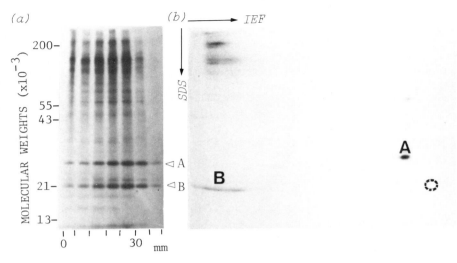

Fig. 2. Protein composition of fast transport as revealed by (a) one- or (b) two-dimensional electrophoresis. Nerves were obtained 3 h after labelling. Gels were 15% in acrylamide for (a) and for the second dimension in (b).

Two polypeptides with molecular weights of 26,000 (protein A) and 21,000 (protein B) are the major components of fast transport in this system, as seen in *Fig. 2a*. On Two-dimensional gel electrophoresis, protein A gave a single spot with a highly acidic isoelectric point of 4.3, while protein B was neutral. The speed of fast transport here is calculated to be 240mm per day.

When the nerve was extracted with isotonic sucrose, more than 90% of the labelled, rapidly transported proteins were found associated with the particulate fraction, consistent with the involvement of membranous structures. Labelled proteins were not solubilised by hypotonic shock, sonication nor EGTA treatment. Part of them including protein A was solubilised with a nonionic detergent Lubrol PX.

B. Homology of Protein A with Calmodulin

The highly acidic nature of protein A together with its low molecular weight suggested that it might be related to a group of acidic, calcium-binding proteins (18). Calmodulin is one of such proteins found ubiquitously, possessing regulatory

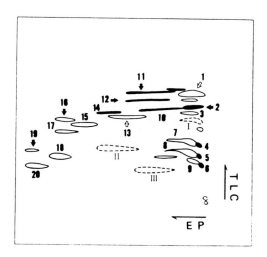

Fig. 3. Comparison between protein A and bovine brain calmodulin by peptide mapping. A mixture of calmodulin and protein A was digested with trypsin. Peptides originating from calmodulin and from protein A were differentially visualised by ninhydrin staining and fluorography, respectively.
Filled spots; radioactively labelled and ninhydrin-positive.
Open spots; ninhydrin-positive, but not radio-labelled.
Dotted spots; radio-labelled, but ninhydrin-negative.
Filled arrows; peptides known to contain methionine.
Open arrows; peptides devoid of methionine (see ref. 16).

functions in a variety of calcium-dependent enzyme systems (19
-21). Calmodulin has considerable sequence homologies to other
members of this group, such as troponin C and brain S-100 protein, and a common calcium-binding sequence has been proposed
for this group of proteins (19, 22). To examine if such sequence homologies exist between protein A and calmodulin, tryptic
peptide mapping was performed. The high methionine content of
calmodulin (9 methionines in the total of 148 amino acids)
makes it feasible to compare the two on the basis of methio-
nine-containing peptides.

Labelled protein A was extracted out of SDS gels, mixed
with a large excess of authentic bovine brain calmodulin and
digested with trypsin. Since the amount of protein A was too
small to give ninhydrin-positive spots on the map peptides
originating from calmodulin and from protein A were distinguished from each other on the same map by ninhydrin staining and
fluorography, respectively.

The results are summarised in *Fig. 3.* Protein A produced

eleven radioactive spots (spots 2, 4-6, 10-12, 14 and I-III),
among which eight had corresponding spots on the ninhydrin
stain, representing peptides common to both protein A and calmodulin. The remaining three (spots I-III) were detected only
by fluorography. Ninhydrin-positive spots that are known to
contain methionine (16) are marked with filled arrows in *Fig.
3*. The fluorographic pattern agrees with these spots, except
for the two spots in the basic region (spots 16 and 19). Both
of these originate from the sequence of Met-Ala-Arg-Lys-Met-
Lys (residue No.72 to 77; ref. 16) which is cleaved to varying
extents, producing weak and variable spots. Lack of radioactivity may have resulted from such a variability. Spots known to
be devoid of methionine (open arrows) do not contain any radioactivity.

It is thus concluded that protein A is chemically related
to calmodulin. It contains additional peptides not found in
calmodulin, accounting for its larger size. In contrast to
calmodulin which is essentially a soluble protein, protein A
is tightly associated with the membrane.

C. Axonal Transport of Calmodulin

Calmodulin has been shown to be concentrated at postsynaptic densities (23-25), but its localisation within the axon is
not known. Iqbal and Ochs have suggested that calmodulin is
transported at a fast rate (26), while Erickson *et al.* (27)
and Brady *et al.* (28) report that most of this protein is
found in the slow transport. We found calmodulin in the slow
transport and not in the fast, as described below.

In our electrophoretic system, calmodulin migrates as a
protein with a molecular weight of 20,000 and pI at 4.1 (*Fig.
2*). No radioactivity was detected at this position with the
proteins of fast transport (*Fig. 2*). In SDS electrophoresis
calmodulin migrates faster in the presence of calcium than in
its absence (23). None of the radioactive bands showed this
anomalous behavior. Another characteristic of calmodulin is
its specific, calcium-dependent binding to phenothiazine derivatives (17). Labelled proteins in the fast transport were
subjected to affinity chromatography on immobilised Fluphenazine to test the presence of calmodulin. Authentic calmodulin
is bound to this column in the presence of calcium, not eluted
under high salt condition, but eluted with EGTA. When Lubrol-
solubilised proteins obtained from the particulate fraction of
nerves containing labelled, rapidly migrating proteins were
applied to the column, 45% of radioactivity was retained.
However, all of it was eluted with 0.5M NaCl, leaving no radioactivity to be eluted with EGTA (in the calmodulin fraction).
Protein A was found exclusively in the retained and salt-

eluted fraction. EGTA-eluted radioactivity was not detected in the soluble fraction either. These experiments indicate that calmodulin is not present among the proteins of fast transport.

When similar affinity chromatography was carried out with proteins of slow transport (obtained one week post injection), a small but definite peak of radioactivity was detected in the EGTA-eluted fraction. The peak contained 0.9% of total radioactivity. Since calmodulin is rich in methionine, the actual amount of calmodulin as protein is smaller, 0.3-0.5% of transported proteins, depending on the methionine content of other major proteins (2-3.4% for tubulin and neurofilament protein compared to 6.1% for calmodulin). This value is comparable to that estimated by Erickson *et al.* using different methods of analysis (27).

Thus it is clear that calmodulin and protein A are transported in different compartments. This result together with our observation that protein A reaches the terminal in 20 h without any detectable changes in molecular weight makes it unlikely that protein A serves as a precursor or a transport form of calmodulin. Rather, it is suggested from its chemical properties to be another member of acidic, calcium-binding proteins, though the physiological activity of the protein has yet to be seen.

ACKNOWLEDGMENTS

We thank Dr. H. Kasai of the Tokyo Metropolitan University for a gift of bovine brain calmodulin, and for his valuable suggestions.

REFERENCES

1. Hoffman, P. N., and Lasek, R. J., *J. Cell Biol.* *66*, 351 (1975).
2. Lasek, R. J., and Hoffman, P. N., *in* "Cell Motility" (R. Goldman, T. Pollard, and J. Rosenbaum, ed.), 1021. Cold Spring Harbor Laboratory, New York, (1976).
3. Mori, H., Komiya, Y., and Kurokawa, M., *J. Cell Biol.* *82*, 174 (1979).
4. Abe, T., Haga, T., and Kurokawa, M., *Biochem. J. 136*, 731 (1973).
5. Di Giamberardino, L., Bennett, G., Koenig, H. L., and Droz, B., *Brain Res. 60*, 147 (1973).
6. Droz, B., Rambourg, A., and Koenig, H. L., *Brain Res. 93*,

1 (1975).

7. Cancalon, P., and Beidler, L. M., *Brain Res. 89*, 225 (1975).

8. Krygier-Brévart, V., Weiss, D. G., Mehl, E., Schubert,P., and Kreutzberg, G. W., *Brain Res. 77*, 97 (1974)

9. Lorenz, T., and Willard, M., *Proc. Natl. Acad. Sci. USA 75*, 505 (1978).

10. Tsukita, S., and Ishikawa, H., *J. Cell Biol. 84*, 513 (1980).

11. Briggs, C. A., and Cooper, J. R., *J. Neurochem. 36*, 1097 (1981).

12. Tashiro, T., Kasai, H., and Kurokawa, M., *Biomedical Res. 1*, 292 (1980).

13. Weber, K., and Osborn, M., *in* "The Proteins (Third Edition)" (H. Neurath and R.L. Hill, ed.), vol. 1, p. 179. Academic Press, New York, (1975).

14. O'Farrell, P. H., *J. biol. Chem. 250*, 4007 (1975).

15. Bonner, W. M., and Laskey, R. A., *Eur. J. Biochem. 46*, 83 (1974).

16. Kasai, H., Kato, Y., Isobe, T., Kawasaki, H.,and Okuyama, *Biomedical Res. 1*, 248 (1980).

17. Charbonneau, H., and Cormier, M. J., *Biochem. Biophys. Res. Commun. 90*, 1039 (1979).

18. Kretsinger, R. H., *Annu. Rev. Biochem. 45*, 239 (1976).

19. Kakiuchi, S., Yamazaki, R., and Nakajima, H., *Proc. Japan Acad. 46*, 587 (1970).

20. Means, A. R., and Dedman, J. R., *Nature 285*, 73 (1980).

21. Cheung, W. Y., *Science 207*, 19 (1980).

22. Dedman, J. R., Jackson, R. L., Schreiber, W. E., and Means, A. R., *J. biol. Chem. 253*, 343 (1978).

23. Grab, D. J., Berzins, K., Cohen, R. S., and Siekevitz,P., *J. biol. Chem. 254*, 8690 (1979).

24. Lin, C.-T., Dedman, J. R., Brinkley, B. R., and Means, A. R., *J. Cell Biol. 85*, 473 (1980).

25. Wood, J. G., Wallace, R. W., Whitaker, J. N., and Cheung, W. Y., *J. Cell Biol. 84*, 66 (1980).

26. Iqbal, Z., and Ochs, S., *J. Neurochem. 31*, 409 (1978).

27. Erickson, P. F., Seamon, K. B., Moore, B. W., Lasher, R. S., and Minier, L. N., *J. Neurochem. 35*, 242 (1980).

28. Brady, S. T., Tytell, M., Heriot, K., and Lasek, R. J., *J. Cell Biol. 89*, 607 (1981).

CHAPTER 38

A PERMEABILIZED MODEL OF PIGMENT PARTICLE TRANSLOCATION:
EVIDENCE FOR THE INVOLVEMENT OF A DYNEIN-LIKE MOLECULE
IN PIGMENT AGGREGATION

Theodore G. Clark
Joel L. Rosenbaum

Department of Biology
Kline Biology Tower
Yale University
New Haven, Connecticut

I. INTRODUCTION

   Organelle transport is known to occur in a wide variety
of eucaryotic cells in conjunction with physiologically important activities. A striking example of this type of movement
is the coordinate migration of pigment granules in dermal
chromatophores. These cells contain thousands of individual
pigment particles which can rapidly aggregate to the cell
center, or become dispersed throughout the cell. Pigment
aggregation and dispersion within chromatophores, underlie
rapid changes in skin coloration. Because changes in the intracellular distribution of pigment granules can be induced
in isolated preparations with appropriate reagents, and can
be easily monitored at the level of the light microscope,
chromatophores provide an extremely convenient system for experimental purposes.
   There is substantial indirect evidence for the involvement
of microtubules in pigment granule translocation, as well as
other types of saltatory particle movement (1-3). One approach toward understanding the role of microtubules in such
movements relies on the strict biochemical analysis of isolated microtubules, with the eventual aim of reconstituting a
functional motile system from purified components. While this
approach has yielded considerable information with regard to
the structure and assembly properties of isolated microtubules, most of what we have learned about the function of micro-

Biological Functions of Microtubules
and Related Structures

433

tubules in cell movement has come from studies of drug effects
on intact cells and from the analysis of reactivated model
systems. The use of drugs and other probes on intact cells is
limited by the permeability barrier of the plasma membrane.
Cell-free model systems, on the other hand, are accessible to
large molecules and can be subjected to selective extraction
and reconstitution experiments of the type used by Gibbons *et
al.* (4-6) to unravel the nature of force production in ciliary
and flagellar beating.

In order to determine how microtubules (and/or other cyto-
structural elements) might be involved in pigment particle
translocation, an effort has been made to develop a melanophore cell model which is permeable to large molecules, and
yet retains the ability to move pigment. Specific immunologi-
cal or biochemical probes could be introduced into such a
system in order to critically identify the components responsible for pigment translocation. Recently, Cande has described conditions for the reactivation of anaphase chromosome
movements in detergent-lysed PtK$_1$ tissue culture cells (7).
We have adopted similar procedures in order to investigate
the molecular basis of pigment transport in the dermal melanophores of *Fundulus heteroclitus*, the common killifish.

We have found the melanophores treated with the detergent
Brij-58, in combination with polyethylene glycol, are capable
of aggregating their pigment in response to adrenalin. During
the time in which cells can respond to the drug, plasma membrane disruption occurs and the cells appear to become accessible to large probes. Vanadate, an inhibitor of ciliary and
flagellar dynein ATPase which does not readily enter cells
(8,9), blocks pigment aggregation in detergent-treated melanophores. Erythro-9-(2-hydroxy-3-nonyl) adenine (EHNA), another
inhibitor of dynein ATPase (10), also blocks pigment aggregation in response to adrenalin suggesting that a dynein-like
molecule may play role in pigment granule translocation.

II. MATERIALS AND METHODS

Detailed materials and methods are to be published elsewhere (11). *Fundulus heteroclitus* were obtained from local
waters and from the Marine Resources Department, Marine Biological Laboratory, Woods Hole, Massachusetts. All experiments
were carried out with melanophores still attached to the
scale. The epidermal tissue lying above the melanophores was
removed by manual dissection prior to experiments. Melanophores were permeabilized using a modification of procedures
described by Cande for the preparation of PtK$_1$ cell models
(7). Experiments were carried out in lysis buffer containing

90 mM PIPES (piperazine-N,N'-bis-2-ethane sulfonic acid) pH
6.95, 2.5 mM MgSO$_4$, and 10 mM EGTA (ethyleneglycol-bis-(β-
aminoethylether)-N,N'-tetraacetic acid). Scales with attached
cells were washed briefly in buffer without detergent, and
then placed into lysis buffer containing 0.05 % Brij-58.
After 1 minute, preparations were transferred into buffer
containing 0.15-0.3 % Brij and 2.5 % polyethylene glycol (MW
20,000), and the incubation in detergent continued for varying
times as described in the text. All manipulations were carried out at room temperature. Control preparations were incubated in lysis buffer without detergent or polyethylene glycol
(PEG). Pigment aggregation and dispersion were induced by
treatment of melanophores with 100 µM adrenalin and 5 mM 3-
isobutyl-1-methylxanthine (IBMX), respectively.

III. RESULTS

Pigment aggregation in detergent-treated melanophores

     Initial attempts to develop a functional cell model of
Fundulus melanophores asked whether cells treated with detergent under conditions similar to those used for the reactivation of anaphase chromosome movements in lysed PtK$_1$ cells (7),
could be made to aggregate their pigment in response to adrenalin. With pigment in the dispersed state, melanophore
preparations were placed into lysis buffer containing the nonionic detergent Brij-58, and at various times afterward,
adrenalin was added. Figure 1 shows the results of a typical
experiment of this kind in which the response of detergent-
treated melanophores, pigment aggregation occurs and proceeds
over the course of several minutes while cells are in the
continuous presence of detergent. At concentrations between
0.15 and 0.3 % Brij, virtually all cells respond to adrenalin
when the drug is added within 4 minutes after initial treatment with detergent. The overall rate and extent of pigment
aggregation in detergent-treated cells is less than that of
control, non-detergent treated preparations, and the response
to adrenalin becomes progressively diminished with increasing
time of incubation in Brij. Nevertheless, at 0.15 % Brij,
individual pigment granules may continue to move 10 minutes or
more after cells are placed into detergent. Figure 2 shows
the migration of pigment granules toward the cell center over
a 28 second time span, beginning 10 minutes after placing melanophores into lysis buffer.

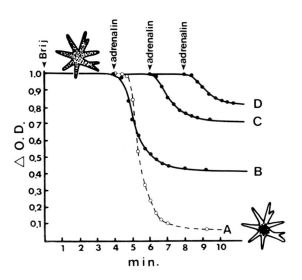

*Fig. 1. Photometric assay of pigment aggregation in
detergent-treated cells. Melanophore preparations were placed into lysis buffer containing 0.05 % Brij at time zero,
and after 1 minute, transferred to buffer containing 0.15 %
Brij and 2.5 % PEG. Preparations were then mounted on a perfusion slide and examined in the light microscope. A region
of the scale was selected and a pinhole aperture enclosing
1-3 cells was inserted between the objective lens and the
photometer. Melanophores were then perfused with 3 ml of
lysis buffer containing adrenalin at 4 minutes (B), 6 minutes
(C), and 8 minutes (D) after initial treatment with detergent.
Control preparations (A) were incubated in lysis buffer
without detergent or PEG. The response to adrenalin was measured as a change in optical density (arbitrary units norma-
lized to an initial value of 1.0) over time. Each time point
represents the average of 6 samples. All samples were taken
from the same animal. In order to prevent slow spontaneous
aggregation of pigment in lysis buffer, IBMX, a pigment dis-
persing agent, was added to all buffers until the addition of
adrenalin.*

Ultrastructural analysis

Examination of detergent-treated melanophores by transmission electron microscopy reveals a progressive disruption
of the plasma membrane during the time in which cells are
responsive to adrenalin. Figure 3a is a representative thin
section through a melanophore cell process from a preparation
fixed 3 minutes after treatment with 0.3 % Brij. Several

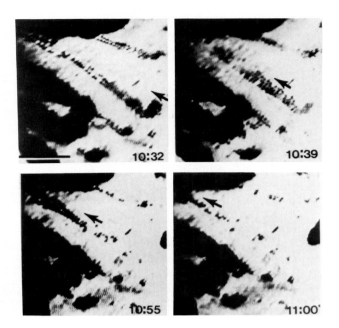

Fig. 2. Light microscope analysis of pigment aggregation in a cell treated with a final concentration of 0.15 % Brij. Migration of pigment toward the cell center (left) can be seen in three separate cell processes. Proximal regions of the cell processes are filled with pigment and appear black. The arrow traces the direction of movement of granules in one cell process beginning 10 minutes 32 seconds following initial treatment with detergent. Time is shown in the lower right of each micrograph. Observations were made using video enhanced contrast differential interference constract microscopy (12). Bar = 10 μm.

holes or interuptions in the plasma membrane can be seen. Although many cytoplasmic microtubules are present in these preparations, some microtubule breakdown may occur as evidenced by curved sheets of photofilaments which are occasionally seen near clusters of intact microtubules (Fig. 3c). With increasing time of detergent treatment, membrane disruption becomes further evident. Figure 3d shows a representative thin section through a melanophore cell process from a preparation which had been treated with 0.15 % Brij for 12 minutes. Large regions of the plasma membrane are completely gone from the preparation.

In order to determine whether detergent-treated cells are

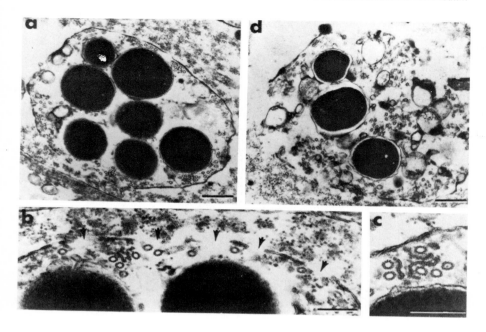

Fig. 3. Electron micrographs of sections taken through melanophore cell processes of detergent-treated preparations. (a-c) Increasing magnifications of a cell process from a preparation treated with a final concentration of 0.3 % Brij for 3 minutes. Arrows in (b) indicate interuptions in the melanophre plasma membrane. (d) Cross-section through a cell process from a preparation incubated in a final concentration of 0.15 % Brij for 12 minutes. Note the absence of plasma membrane from large regions of the cell process, and the distended appearance of the membranes surrounding the pigment granules. Bar a,d = 0.2 µm; b,c = 0.1 µm.

accessible to large molecules, cells were incubated in lysis buffer containing ferritin for 3.5 minutes and then transferred to the same medium without detergent from an additional 8 minutes prior to fixation and preparation for electron microscopy. A melanophore cell process from a sample which had been treated in this way is shown at high magnification in Fig. 4. Ferritin particles are clearly visible on the cytoplasmic side of the plasma membrane. While ferritin could be visualized within the cytoplasm of many of the cell processes of detergent-treated melanophores, no free ferritin was seen within the interior of non-detergent treated controls.

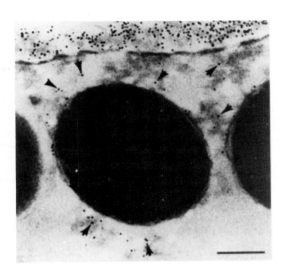

*Fig. 4. Electron micrograph of a detergent-treated melano-
phore incubated in ferritin. The preparation was incubated in
lysis buffer containing a final concentration of 0.15 % Brij
and ferritin at 35 mg/ml before being transferred to buffer
without detergent as described in the text. Note the high
concentration of electron dense ferritin particles in the extracellular space (top), and the free ferritin (arrows) present in the cytoplasm. Bar = 0.1 µm.*

Immunofluorescence staining of cytoplasmic microtubules

As a further test of the accessibility of detergent-treated melanophores to molecular probes, cells were incubated in
lysis buffer with or without detergent, then fixed and treated
with rat monoclonal antibodies against tubulin, followed by
FITC-labelled goat anti-rat IgG. Detergent-treated cells show
a bright, radial pattern of microtubule staining when examined
in the fluorescence microscope (Fig. 5b). Roughly 70 % of the
melanophores in detergent-treated preparations can be stained.
Non-detergent treated control preparations on the other hand,
show a complete absence of microtubule staining (Fig. 5a).

Inhibition of pigment aggregation by vanadate

Vanadate has been shown to inhibit dynein ATPase activity
and to block reactivation of ciliary and flagellar axonemal
beating (8,9). To determine the effect of vanadate on pigment

particle translocation, melanophore preparations were incubated in detergent solutions containing vanadate, then treated
with adrenalin and assayed photometrically for their response
to the drug. Figure 6a shows the results of a typical experiment in which cells were incubated in lysis buffer containing
100 μM vanadate, and then treated with adrenalin 4 minutes
after being placed into detergent. A significant inhibition
of the overall rate and the extent of pigment aggregation can
be seen in melanophore preparations exposed to vanadate.
Vanadate at the same concentration has virtually no effect on
the response of non-detergent treated cells to adrenalin as
seen in Fig. 6b.

Effect of EHNA on pigment aggregation

Erythro-9-(2-hydroxy-3-nonyl) adenine, another inhibitor
of dynein ATPase (10), was examined for its effect on pigment
aggregation. Melanophore preparations were incubated in
detergent solutions containing EHNA, and the ability of cells
to respond to adrenalin was determined. As seen in Fig. 6c,
EHNA, at a concentration of 2 mM, almost completely abolishes
pigment aggregation. EHNA was found to inhibit pigment aggregation in intact melanophores as well, and is completely

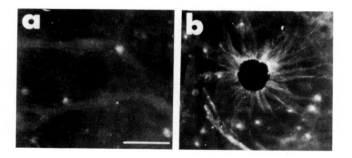

*Fig. 5. Immunofluorescence light micrographs of melano-
phores from a control, non-detergent treated preparation (a),
and a preparation treated with lysis buffer containing a
final concentration of 0.15 % Brij for 2.5 minutes (b). Cells
were fixed and stained with rat monoclonal antibodies against
tubulin and FITC-labelled goat anti-rat IgG. Prior to the
experiment, cells were treated with adrenalin to induce pigment aggregation. Cell bodies appear as a dark spot at the
center of each micrograph. Bar = 50 μm.*

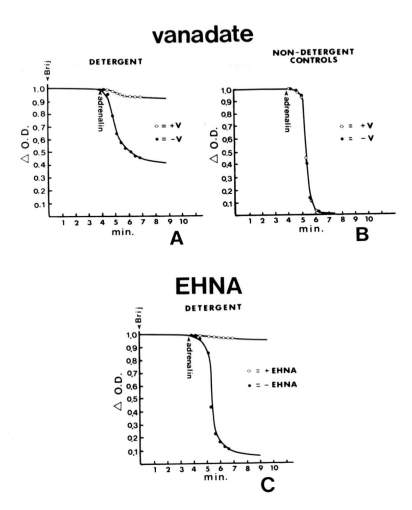

Fig. 6. *Vanadate and EHNA effects on pigment aggregation as assayed photometrically. (A) Melanophore preparations incubated in lysis buffer containing a final concentration of 0.3 % Brij. (B) Preparations incubated in lysis buffer without detergent. In (A) and (B), vanadate concentrations were 100 μM. Time points represent the average of 10 samples. (C) Preparations incubated in lysis buffer containing a final concentration of 0.15 % Brij. EHNA concentration was 2 mM. Time points represent the average of 6 samples.*

reversible in its effect (data not shown).

III. DISCUSSION

     After treatment with the detergent Brij-58, in combination
with polyethylene glycol, melanophore preparations are capable
of aggregating their pigment in response to adrenalin, and
pigment movement will continue for 10 minutes or more after
the addition of detergent. On the basis of several criteria,
it appears that cell lysis has occured under the detergent
conditions used. Electron microscopic examination of detergent-treated cells reveals a progressive disruption of the
melanophore plasma membrane during the time in which pigment
aggregation can occur. Brij-treated cells are accessible to
ferritin, a large, electron dense probe which is effectively
excluded from non-detergent treated controls. In cells incubated in detergent, fixed, and treated with rat monoclonal
antibodies against tubulin followed by FITC-labelled goat
anti-rat IgG, a characteristic radial pattern of microtubule
straining can be visualized by fluorescence microscopy.
Control preparations treated identically, but without detergent, do not stain. Finally, as assayed photometrically,
vanadate (100 μM), an inhibitor of ciliary and flagellar
dynein ATPase, blocks pigment aggregation in response to adrenalin in detergent-treated preparations, but has no effect on
intact cells. When taken together, it is believed these
criteria argue strongly that pigment aggregation can occur in
permeabilized cells.
     There is substantial indirect evidence that microtubules
may play a role in pigment granule movement. Pigment aggregation in several types of chromatophores, is inhibited by
drugs which interfere with microtubule assembly (2,3). Microtubules in chromatophores are generally organized in a radial
pattern extending from the cell center outward toward the cell
periphery and thus may provide an orienting framework along
which pigment particles travel (2,3,13). On the basis of
these observations, it has been argued that microtubules in
some way generate the forces underlying particle movements.
Although in certain cases, a significant decrease in microtubule number appears to accompany pigment aggregation (13,
14), the density and distribution of microtubules in the peripheral cell processes of *Fundulus* melanophores remains constant in the dispersed and aggregated states, raising the
possibility that pigment granules slide along fixed microtubules (15). An interaction between pigment granules and
microtubules through a force-transducing element such as
dynein would provide a mechanism by which such sliding could
occur. The evidence presented here suggests that a dynein-
like molecule may, in fact, play a role in pigment aggregation. Vanadate and EHNA have been shown to inhibit dynein

ATPase activity and to block ATP-reactivated axonemal beating (8-10). Both inhibitors block pigment aggregation in melanophore preparations from *Fundulus*. However, since both vanadate and EHNA are known to inhibit enzyme activities other than dynein (16-18), caution is necessary to attributing the effects of these inhibitors to a specific interaction with a dynein-like ATPase.

The results of vanadate and EHNA experiments described here are consistent with very recent reports on the effects of these agents on particle movements in other cell types. Vanadate appears to inhibit saltations of pigment granules in digitonin-treated erythrophores of the squirrelfish, *Holocentrus ascensionis* (19), and to block pigment aggregation when microinjected directly into the cytoplasm (20). EHNA stops shuttling of pigment granules in *Holocentrus* erythrophores as well (20). In addition, vanadate inhibits saltatory particle movements in detergent-treated human skin fibroblasts (21), and both vanadate and EHNA inhibit anaphase chromosome movements in lysed PtK$_1$ cell models (7,22). These results suggest that dynein, or a dynein-like ATPase, may play a role in a variety of cytoplasmic microtubule-based motility phenomena.

Although direct evidence for the existence of dyenin-like molecules in the cytoplasm of eucaryotic cells remains equivocal (23-26), it is hoped that the use of permeabilized model systems will provide further insight into this problem.

ACKNOWLEDGMENTS

We are grateful to Dr. Robert Allen, Department of Biological Sciences, Dartmouth College, Hanover, New Hampshire, for assisting us in the use of video enhanced contrast microscopy, and to Dr. John Kilmartin, MRC, Cambridge, England, for his generous gift of rat monoclonal antibody preparations.

REFERENCES

1. Rebhun, L. I., *Int. Rev. Cytol.* 32, 93-137 (1972).
2. Dustin, P., *in* "Microtubules", pp. 266-270. Springer-Verlag, Berlin (1978).
3. Roberts, K., and Hyams, J. S., *in* "Microtubules", pp. 487-530. Academic Press, New York, (1979).
4. Gibbons, I. R., *Proc. Nat. Acad. Sci. USA.* 50, 1002-1010 (1963).
5. Gibbons, B. H., and Gibbons, I. R., *J. Cell Sci.* 13, 337-357 (1973).

6.   Gibbons, B. H., and Gibbons, I. R., *J. Biol. Chem.* 254, 197–210 (1979).

7.   Cande, W. Z., and Wolniak, S. M., *J. Cell Biol.* 79, 573–580 (1978).

8.   Gibbons, I. R., Coson, M. P., Evans, J. A., Gibbons, B. H. Houck, B., Martinson, K. H., Sale, W. S., and Tang, W.-J. Y., *Proc. Nat. Acad. Sci. USA.* 75, 2220–2224 (1978).

9.   Kobayashi, T., Martensen, T., Nath, J., and Flavin, M., *Biochem. Biophys. Res. Commun.* 81, 1313–1318 (1978).

10. Bouchard, P., Penningroth, S. M., Cheung, A., Gagnon, C., and Bardin, C. W., *Proc. Nat. Acad. Sci. USA.* 78, 1033–1036 (1981).

11. Clark, T. G., and Rosenbaum, J. L., Manuscript submitted (1982).

12. Allen, R. D., Allen, N. S., and Travis, J. L., *Cell Motil.* 1, 291–302 (1981).

13. Porter, K. R., *CIBA Found. Symp.* 14, 149–166 (1973).

14. Schliwa, M., and Euteneur, U., *J. Supramol. Struc.* 8, 177–190 (1978).

15. Murphy, D. B., and Tilney, L. G., *J. Cell Biol.* 61, 757–779 (1974).

16. Simons, T. J. B., *Nature* 281, 337–338 (1979).

17. Schaeffer, H. J., and Schwender, L. F., *J. Med. Chem.* 6–8 (1974).

18. Pike, M. C., Kredich, N. M., and Snyderman, R., *Proc. Nat. Acad. Sci. USA.* 75, 3928–3932 (1978).

19. Stearns, M., and Ochs, R. L., *J. Cell Biol.* 91, 416a (1981).

20. Beckerle, M. C., and Porter, K. R., *J. Cell Biol.* 91, 302a (1981).

21. Forman, D. S., *Exp. Cell Res.* in press (1982).

22. Cande, W. Z., *J. Cell Biol.* 91, 320a (1981).

23. Mohri, H., Mohri, T., Mabuchi, I., Yazaki, I., Sakai, H., and Ogawa, K., *Dev., Growth & Differ.* 18, 211–219 (1976).

24. Zieve, G. W., and McIntosh, J. R., *J. Cell Sci.* 48, 241–257 (1981).

25. Sakai, H., Mabuchi, I., Shimoda, S., Kuriyama, R., Ogawa, K., and Mohri, H., *Dev., Growth & Differ.* 18, 211–219 (1976).

26. Pratt, M. M., Otter, T., and Salmon, E. D., *J. Cell Biol.* 86, 738–745 (1980).

Author Index

Numerals refer to chapter numbers.

Subject Index

Numerals refer to chapter numbers.

Dissociation constant, *continued*
tubulin–GDP, 6
tubulin–DTP, 6
tubulin–vinblastine complex, 7
DNA, 4, 32
hybridization, 4
Saccharomyces cerevisiae, 4
Schizosaccharomyces pombe, 4
Doryteuthis bleekeri, 35
Doublet microtubules
dynein, 12, 13, 14
sperm flagella, 14
Dynein, 12, 13, 14, 17, 38
arms on doublet microtubules, 13, 17
cytoplasmic, 14
electron microscopy, 13
latent activity dynein-1, 13, 14
sea urchin sperm flagella, 13, 14, 17
Dynein arm
cross-bridge cycle of, 17
force generated by, 17
Dynein heavy chains
binding to doublet microtubules, 13
separation of A_α and A_β, 13

E

Echinosphaerium akamae, 10, 11
Elastase, 17
Electron dense plaque, 34
Epidermal tendon cell, 34
Equilibrium gel filtration, 8
Erythro-9-(2-hydroxy-3-nonyl)adenine
(EHNA), 38
Ethylphenylcarbamate–Sepharose, 1

F

F-Actin, 2, 5
elastic modulus of, 5
Falling ball viscometer, 25, 26, 28
Fascin, 24
Ferritin, 38
Fertilization, 19, 24
Flagellar movement, 15, 16, 17, 18
asymmetric beat, 18
Chlamydomonas, 18
initiation by cAMP, 15
inhibition by potassium ions, 15
microtubule sliding, 17
reactivation of, 16
symmetric beat, 18

Flagellar axoneme, 14, 15, 16, 17, 18
disintegration by trypsin, 15
microtubule sliding in, 17
Flexural rigidity
actin filament, 5
singlet microtubule, 5
Fluorescent staining
calmodulin, 19
MAPs, 33
tubulin, 32, 33
Fluphenazine–Sepharose 6B, 37
Force–velocity relation, 17
Freeze-etch replica, 31
Fundulus heteroclitus, 38

H

Half spindle, 22, 23
Heavy meromyosin (HMM), 24, 34
Heavy meromyosin S-1, 34
Heavy water (D_2O)
effect on mitosis, 20
effect on microtubule assembly, 20
Hemicentrotus pulcherrimus, 16, 17, 20, 24

G

GDP
bound to microtubules, 6
interaction with tubulin, 6
Gelation, 26, 28
Gerbil fibroma cell, 33
Glia, 3
Glial fibrillary acidic (GFA) protein, 29
Grasshopper spermatocyte, 21
GTP
exchangeable site of tubulin, 6
hydrolysis during microtubule treadmilling, 6
nonexchangeable site of tubulin, 6
GTP-Regenerating system, 7
Guanylate cyclase, 12

I

IBMX, 36
Immunoelectron microscopy, 12
Immunofluorescence staining
tubulin, 32, 38
Initial segment, 34
Inner arm, 14
Interconnecting strands, microtubules and
neurofilaments, 31